Biology in Transition

The Life and Lectures of Arthur Milnes Marshall

Curated and annotated by
Martin Luck

Pelagic Publishing | www.pelagicpublishing.com

Published by Pelagic Publishing
www.pelagicpublishing.com
PO Box 725, Exeter EX1 9QU, UK
www.pelagicpublishing.com

Biology in Transition: The Life and Lectures of Arthur Milnes Marshall

ISBN 978-1-78427-166-4 Hardback
ISBN 978-1-78427-167-1 ePub
ISBN 978-1-78427-168-8 PDF

British Library Cataloguing in Publication Data
A catalogue record for this book is available from the British Library

Cover images:

Arthur Milnes Marshall with kind permission of the Centre for Heritage Imaging and Collection Care at the John Rylands Library, University of Manchester.

Charles Darwin from Darwin F, *Charles Darwin*, 2nd edn (London: John Murray, 1902).

Outline sketch of the North Face of Scafell, as seen from the Pulpit Rock on Scafell Pike, copied from Dr Dixon's photograph. A M M April 1893 from the Wasdale Head Climbing Book,1884–1919, held at Kendal among the Archives of the Fell and Rock Climbing Club of the English Lake District, and reproduced with their kind permission.

DEDICATION
For Jacob and Toby, that one day Charles Darwin may be their hero too, *"for the greatness of his services to mankind and his contributions to human knowledge; and love for the truthfulness, the patient endurance in suffering, and the gentle courtesy of his life"*.

CONTENTS

FOREWORD
MATTHEW COBB

These lectures by Arthur Milnes Marshall, my predecessor as Professor of Zoology at the University of Manchester, provide an extraordinary glimpse into the scientific world at the close of the 19th century, and give a telling example of how science was popularized at the end of the Victorian era. A specialist in vertebrate development and anatomy, Marshall was appointed Professor of Zoology at Owens College in 1879, at the young age of 27, and rapidly gained a reputation as a great teacher and an excellent public speaker. The lectures reproduced here reveal why. These were not specialist talks to a select group of colleagues, but examples of what we would now call 'outreach', given at a time when the public had an insatiable appetite for science and self-improvement. Expressing himself vividly but simply, Marshall provided concrete examples to back up his arguments, patiently explaining complex ideas to non-specialists.

Some of the lectures were given to keen amateurs, such as the Manchester Microscopical Society or the Biological Section of the British Association. But several talks were given at New Islington Hall in Ancoats—a part of Manchester just east of the city centre that was criss-crossed with canals and warehouses, and at the time was one of the poorest in the country. Today the area is the focus of intense economic and cultural redevelopment—the site of the Hall is now covered with smart new town houses. The Sunday Lectures to Men and Women at which Marshall spoke were begun in 1884 by the Ancoats Brotherhood, which was set up by two social reformers called Charles Rowley and Thomas Horsfall. Sylvia Pankhurst attended some of these Sunday lectures and later recalled that they 'brought to that factory-blighted district examples of the best things of the time in music, art and science'. Marshall's lectures certainly fit that description.

In many respects, Marshall's interests and his understanding of biology appear profoundly modern. His defence of Darwinism is determined and precise, while his emphasis on how organisms develop from a single cell to an adult form reveals an interest in the role of developmental effects in animal evolution that is now the focus of some of the most exciting research in biology. He even foreshadowed our current conservation concerns, complaining about the 'terrible destruction' of the natural world, in particular the disappearance of elephants and whales, all caused by the actions of 'a ruthlessly advancing civilization'. But in two key respects Marshall was trapped in his time, understandably unable to peer into the future that was mere years away, despite showing remarkable insight.

Marshall was convinced that understanding how an organism develops would reveal something about its evolutionary past. Nowadays, what Marshall called 'Recapitulation Theory' is generally associated with Ernst Haeckel's 'biogenetic law', according to which 'ontogeny recapitulates phylogeny'—in other words, each organism necessarily goes through the steps of its evolutionary past as it grows. This is not actually true, and Haeckel's evidence, widely reproduced in engravings from his papers, was inaccurate and had been fudged to fit his theory. But while there is no 'law', it is the case that development provides evidence for our evolutionary past and Marshall's lectures, in particular the one on Animal Pedigrees, show strikingly how that embryonic evidence can reveal deep evolutionary links between apparently distinct lineages.

Secondly, like all his contemporaries, Marshall was understandably confused by what he described as the 'bewildering problem of heredity'. In his lecture on Inheritance, Marshall describes the two major hypotheses of the time—Darwin's theory of pangenesis, according to which every part of the body contains hereditary particles that are affected by experience and are passed on to the next generation, and August Weismann's more recent suggestion that in animals the sex cells—egg and sperm—form a separate lineage from those that form the body, and are solely responsible for heredity.

Marshall was doubtful about Darwin's idea, mainly because of the complexity of getting all those particles to find their way to the egg and sperm, and the lack of experimental evidence to support it. He was far more impressed by Weismann's theory and the empirical work that underpinned it, but there was no decisive proof. Marshall concluded that insight into the nature of inheritance might come only once scientists had some clue about the origin of life.

This turned out not to be the case. Unknown to Marshall, one of the greatest conceptual breakthroughs in biology had taken place decades earlier in Moravia (part of the Austrian empire), as a series of thinkers grappled with they termed 'the genetic laws of nature'—the nature of heredity. The key insight came in 1837, when the head of the monastery at Brno, the Abbé Napp, asked his colleagues 'What is inherited, and how?' To answer this question, Napp encouraged one of his protégés, Gregor Mendel, to explore what happened when two kinds of organisms were crossed, or hybridized.

Mendel's studies on pea plants were published in 1866, to little effect. Marshall, like every other 19th-century scientist, either never heard of these findings or did not appreciate their significance. The answer to Marshall's understandable uncertainty about the nature of heredity would become apparent a few years later, after Marshall's death. Mendel's work was simultaneously rediscovered and replicated in 1900, and within 15 years the newly named genes were shown to be located on chromosomes, which been identified shortly before Marshall gave his lectures, and are the subject of some intriguing discussion in the lecture on Some Recent Developments of the Cell Theory.

In one respect, some of Marshall's arguments could easily be brought forward over a century, with little amendment. Opponents of evolution by natural selection regularly come up with what they consider to be novel, supposedly decisive arguments without appreciating that scientists have repeatedly rebutted these criticisms down the decades. Marshall's lecture on Objections to the Darwinian Theory deals with many of these, politely but systematically demolishing them. Among the criticisms Marshall explores are the 19th-century version of today's oft-repeated jibe: 'If humans evolved from apes, how come there are still apes?' In Marshall's time this criticism was called 'the persistence of lowly organized forms alongside more highly organized ones', and he gave it short shrift.

In his lecture on The Colours of Animals and Plants, Marshall mentions one of the iconic examples of natural selection in action—the appearance of dark forms of the Peppered Moth, *Biston betularia*, in the Greater Manchester region following the pollution of the area with coal smoke from what Blake memorably called 'dark satanic mills'. Marshall's interpretation of the significance of this event was rather different to our modern understanding, all of which is now supported by experimental and molecular evidence.

The moth was originally present in a pale form with speckled black markings that camouflaged it from predatory birds when it settled on the bark of trees such as silver birch. As the trees darkened with soot through industrial activity in the first part of the 19th century, a darker form arose, which was camouflaged against the dark trees (this form was first recorded in Manchester in 1848; genetic evidence suggests that it appeared in 1819). In the 1950s a combination of the Clean Air Act and deindustrialisation led to the slow disappearance of soot-stained trees, and the lighter form became predominant once more. Birds were the agents of natural selection, differentially eating first the light form when the trees were dark, then the dark form when the trees lightened (this too has been demonstrated experimentally).

For Marshall, the change in the colour of the Peppered Moth—and of other butterflies and moths—had an entirely environmental cause. They turned dark not because of any change in the frequency of heredity factors in the population, but because of the soot they ingested as caterpillars. Instead of being an example of natural selection, Marshall presented the Peppered Moth as a case of 'the direct action of the environment' on the colour of an organism. He does not seem to have wondered either about the consequences of the moth being a different colour or about whether the change was inherited or acquired. This insight into how a key example of natural selection was seen at the time is fascinating and will undoubtedly prompt historians to delve deeper into Victorian views about examples of natural selection.

Marshall's name is now forgotten by scientists, historians, and even Mancunians—I am ashamed to say I had not heard of him until Professor Luck invited me to write this foreword. The republication of these lectures will bring his work to a 21st-century audience, and will provide a rich source for historians of science and of culture who will undoubtedly be keen to explore how scientific ideas were communicated in the latter years of the 19th century. Scientists working in evolution, anatomy, and development will all be intrigued by Marshall's views, both those that are still held today, and those that have been superseded. These lectures will also interest and inform the general reader, for Marshall's easy style conveys his material in a clear and engaging way, while Martin Luck's excellent annotations provide invaluable context and clarification. Above all, teachers, lecturers and those involved in science communication will be inspired by Marshall's commitment to explaining science to the general public.

Marshall's tragically early death deprived us of someone who made a major contribution to the public understanding of science. Had he lived, who knows how he might have influenced both biology and the perception of science, in particular in Manchester. The republication of these brilliant and fascinating lectures marks a suitable memorial to Marshall's life and work. Over a century after they were first delivered, they still deserve to be widely read.

HISTORY BY SERENDIPITY

Sometime in the very early years of the 21st century, my department at the University of Nottingham acquired a small library belonging to Sir John Hammond, FRS (1889–1964), veterinarian, physiologist, and pioneer of artificial insemination in cattle. Hammond was a mentor of the late Professor Eric Lamming, a long-serving head of the department, and his *Hammond's Farm Animals* was a standard text. Alumni and other animal scientists will be familiar with both these names.

The library languished on a trolley for several years, no one being quite sure what to do with it. After rescue from a flooded laboratory, in which some books sustained water marks, it came to rest in a glass-fronted cabinet against a coffee-room wall. As a department of pioneering research and education rather than history, we tolerated it more for its academic associations than its contents.

The collection consists largely in obsolete texts and conference reports on animal production, agriculture, and reproductive physiology, together with some of Hammond's own experimental notebooks. Among the more notable volumes are early editions of D'Arcy Thompson's *On Growth and Form*, Samuel Brody's *Bioenergetics and Growth* and other minor science classics, but these are hardly collectors' items.

On closer inspection one afternoon, I noticed a couple of dark red-bound volumes with rough-cut, brown-stained pages and the name Marshall on the spine. To any reproductive biologist that surname bespeaks *Marshall's Physiology of Reproduction*, the legacy treatise of Cambridge physiologist FHA Marshall, FRS, and for many years the most comprehensive authority on the topic. Lamming himself edited and co-authored its final (4th) edition in 1994.

The two red books turned out not to be by FHA Marshall but to contain a series of lectures by one Arthur Milnes Marshall (AMM), edited by his brother CF Marshall (CFM) and published in 1894. I wondered if there was a family connection with FHA but enquiries have failed to reveal any. Both volumes have Hammond's signature boldly inscribed inside the front cover and dated 1949. The names of several previous owners are there too, so these books have travelled.

Marshall's lectures proved to be an absorbing read. They deal with evolution, embryology, and related biological matters, with Charles Darwin's name and those of other important 19th- century biologists distributed liberally through the text. AMM was clearly a knowledgeable and insightful scientist as well as a gifted author and teacher. He was evidently capable of explaining the biological debates of his time to specialist and non-specialist audiences alike.

Marshall's back story (included here in a Biography) also proved unexpectedly intriguing, especially for anyone who enjoys tramping the Lakeland fells. Crucially for the history of science, his premature death on the last day of 1893, just 11 years after Darwin's and only 34 years after the publication of *On the Origin of Species*, sets a decisive chronological fix on the state of biological knowledge at the end of the 19th century. The bereaved CFM must have felt something of this as he set about gathering his deceased brother's lectures and preparing them for publication. Reading them now, we can see both the transformative effect of Darwin's intellectual legacy and *exactly* how far our science has come in the subsequent 125 years.

Several of the original lectures were published separately during AMM's lifetime and can still be found in journal archives. Copies of the books, over a century out of print and never getting beyond a first edition, may be extracted with some effort from the world's library system and odd copies are on sale in antiquarian bookshops. But who has really heard of AMM (or his brothers) and who, after all, would be searching for them?

My principal aim in re-presenting these time-stranded documents, together with a little contextual commentary, is to illuminate a transitional period in the history of biology. Evolution by natural selection was the established doctrine but genes were undefined. Microscopy was flourishing and cell science was finding its feet but molecular science was yet to come. Embryology was suggesting recapitulation, but ancestry, inheritance, and missing links awaited liberation from theoreticians and the stones of palaeontology. Most revealing, perhaps, is the complete absence of Gregor Mendel's name from the lectures but the extensive incorporation of August Weismann's developing theories.

We can also reflect, perhaps sadly, on the abruptly truncated life of an inspirational teacher and scientist. Had he continued his work into the 20th century, he would surely have grasped the new biology and eagerly communicated its wonders to his students and to the public audiences in whom he took such a generous interest. This, too, underscores the uniqueness of the time when the lectures were written: had Marshall lived but a few years longer, they would be of no interest at all.

Old academic books are dusty and easily ignored but their faded pages sometimes illuminate, unexpectedly, one's own position along the time line of knowledge. The trail back through Hammond and others in whose hands these two remarkable volumes have rested allows me the conceit of imagining a connection to the great biologists of the 19th century, which I hope the reader will forgive.

Martin Luck
University of Nottingham
2018

GENERAL NOTE

The 21 lectures by AMM are presented here in the order in which they originally appeared in CFM's two-volume collection, renumbered as a single sequence. Each volume set is preceded by its title page and by CFM's preface and table of contents. Between the two sets I have inserted a book review by AMM which particularly illuminates his views on certain biological questions.

The figures in the second set of lectures are photographs of the originals. The diagrams in Lectures 10, 15 and 19, and tables appearing elsewhere in the text, have been re-created to match the originals as closely as possible.

My comments are presented as contextual introductions to each volume and lecture and as footnotes, the latter numbered as a single sequence to facilitate cross-referencing. Footnotes preceded by {M} are those which appeared in the original texts, assumed to be by CFM. References to the Hammond Library (HL) copies of the Marshall lectures are explained in the Apology above.

Authors named in the text are identified (with a very few exceptions) in the List of Authorities which follows the lectures, along with their dates and a brief indication of their importance in the history of biology.

ACKNOWLEDGEMENTS

I am grateful to archivists James Peters, Suzanne Fagan, and Henry McGhie (University of Manchester), Claudia di Somma and Carmela Scotti (Zoological Station Anton Dohrn, Naples), Rosemary Clarkson (Darwin Correspondence Project, Cambridge), Katherine Harrington (Royal Society of London), Christopher Hilton (Wellcome Library), Kelda Roe (Keswick), Max Clark (Kendal) and Geoff Burns (Library of Birmingham) for their expertise and guidance; to Chris Sherwin (Fell and Rock Climbing Club) for access to climbing records, photographs, and other information; to Mark Francis, Nick Hopwood, Brenda Luck, John van Wyhe, and Julian Wiseman for helpful correspondence; to Mark Bentley (University of Nottingham) and James Robinson (University of Manchester) for skilful document photography; to Brigitte Graf and Robin for generous hospitality during visits to Manchester; to Janine for assistance with document checking and for tolerating my obsession; and to Alice and Chris for encouraging my unreliable knees across the screes of Scafell.

Thank you all.

Note regarding sources

Lecture texts were obtained from copyright-free, electronic versions held by the University of Edinburgh, accessed through the Wellcome Library. The associated illustrations were photographed directly from the Hammond Library copies or my personal copies of the original volumes published in 1894.

The cover photo of AMM was supplied, with kind permission, by the Centre for Heritage Imaging and Collection Care at the John Rylands Library, University of Manchester. That of Charles Darwin is from my copy of Darwin F, *Charles Darwin*, 2nd edn (London: John Murray, 1902).

The background sketch of Scafell is from the Wasdale Head Climbing Book, 1884–1919, held at Kendal among the Archives of the Fell and Rock Climbing Club of the English Lake District, and reproduced with their kind permission. The original bears the inscription "*Outline sketch of the North Face of Scafell, as seen from the Pulpit Rock on Scafell Pike, copied from Dr Dixon's photograph. A M M April 1893.*"

VOLUME 1

BIOLOGICAL LECTURES AND ADDRESSES

The 21 lectures in this collection were delivered by Marshall over a relatively short period of time, between 1879, when he was appointed to the newly established Chair of Zoology at Owens College Manchester, and 1893, the year of his death at the age of 41 years.

Owens College was the forerunner of the University of Manchester. It was founded in 1851 and joined the federal Victoria University in 1880, prior to the granting of independent university status in 1904. One of AMM's responsibilities was to take over the teaching of zoology from the Professor of Natural History, WC Williamson, so that the latter could concentrate on botany.

The lectures of the first volume span the whole of that period and were delivered to several different types of audience: students, the public and members of scientific societies. The presentational styles are appropriately varied but each shows a combination of accessibility, detail and conceptual explanation, with an abundance of examples.

An interesting lexicographical curiosity emerges from these lectures: the use of the word *genetic(s)* (Lectures 1, 2 and 3; also in conjugated form in lectures 9, 10, 11 and 13). Many modern writers (for example, Dronamraju K, *Popularizing Science*, Oxford, Oxford University Press, 2017; Blackman H, *Studies in the History and Philosophy of Biology & Biomedicine* 2004; **35**: 93–117; see also *Wikipedia*) declare or imply that the word was coined in 1905 by William Bateson, in a letter to Adam Sedgwick, as a name for the new science of heredity initiated by the work of Gregor Mendel (1822–1884). (The first use of the word *gene* for a unit of inheritance is attributed the Danish botanist Wilhelm Johansen (1857–1927) in 1909.)

As these lectures show, *genetic* was in use at least a quarter of a century earlier. AMM uses it to mean *relationship through lineage* or *pedigree* (reflecting the Greek root *genos* = origin), rather than as the name of a science or a process, but the fact of its pre-Batesonian coinage deserves to be more widely recognised. Unfortunately, it is such a familiar word to us in its Batesonian and post-Batesonian (molecular) guise that it is difficult to read it without a modern interpretation.

Bateson was an early advocate of Mendel, whose experiments demonstrating the particulate nature of inheritance, carried out between 1856 and 1863 and published in German in 1866, were rediscovered at the very start of the 20th Century (supposedly by Correns, de Vries and von Tschermak, working independently; see Kampourakis, *Science & Education* 2013; **22** 293–324). Although AMM's death preceded their rediscovery, is it possible that he had read Mendel's original paper and knew about them? We can safely answer "No" to this question, for at least three reasons:

1. AMM was alive to the latest developments in biology, yet neither Mendel's name nor the laws (of segregation and of independent assortment) attributed to him appear anywhere in his writings.

2. AMM discusses the *gemmule*, the hypothetical entity invoked by Darwin in his pangenesis theory to explain inheritance (Lectures 7 and 12). It is generally accepted that Darwin was unaware of Mendel's work or at least of its significance; had he been thus aware, he would undoubtedly have abandoned his own provisional mechanism. (Gemmules with a completely different meaning feature in Lectures 6 and 11.)

3. AMM was familiar with August Weismann's work on inheritance and critically evaluates his theories (Lectures 7, 9, 11, 12 and 13); Weismann distinguished between somatic and reproductive cells but, like Darwin, did not know how the process of inheritance worked. Mendel's work supplied the crucial clues that enabled 20th-century genetic science to emerge and thrive, completely overtaking Weismann's insightful but mechanistically bewildered ideas.

Had AMM lived but a decade longer, Mendel's work would surely have intrigued and excited him. He would have recognised its significance and been the first to explain it to a wide audience.

BIOLOGICAL LECTURES
AND ADDRESSES

DELIVERED BY THE LATE

ARTHUR MILNES MARSHALL

M.A., M.D., D.Sc., F.R.S.

PROFESSOR OF ZOOLOGY IN OWENS COLLEGE ; LATE FELLOW OF

ST. JOHN'S COLLEGE, CAMBRIDGE

EDITED BY
C. F. MARSHALL

M.D., B.Sc., F.R.C.S.

LONDON
DAVID NUTT, 270–271, STRAND
1894

PREFACE

THE majority of the lectures and addresses collected together in this volume have already been printed in the Transactions of several Societies—viz., the lecture on Animal Pedigrees, published in the *Midland Naturalist;* the Presidential address to the Biological Section of the British Association; and several reprinted from the *Transactions of the Manchester Microscopical Society.* Of these printed addresses I have reproduced as many as possible without involving too much repetition. In the case of the British Association address however it will be found that many of the points discussed are dealt with in other addresses, especially in the lecture on Animal Pedigrees, which is indeed based on that address. It appeared however desirable to include the British Association address, even at the risk of repetition, on account of the importance of its scientific value, and for this reason I have placed it at the end of the series, the others being arranged chronologically.

With regard to the lectures in manuscript hitherto unpublished, I have selected a few of those which appeared to be of most interest, and in this I have of necessity been obliged to confine myself to those which were most fully written out. Where amplification was required I have endeavoured, as far as possible, to do this in words which, from my own personal knowledge, I believe would have been used.

The lectures on the Darwinian Theory, which form a distinct course by themselves, will be published as soon as possible in a separate volume, together with other series of lectures, if there appears to be a sufficient demand for them.

I must express my thanks to the Committees of the Manchester Microscopical Society, the Birmingham Natural History Society, and the British Association for permission to reproduce the addresses printed in their Transactions.

I am under great obligations to Professor G. B. Howes for his kindness in reading the proofs, and for supervising the technical points. My thanks are also due to Professor Ray Lankester for valuable suggestions, to my brother Mr. P. E. Marshall for correcting the proofs, and to Dr. C. H. Hurst for assistance on several points.

C. F. MARSHALL
LONDON, *April* 1894.

CONTENTS

11. SOME RECENT EMBRYOLOGICAL INVESTIGATIONS.

An address delivered on the occasion of the Annual Conversazione of the Manchester Microscopical Society, January 21st, 1893. Reprinted from the Society's Transactions.

12. DEATH.

The President's address delivered at the Manchester Microscopical Society, February 2nd, 1893. Reprinted from the Society's Transactions.

13. THE RECAPITULATION THEORY.

The President's address to the Biological Section of the British Association delivered at Leeds, September 1890. Reprinted from the Transactions of the Association.

LECTURE 1

THE MODERN STUDY OF ZOOLOGY

Most of the students AMM taught on his zoology courses were studying for medical rather than zoology degrees and for that reason few of them joined him in practical laboratory research. Given the context, this lecture can be seen as serving at least two important purposes. It is clearly a calling card—a presentation of academic credentials—by a new professor, keen to offer his perspective on the current status of his subject. We might even view it as amounting to an inaugural lecture, although whether some other presentation to the College served that formal role is not recorded.

Its other clear purpose is to introduce students to the subject they are about to study. As medical students, he could reasonably assume that the majority had an appreciation of general science and were capable of absorbing theoretically challenging material. His objective is to provide them with a conceptual framework for the studies they are about to undertake: to set foundations and identify the direction in which zoology was moving.

Regarding AMM's didactic style, one notes the accessible starting point, the use of extended analogy and metaphor to communicate difficult concepts, the occasional use of ironic humour, and the frequent recourse to established authorities for justification. He skilfully draws lessons from history, including from the work of his hero Darwin, and is not afraid to point out where previous interpretations faltered or have been superseded. He establishes his academic authority by demonstrating the extent of his knowledge and his well-informed appreciation of current and past debates.

Many topics discussed in this lecture form the kernels of later, more detailed lectures (indicated in comments). Thus it is an introductory lecture in every sense.

LECTURE I

THE MODERN STUDY OF ZOOLOGY

THE man of business knows full well—at times too well—the importance of periodical stock-taking; of comparing his actual position with his estimated one, of ascertaining exactly how he stands, of assuring himself that his affairs are in a sound and healthy condition, and that the gain on the year's transactions is a real one. The man who neglects such precautions is apt, sooner or later, to find himself in difficulties: his latest transaction proves a failure, and on attempting to fall back on his former position and start afresh, he finds the ground cut away from beneath him, his reserve fund mysteriously vanished, and his affairs in hopeless confusion.

As in business, so in science, it is well to have periodical stock-takings. Scientific facts accumulate rapidly, and give rise to theories with almost equal rapidity. These theories are often wonderfully enticing, and one is apt to pass from one to another, from theory to theory, without taking care to establish each before passing on to the next, without assuring oneself that the foundation on which one is building is secure. Then comes the crash; the last theory breaks down utterly, and on attempting to retrace our steps to firm ground and start anew, we may find too late that one of the cards, possibly at the very foundation of the pagoda, is either faultily placed or in itself defective, and that this blemish—easily remedied if detected in time—has, neglected, caused the collapse of the whole structure on whose erection so much skill and perseverance have been spent.

Thus men of science find it well occasionally to take stock, to look back for the moment instead of forward, to assure themselves that their operations since the last stock-taking have really resulted in a gain, and to define accurately the nature and extent of that gain.

Science has been aptly compared to a globe, similar to our own earth—a globe with a solid hard crust bounded by an irregular surface. The solid crust represents ascertained facts, facts that have been confirmed and stowed away in their proper places; the irregularity of its surface indicates the unequal accumulation of facts in the various branches of knowledge. The atmosphere by which the whole globe is invested represents the world of speculation, of theories—an atmosphere heavily laden with germs and particles of truth, but germs as yet immature, particles whose position relative to the solid crust is not yet a fixed and determined one.

Our process of stock-taking consists in defining the boundary line between the crust and the atmosphere, between earth and air; such a process becomes periodically necessary because the contour of the surface is constantly changing; particles are continually being added to the crust, while those whose places are already determined are liable by reason of these additions to have their relative positions and impor-tance altered. Thus what was at one time a lofty peak, a startling though established generalisation, may become overshadowed by the formation of a far loftier one by its side, of which the original peak becomes but an insignificant shoulder whose original importance is soon forgotten.

I propose, then, in the present paper to take stock of our zoological knowledge, to attempt to define the actual position and aims of zoological thought, the steps by which this position has been attained, and the methods by which it is hoped to achieve these ends.

Such a process is of special and peculiar interest as applied to zoology, firstly, by reason of the great and rapid accumulation of facts that has occurred of late years; secondly, because of the far-reaching and fiercely contested theories to which these facts have given birth; and, thirdly, because the study of zoology includes the study of man, so that generalisations concerning the rest of the animal kingdom must apply also to man himself. For these reasons, and more especially for the third one, the theories and generalisations of zoology are always subjected to rigid and jealous scrutiny, not only by those who make zoology a special study, but by the world at large.[1]

In order to know clearly with what we are dealing we may, with Professor Huxley,[2] define zoology as "the whole doctrine of animal life," as being in fact, if such marked alliteration may be excused, all about animals. Now, from very remote times indeed there have existed not only names for different animals, but also collective names for groups of animals agreeing with one another in certain respects but differing widely amongst themselves in others; collective names such as fish, under which head a great number of animals are commonly included, some of which, such as the whale, are not fish at all; or birds, including forms as diverse as a starling and a stork, a humming bird and an ostrich. The introduction of such collective names marks the earliest attempts at zoological classification.

Of such classifications we meet with examples in the Old Testament. Thus we read of Solomon that *"he spake of trees, from the cedar tree that is in Lebanon even unto the hyssop that springeth out of the wall he spake also of beasts, and of fowl, and of creeping things, and of fishes."*[3] The object of the writer in the above passage is manifestly to bring into prominence the extent of Solomon's knowledge, and we are certainly led to believe that Solomon had made a personal study of the several groups of animals mentioned—i.e., that he was a zoologist. The passage quoted bears evidence in itself that the four groups named were intended to include the whole of the animal kingdom; but any doubt on this point is removed by the fact that in other parts of the Old Testament the animal kingdom is distinctly divided into these same four groups.[4] We are therefore justified in speaking of this as a zoological classification.[5]

---

1   This long introduction, concerning the accumulation of knowledge and the importance of periodically reviewing accepted facts, indicates that AMM felt the need to remind his audience about the process of science in an age of rapid change. He employs several metaphors, sometimes stretched and testing to the reader. This device, found throughout the Lectures, was probably characteristic of his teaching style.

2   Thomas Henry Huxley, in *A Lobster; or, The Study of Zoology.* Lay Sermons, Collected Essays VII (1861).

3   {M} I Kings iv. 33.

4   {M} *e.g.* Deuteronomy iv. 17, 18.

5   The biblical quote is clearly being used as factual evidence of the human tendency to order and classify things, rather than to claim religious authority for a historical fact or point of view.

If we examine this classification more closely we see that the habits of the different animals, and more especially the media in which they live, are made the basis on which the several divisions are founded. Thus beasts include terrestrial animals, animals living on dry land; fowl are those animals that possess the power of flight and so are enabled to live as denizens of the air; fishes are animals adapted for living in water; while creeping things probably included what we now call insects, and any other small forms that could not be referred readily to either of the other groups. Such a system may be spoken of as a classification by distribution. It is one of easy application, and so far a convenient one, but inasmuch as it takes no account whatever of structural and physiological resemblances and differences between the several animals with which it deals it must be regarded as an exceedingly primitive one.

The next classification of any great importance that we meet with is that given by Aristotle, perhaps the greatest and most truly scientific man in the highest sense of the word that the world has ever known. Aristotle, like Solomon, is better known in connection with other branches of knowledge than zoology; still he devoted much attention to the study of animals, and placed zoology on a far more scientific basis than his predecessors had done. It would appear that Aristotle never drew up a formal scheme of classification; the system commonly ascribed to him, which is in reality compiled from his various writings and was never given by him in its modern form, is as follows,[6] the modern equivalents of the several groups being indicated in the right hand column.

A.   Animals with red blood and a backbone . . . . *Vertebrata*

    I.   Provided with four legs

        (a) Viviparous. . . . . *Mammalia*

        (b) Oviparous . . . . *Reptilia*

    2.   Provided with two legs and two wings . . . . . . *Aves*

    3.   Devoid of legs, but provided with fins . . . . . . *Pisces, Cetacea*

B.   Animals without red blood and with no backbone . . *Invertebrata*

    I.   Soft externally . . . . *Mollusca*

    2.   Soft internally, hard externally . . . . *Crustacea, Testacea, Insecta*

Such a classification is manifestly based on a totally different system to that of Solomon; the several groups are now characterised not by their habits or the media in which they live, but by resemblances and differences in anatomical structure. The branch of zoology that treats of the structure of animals is called Morphology; hence Aristotle's classification may be contrasted with that of Solomon as being not a classification by distribution, but a morphological classification. Inasmuch as the latter springs from a closer and more accurate acquaintance with animals than the former, it is a better and more scientific one and may be taken as marking a distinct and very important advance in the study of zoology.[7]

---

6   {M} *Vide* Claus: "*Grundzuge der Zoologie*": French Translation by Moquin-Tandon. Note B., p. 1099.

7   The move from *habits and media* to morphology as the basis for animal classification

The next writer on zoology of any great importance is the elder Pliny, who lost his life A.D. 79, at the celebrated eruption of Mount Vesuvius by which Pompeii and Herculaneum were destroyed. Pliny was to a far greater extent than Aristotle a professed zoologist, and left a voluminous work on natural history in thirty-seven books. He divided the animal kingdom into four main groups, which he named as follows:

1.   Animalia terrestria;
2.   Animalia aquatilia;
3.   Volucres;
4.   Animalia insecta;

i.e., he classified animals according as they lived on the ground, in the water, or in the air; dividing them into terrestrial, aquatic, and volatile,[8] with a distinct class for those animals, such as insects, which do not belong to any element exclusively.

This is clearly a classification by distribution, and therefore differs totally from that of Aristotle, while it agrees in principle with that of Solomon. This agreement, however, is not only in principle; if the two schemes of classification be compared it will be seen that they are really identical, a point of some interest. Thus, Pliny's *Animalia terrestria* are the same as the beasts of Solomon; the *Animalia aquatilia* as the fishes; while *Volucres* are obviously equivalent to fowl, and *Animalia insecta* to creeping things. The sole difference between the two systems is in the order in which the several groups are arranged. It would, therefore, appear, that while Aristotle was a long way in advance of any of his predecessors, Pliny, who lived more than 400 years after Aristotle, not only made no advance, but even fell back on the very empirical classification that was in use in the days or Solomon, 1100 years previously, and that had probably been in use for a still longer time.

As Pliny is a writer who owed a considerable part of his reputation to his work on natural history, it may not be inappropriate here to quote the criticism passed on him many centuries after by Cuvier, in order to support my statement that Pliny, instead of placing zoology on a more scientific basis, in reality did it incalculable damage, and threw it back as a science to the condition in which it had been before Aristotle's time.[9] Cuvier's words are as follows:[10]—"*In general, he is only a compiler, and, indeed, for the most part, a compiler who has not himself any idea of the subjects on which he collects the testimony of others, and therefore cannot appreciate the truth of their testimonies, nor even always understand what they mean. In short, he is an author devoid*

---

was indeed a significant advance. In the present age we take a further step, basing taxonomy on genetic relationships and other demonstrable links to common ancestors. Later in the lecture AMM distinguishes between morphological classifications made by similarity (*definition*) and those based on difference (*type*). This progressive development of perspective is a reminder, as relevant now as then, that all classifications are arbitrary human constructs, contrived for whichever purpose is most informative or most useful in the moment.

8   = aerial.

9   Thus Pliny failed to take stock of accumulated knowledge and its veracity, in the manner which this Lecture proposes as essential to science.

10   {M} "*Bibliographie Universelle*," xxxv.

of criticism, who, after having spent a great deal of time in making extracts, has ranged them under certain chapters, to which he has added reflections that have no reference to science properly so called, but display alternately either the most superstitious credulity or the declamations of a discontented philosophy, which finds fault continually with mankind, with nature, and with the gods themselves."

Pliny's influence on zoological thought, though most pernicious, was sufficiently great to completely outweigh his illustrious Greek predecessor; and to this must, I think, be ascribed in great part the almost complete gap in zoological literature of any value that extends from the time of the Roman zoologist to about the sixteenth century. It was not, indeed, until nearly the middle of the eighteenth century that a system of zoological classification of any permanent value was proposed. For this we are indebted to the great Swedish naturalist Linnaeus, the founder of modern natural history as he has been well called.

The system of classification proposed by Linnaeus was, like that of Aristotle, a morphological one, based on resemblances and differences of structure in the several animals and groups of animals. He divided the whole animal kingdom into six classes, defined as follows:-

A.   Cor biloculare biauritum, sanguine calido rubro.

    1.   Viviparis.   *Mammalia.*

    2.   Oviparis.   *Aves.*

B.   Cor uniloculare unitarium, sanguine frigido rubro.

    1.   Pulmone arbitrario.   *Amphibia.*

    2.   Branchiis externis.   *Pisces.*

C.   Cor uniloculare inauritum, sanie frigida alba.

    1.   Antennatis.   *Insecta.*

    2.   Tentaculatis.   *Vermes.*

Of morphological classifications there are two principal varieties, classification by definition and classification by type. The Linnaean classification is a typical example of the former of these. In it the whole animal kingdom is divided up into groups of convenient size, each characterised by the presence or absence of some one, two or more easily recognisable features; stress being laid on the differences between the several groups, rather than on the resemblances between the several animals included in each individual group. An illustration will perhaps serve to give a clearer idea of what is meant. Take a piece of paper and make a number of dots on it in a perfectly irregular manner. We want to classify these dots, to arrange them in groups: if we were to classify them by definition, we should divide the paper by means of lines passing between the dots into a number of compartments of convenient size, to which we should give distinctive names; we should then define the position of any one dot by simply saying in which division it was. As our whole paper is divided up, every dot must fall into some one or other of these divisions, so that our classification is at any rate a simple and a convenient one.[11]

---

11   This illustration, and to a certain extent the next, are initially difficult to follow and appear tangential to the biological point being made. The dots, or the people, are

Or, again, imagine a map of England in which the county boundaries are laid down, but all the towns and villages are left out; such a map would give us a classification of the inhabitants of England, and a classification by definition. Stress is laid simply on the boundary lines between the several divisions, and the sole interest attaching to any particular individual consists in the question on which side of a given arbitrary line he happens to reside. It follows also that in such a scheme those individuals who reside in the centres of the several counties are subordinate in interest to those near the margins of the counties, since about these latter there may be doubt as to which division they should be referred to, while such doubt can hardly exist in the case of the former. Such a map might be very useful, and for purposes of minor importance, such as a parliamentary election or a cricket match, might contain all the information necessary, the sole interest consisting in which side of an artificially drawn line a given individual happened to live.

As the knowledge of anatomy advanced; as zoologists became gradually acquainted with the structure of a larger and continually increasing number of animals; as the microscope in the hands of Malpighi, Swammerdam, and their successors gradually revealed the details of minute structure and rendered possible a correct appreciation of the anatomy of animals previously too small to be investigated, it was gradually realised that the Linnaean system, with its hard and fast lines of division, no longer represented the actual state of our knowledge, and classification by definition gradually gave way to the second form of morphological classification—classification by type.

We may explain the difference between the two by means of our former illustrations: thus, to take our first case, we no longer divide our paper by artificial lines, we now look to the dots themselves; we find that the dots are not always the same distance apart, that many of them fall naturally into groups of various sizes; each well-marked group we give a name to, and the central member of the group round which the others seem to be arranged we call the type of the group: of the remaining dots, some are so close to our big groups that we include them with these, others form distinct smaller groups of their own, whilst some solitary ones stand quite apart and isolated from all the rest.[12] Or we may, to take our second instance, illustrate classification by type by a map of England, in which the county boundaries are left out, but all the towns and villages marked.[13] Here we have large centres such as London or Manchester, containing large numbers of inhabitants, and representing distinct types; smaller centres lying immediately round them, not definitely connected with them as yet, but destined ultimately to be so, other small centres lying at a distance from the large ones, and constituting distinct types, and, finally, isolated houses representing species of animals widely separated from their fellows, and forming for the time at any rate small but distinct types of their own.

---

being classified by location and defined on that basis. One can probably think of more accessible analogies.

12   Here, the dot illustration starts to make (biological) sense for it suggests variance around the mean.

13   The geographical illustration now also becomes instructive for it links with the concept of degrees of separation and similarity.

The distinguishing characteristics of classification by type, and especially the points in which it contrasts most strongly with classification by definition, have been admirably stated by the late Master of Trinity College in the following words:— "*The class is steadily fixed, though not precisely limited; it is given though not circumscribed; it is determined, not by a boundary line without, but by a central point within; not by what it strictly excludes, but by what it eminently includes; by an example, not by a precept; in short, instead of a definition we have a type for our director. A type is an example of any class, for instance, a species of a genus, which is considered as eminently possessing the characters of the class. All the species which have a greater affinity with the type-species than with any others form the genus, and are ranged about it, deviating from it in various directions and different degrees.*"[14]

Such a classification represents the real affinities of animals much more truthfully than classification by definition. The sharp boundary lines, of which nature knows nothing, and which formed the main feature of the older system, are here swept away; the resemblances of animals are made of more weight than their differences; and no attempt is made to define the limits of the several groups.

This doctrine of animal types[15] was first brought forward prominently by Cuvier and Von Baer at the commencement of the present century. Cuvier, in his latest system of classification, distinguished four leading types or plans of structure in the animal kingdom, to one or other of which all animals could, according to him, be referred. Mainly owing to the weight of Cuvier's authority this doctrine of types made considerable progress during the first half of the present century; it never, however, wholly replaced classification by definition, and probably never would have done so; for, in the first place, the essence of classification is convenience, and classification by definition is far more convenient for the ordinary purpose of a zoologist than classification by type;[16] and, secondly, although the idea of types expressed a great and important truth, yet it was but the partial expression of a still greater one, which, when fully developed, was destined to completely overthrow all former attempts, and to reveal the only true and unassailable basis of classification. The gradual rise of this new doctrine[17] we have now to notice briefly.

About the commencement of the present century two new influences began to make themselves felt in zoology, two new branches that were afterwards to exert great influence on zoological thought began for the first time to receive serious attention. These were Palaeontology and Embryology.

---

14   {M} Whewell, "*The Philosophy of the Inductive Sciences,*" vol. i. pp. 476–7.

15   *Animal types*, as referred to here, must not be confused with the *type specimens* that Linnaeus used as reference material for the identification of the species he named and catalogued. Indeed, the distinction illustrates precisely the difference between classification by type (possession of common features) and classification by definition (exclusion of variants).

16   The validity of this practical point is illustrated by the identification keys used by field biologists.

17   The new doctrine, as we shall see in later lectures, concerns the ancestral links between existing species, as evidenced by embryology and the finding of fossil relatives. As a view of the biological world, it aligns precisely with contemporary taxonomy based on genetic inheritance and common ancestry and depicted in clade diagrams.

Palaeontology, the investigation of extinct animal forms, of those animals and portions of animals known to us only through their fossil remains, was first studied systematically and raised to the rank of a science by Cuvier. Previous to his time fossils had not received serious attention; even their animal origin was far from being commonly recognised, and the most absurd ideas were in vogue as to their nature and origin; some supposing them to be mere freaks of nature, others that they were models used by the Creator when he was preparing to stock the earth with animals.

Cuvier did not confine himself to demonstrating that these fossil remains must have proceeded from animals that once lived on the surface of the earth; he studied the distribution of fossils in the different geological strata with great care, and was led to form generalisations of extreme value and interest. The most important of these conclusions are contained in his "Theory of the Earth,"[18] and are to the following effect:—In the oldest strata of all there are no fossil remains at all; organised beings were not all created at the same time, but at different times, probably very remote from one another; the fossil remains of the recent strata approach far nearer to the existing forms of animals than do those of the older strata; finally, of the highest forms of animal life—man and the quadrumana[19]—there are no fossil remains whatever.[20]

From these conclusions, the importance of which it is impossible to overrate, Cuvier was led to found his doctrine of *Catastrophism,* according to which there have been periodical annihilations at long intervals of time of all the animals living on the earth at the time; each cataclysm being followed by the creation of a totally new set of animals,[21] which though agreeing in many points with their predecessors, yet presented many marked differences from them.

---

18   {M} First published in 1798 as the preliminary discourse to the "Recherches sur les Ossemens Fossiles;" republished separately, with many editions, in 1825.

19   = "four-handed" animals.

20   This is not entirely correct. The first Neanderthal fossil was reported in 1829 and further discoveries were made throughout the 19th century. William Buckland discovered the Red Lady of Paviland in 1823 and other humanoid fossils were known, if not understood. The Lucy fossil was discovered in East Africa in 1974.

21   The periodic occurrence of catastrophic annihilations of life is now well established, the most famous being that which followed the meteor impact at the Chicxulub crater some 66 million years ago and which is said to have wiped out 75% of life including the large dinosaurs. Where we now dissent from the view attributed to Cuvier is in the belief that all life was wiped out and a fresh set created. We see these events rather as survival bottlenecks and periods of intense species selection. However, the Hammond Library (HL) copy of *Biological Lectures* contains the following handwritten marginal annotations by a previous owner (probably Wynfrid LH Duckworth, 1870–1956, an anatomist and Master of Jesus College, Cambridge): "*No, Cuvier did not postulate this. He postulated an immigration from adjacent land of comparable animals.*" This sounds much more like our contemporary interpretation. The annotations continue: "*Cuvier postulated 4 types as capable of being actually demonstrated. These he said represent the possibilities of nature in the direction of forming animals. He was not primarily concerned with the immutability of species. He discussed the significance of variations as linking them to causes such as varying heat etc. acting on embryos.*"

Cuvier's doctrines, however, did not meet with general acceptance among geologists, and the publication of the first edition of "The Principles of Geology," by Sir Charles Lyell, in 1830 two years before Cuvier's death, may be said to mark the complete overthrow of the doctrine of catastrophism so far as the changes that have taken place in the earth's crust are concerned. A closer study of what is at present occurring on the earth's surface showed that there are now in action forces amply sufficient, given time enough, to produce changes as great as any of which we have geological record; that the elevation of great mountain chains is not due to the sudden action of immeasurably great forces but to the long continued action of apparently insignificant ones; and that there is not only no evidence whatever of the occurrence of the supposed catastrophic periods, but that all the evidence on the point tends to prove that such periods never have occurred.

Though catastrophism thus received its deathblow so far as the crust of the earth was concerned, men still hesitated to apply the same reasoning to the fossil remains of animals, and in spite of the geological evidence the doctrine of catastrophism, *i.e.*, of periodical annihilations and re-creations, continued to meet with acceptance so far as these fossil remains were concerned.

All this time there was steadily developing and gradually acquiring definite shape a doctrine destined ultimately to overthrow Cuvier's theories concerning fossils as completely as the geologists had done those dealing with the earth's crust. This was the doctrine of the Mutability of Species.

Cuvier, as we have seen, maintained that species were all due to separate acts of creation; the new doctrine maintained that species were not immutable, but that one species might give rise to two or more new ones. The actual birth of this doctrine is involved in some obscurity; it is not quite clear when it first arose, or to whom the credit of its origination is due. It was clearly recognised and advocated by the illustrious Goethe in 1796, but whether this is the date of its birth is not clear.

Its greatest advocates were Lamarck and St. Hilaire,[22] its greatest opponent Cuvier, and long and bitter was the struggle. Though the two former, and more especially Lamarck, worked out the doctrine in the most elaborate manner, yet they were unable to point out the causes at work in the supposed transformation of species; they

---

22   Cuvier's argument with Lamarck and St. Hilaire, referred to in this paragraph, came to a head at a meeting of the French Academy of Sciences on 19 July 1830. Cuvier, evidently exploiting his international reputation, was able to silence his critics and prevented the mutability of species from becoming the accepted doctrine. His cause may have been helped by the fact that immutability left the biblical creation story unchallenged. There is more on this in Lecture 2. Duckworth's annotation of the HL text provides nuance: "*St. Hilaire was not primarily concerned to establish the mutability of species. He was concerned to establish Unity of Type for all animals: one type as against the 4 postulated by Cuvier: & it was on this that Cuvier scored.*" Curiously, modern biology accepts a form of *inheritance of acquired characteristics* in the guise of the epigenetic control of gene expression and cellular differentiation and other sources of adaptive variation. This is argued to broaden Darwinian evolutionary theory from its *modern synthesis* into an *extended evolutionary synthesis* (see Laland *et al.*, *Proceedings of the Royal Society B* 2015; **282**: 20151019, http://dx.doi.org/10.1098/rspb.2015.1019). In that context, one is also struck by Duckworth's earlier comment on Cuvier (above)

were unable to show why species should become modified into other species, and so, the *onus probandi* lying with them, victory in the eyes of the world rested with Cuvier. So complete was this victory considered at the time that for nearly thirty years after his death Cuvier's authority was sufficient to keep this new doctrine in abeyance.

At length came the most eventful epoch in the history of zoology, the simultaneous announcement by two independent investigators, Charles Darwin and Alfred Russel Wallace,[23] of the doctrine of Natural Selection, at the meeting of the Linnaean Society on July 1st, 1858. This doctrine effected for the animal world exactly what the geologist had already done for the earth's crust; it showed that there are now in operation causes sufficient, given time enough, to produce all the changes requisite to convert the extinct fossil species into those now living on the earth, causes that must have been in operation since life first dawned on the earth, causes that must inevitably have led to the passage of species into species.

In this way a complete and consistent theory of the history of life on the earth was at length obtained—not only what had actually occurred, but how and why it had occurred. And now at length the true meaning of the laws of Cuvier regarding the distribution of fossil remains was seen, those laws which had led him to form his erroneous theory of catastrophism. For instance, it was seen now that the reason why the fossils of recent beds resemble existing forms more closely than do those of the older beds is simply that there has been a continuous process of evolution; that the fossils of the recent beds are the ancestors of the now living forms, the descendants of the fossils of the older beds, and thus occupy an intermediate position genealogically; in which case their intermediate position structurally becomes at once intelligible.

Recognising these facts, attempts were soon made to reconstruct the path of descent, to trace out the pedigree of some given species. This was first accomplished with any degree of success in the case of the horse. The horse, the zebra, and the ass stand alone among mammalia in possessing but one complete functional toe on each limb; they must either have been specially created as such, or must have been derived from some more typical mammalian form. Owing to their large size, the fact that bones are fairly easily preserved as fossils, and the great time that must have elapsed during the gradual transformation of some typical mammal to the highly specialised horse, it is only reasonable to expect that some direct evidence of this transformation should be forthcoming. Consequently this furnishes a very good test case. Without entering into details, which would be unnecessary in what is now so familiar an instance, it will suffice to state that a series of fossil forms is now known furnishing a complete gradation from older tertiary forms with four or five toes on each foot, through newer tertiary forms to the horse with its one functional toe on each limb; that, in other words, the pedigree of the horse has been completely and satisfactorily worked out.

---

about variation caused by the effect of heat on embryos: a number of epigenetic effects are known to be exerted by maternal conditions during pregnancy, although of course these do not permanently alter the DNA sequence.

23   AMM's account of Darwin's life and work, including a well-balanced acknowledgement of the role played by Wallace, appears in Lecture 21. Darwin and Wallace take their place amongst other historical luminaries in Lecture 2.

Another familiar but striking example is afforded by birds. These are very highly specialised forms, and stand apart from other vertebrates in a number of anatomical points. We are now acquainted with a large number of fossil forms serving to connect birds with reptiles, and showing the several gradations by which reptiles gradually lost their teeth, acquired wings and feathers, and became birds.[24]

If then it is possible by the aid of fossil remains to reconstruct the pedigree of some particular form or group of forms, why should it not be possible for all animals? As the possibility of this reconstruction gradually dawned on man, so it was slowly realised that such a reconstruction, such a pedigree, would in itself be a perfect system of classification, and the only really natural one. Such a classification may be spoken of as a genetic or genealogical classification.

The idea of such a classification is familiar to all of us through the medium of our own special genealogies: these generally take the form of a tree in which the stem represents our earliest ancestor, who, in this country at least, usually takes the form of some impoverished adventurer, whom we should probably be intensely ashamed of could we see him in the flesh, whose sole virtue lies in the fact that he "came over with the Conqueror," and whose sole possessions of any importance appear to have been a crest, a motto, and a coat of arms; the primary branches of the stem represent the offspring of this all-important ancestor; the secondary branches their offspring, and so on, each branch denoting a generation. Some of the branches die off, stop, and become extinct; others persist and thrive: the ultimate branchlets of these last bear the leaves, which are the actually living representatives of the family, on the topmost of which we inscribe our own name.[25]

A genetic classification of the animal kingdom takes a similar form; the stem represents the earliest and most primitive form of animal life, the branches the several forms derived successively from this and from one another, while the terminal leaves represent the actually existing forms. If we draw horizontal lines across our tree, these may be taken to represent the different geological ages; then the branches cut by any one line or plane will represent the life on the earth at that period. Some of the branches never reach the top of the tree; these represent forms of life that attained their full development in some of the earlier epochs and then became extinct.

Such a genetic classification is at once felt to be the only really natural one, to be in fact an ideal classification. It is simply an embodiment of fact, not an expression of

---

24   We would now qualify this interpretation by identifying *dinosaurs* as a diverse, multi-morphic range of vertebrates having close common ancestry with modern birds. Feathers and egg-laying evolved millions of years before flight, and several feathered dinosaurs with and without teeth are known. The connection between dinosaurs and reptiles is a much more general one, in the sense that dinosaurs had many characteristics that we see in modern reptiles. Somewhere amongst the very early dinosaurs was the common ancestor of reptiles, birds and mammals, presumably an amniote. The role of *Archaeopteryx* in the interpretation of avian origins is dealt with by AMM in Lectures 6, 16 and 20, including references to T.H. Huxley who was prominent in proposing an evolutionary link between dinosaurs and modern birds. A picture of the *Archaeopteryx* fossil appears as the frontispiece to the second volume of lectures.

25   Pedigrees are discussed again, in greater depth and still with engaging ironic humour, in Lecture 10.

opinion: it is therefore absolutely unassailable, and what is more, true for all time.[26] The recognition of the possibility of such a classification is the distinguishing feature of modern zoology, and the determination of such genetic histories or phylogenies of the several groups of animals is the goal towards the attainment of which the efforts of zoologists are directed.

Having traced the way in which this idea arose, how increase of knowledge and perfection of methods caused classification by definition to replace the crude ideas embodied in the classifications adopted by Solomon and Pliny, and then in its turn to give way to classification by type, and having seen how finally the doctrine of natural selection rendered possible the conception of a genetic classification, it may not be out of place to devote a few words to considering the methods by which it is proposed to attain this end. These are as follows:

I.   The accumulation of anatomical facts concerning as large a number of animals as possible; this requires no further explanation.

2.   The systematic study of the geographical distribution of the different animals and groups of animals, both recent and extinct. This, which is a comparatively new branch of zoology, has already yielded results of immense importance, and promises to yield others of even greater value in the future.

3.   Palaeontology, or the study of the anatomy of extinct animals and their relations to existing forms. We have seen above that it is to palaeontology that we owe the first successful attempt to reconstruct the pedigree of some one given form; and inasmuch as the connecting links between the several groups of animals are almost necessarily extinct, it is clear that the evidence yielded by those extinct forms must be invaluable. Still it must be borne in mind that it is only certain animals, and of these animals only certain parts, that are capable of being preserved as fossils, so that palaeontology can only help us as regards certain groups of the animal kingdom, and even concerning these its evidence must necessarily be very imperfect and fragmentary. Indeed, had we only these three methods to aid us, the attempt to reconstruct the pedigree of the animal kingdom could only be successful to a very partial and limited extent. Fortunately there is another method which supplies us with evidence of the very kind we most want, and which bids fair to far outweigh in importance the other three methods. This is:

4.   Embryology,[27] or the study of the actual development of existing forms, the several changes which they undergo during their gradual evolution from the ovum. This, which is the second of the two new influences referred to in a preceding page, is, like palaeontology, a young science, but one that has thriven mightily of late years. It was not till towards the middle of the present century that it began, in the hands

---

26   AMM is being a little overconfident here. We can accept this bold statement only if we are prepared also to accept the correctness and completeness of the information available. For as long as the data is incomplete or uncertain, as it inevitably will be in the paleontological domain, any genetic (pedigree-related) classification remains an interpretation. The third of the *methods* in the subsequent text acknowledges this limitation.

27   One of AMM's specialisms and the subject of Lectures 3, 11 and 17.

of von Baer, to assume its modern form, and it is only of recent years that it has revealed its enormous powers as an instrument for the solution of the hard problems of phylogeny. Von Baer propounded the doctrine of animal types quite independently of Cuvier; and he went further than his illustrious contemporary, for he showed that not only as far as structure was concerned might animals be arranged under certain definite types, but that each type had its own mode of development, and that all the animals included in any one type agreed with one another in the fundamental features of their development. Thus the several members of the type of animals known as *Vertebrata* all develop in a fundamentally similar manner; in their earlier stages they resemble one another so closely that it is often no easy matter to distinguish one from another, the characteristic differences not appearing till the later stages of development. A good example of this is given by von Baer himself, as follows *"In my possession are two little embryos in spirit, whose names I have omitted to attach, and at present I am quite unable to say to what class they belong. They may be lizards, or small birds, or very young mammals, so complete is the similarity in mode of formation of the head and trunk in these animals. The extremities, however, are still absent in these embryos; but even if they had existed in the earliest stages of their development we should learn nothing, for the feet of lizards and mammals, the wings and feet of birds, no less than the hands and feet of men, all arise from the same fundamental form."*

As soon as the idea of the mutability of species was fairly grasped, the reason of these embryological resemblances became apparent, and it was seen that if two species or groups of animals develop in the same way this is due to their being genetically allied to one another.

Further consideration of this question led ultimately by a series of steps, which space forbids me to notice, to the promulgation of what is known as the recapitulation hypothesis.[28] This which is found lurking in the works of von Baer,[29] but first received definite expression from Fritz Müller,[30] is commonly stated thus: *The development of the individual is an epitome of the development of the species*—i.e., the actual changes through which an animal passes in its development from the egg to the adult represent in a condensed form its genealogical history or pedigree.

Thus embryology provides us with a completely new and very accessible clue to the determination of animal genealogies—a clue of such extreme value that we cease to wonder at the immense importance attached to it by the modern zoologist, or at the time and perseverance that are expended in attempts to penetrate its mysteries. For the clue thus afforded is the one on which we have chiefly to rely in our efforts to unravel the knotty problems of genealogy, and it is a clue in the following up of which extreme caution is necessary. If the recapitulation hypothesis were strictly true, if the ontogenetic history of an individual were an accurate and complete reflection of its specific history or phylogeny, then our task would be a comparatively light one. But it is not so. As Fritz Müller has pointed out with admirable clearness[31] the ontogeny of an individual is not a simple abbreviated history of its phylogeny, but a history

---

28   Dealt with at length, with evidence, in Lectures 10 and 13.

29   In fact, von Baer came to reject recapitulation in favour of a theory in which more complex animals develop by diversion from simpler, ancestral forms.

30   See footnote 31.

31   {M} *Für Darwin* (English translation), chap. Xi., pp. 110 *et seq*

obscured and falsified by a variety of causes, the most important of which are the tendency to shorten the processes of development, thereby causing the omission of stages which may be of extreme historical significance; and, secondly, modifications introduced by natural selection into the ontogenetic history in consequence of the struggle for existence which free-living larvae have, in common with adult forms, to undergo.[32]

Thus, in spite of the powerful aid afforded us by embryology, our problem is still one of extreme difficulty, requiring for its solution great patience, great manipulative skill, and, above all, the faculty of distinguishing accurately between essentials and accidentals. A good beginning has already been made, mainly through the energy of zoologists in other countries than our own. England has not yet contributed her fair share towards the solution of a problem the conception of which was first rendered possible by the magnificent labours of one of the greatest of her children, Charles Darwin.[33]

---

32   This second cause is actually difficult to reconcile with the principle of natural selection if it has to apply to embryos as well as to *free-living larvae*: it is hard to see how *the struggle for existence* could impinge on an embryo, other than by the presence of a trait lethal to its continued development or birth. AMM is clearly attributing this suggestion to Müller but does not discuss the limit to its application.

33   The appeal to nationalism which begins the sentence seems anachronistic now but must have been intended to encourage the audience to greater things. It may also appear excessively modest given the influence of Wallace, Owen, Huxley, Lyell, Galton ... However, later lectures (especially 9) show how European scientists had led the development of cell theory and dominated studies on the mechanisms of inheritance through most of the 19th century. It cannot be chance that Darwin's name appears as the final words of the lecture. It is a dramatic flourish which one can imagine being delivered with a thump of the lectern or some other theatrical gesture. It is clearly designed to lead the audience along the correct path of understanding. If ever an exemplar were needed of *great patience, great manipulative skill, and, above all, the faculty of distinguishing accurately between essentials and accidentals*, it would surely be Darwin.

LECTURE 2

THE INFLUENCE OF ENVIRONMENT ON THE STRUCTURE AND HABITS OF ANIMALS

L
ike the first lecture, the audience for this lecture may have been largely students, although perhaps from a wider range of backgrounds and stages of study. There were several student debating societies and it is not clear from AMM's papers in the Manchester University archives exactly which this was. *"The Men's Union originated in a discussion club, which expanded after 1861 to include a Debating Society and a range of subject-specific societies. It ran a clubhouse and later a magazine"* (www.manchester. ac.uk/discover/history-heritage/history/buildings/students-union/).

This lecture contains the first of many references to August Weismann, one of the towering figures of 19th-century biology. Weismann consolidated and advanced biological thought on many fronts including insect development, embryology, cell theory, germ layers, recapitulation, reproduction, heredity, Darwinian evolution (he wrote a biography of Darwin), freshwater invertebrates and lake ecology. He was an exact contemporary and close friend of Ernst Haeckel, although their ideas were not always congruent. His long active life—some 60 years of observation, experiment and theorising—and many written works provide a line of continuity from von Baer and others scientists active in the first half of the century, through Darwin, Kölliker, Müller and the Naples research centre in the second half, and on into the fin-de-siècle, pre-Mendel, pre-genetic period during which AMM was writing. He continued to publish into the first decade of the next century, until his academic retirement in 1911.

Weismann was a prolific letter writer, corresponding and often vehemently arguing with many of the significant figures cited by AMM in these Lectures. There is no evidence of direct communication with AMM but his influence is not in doubt. He attended the 1887 meeting of the British Association for the Advancement of Science, held in Manchester, for which AMM was the local secretary. He spoke on polar bodies and reduction division and debated (against) the inheritance of acquired characteristics. For more on Weismann and his connections, see Churchill FB, *August Weismann: Development, Heredity, and Evolution* (Harvard University Press, 2015).

AMM's championship of Darwin, although not the principal subject of this lecture, is emphasized by the way he dissociates himself from Herbert Spencer's supposed Lamarckian beliefs in the antepenultimate paragraph. However, he makes use of Spencer's idea of "correspondence" between an organism's characteristics and its environment and the implications this has for survival. As suggested in the comments at this point, there was a wider context to this discussion—that of what life actually is and how it might be defined. AMM avoids that thorny issue.

THE INFLUENCE OF ENVIRONMENT ON THE STRUCTURE AND HABITS OF ANIMALS

I PROPOSE to ask your attention for a short time to the consideration of the dependence of any body, living or dead, organised or unorganised, organic or inorganic, on its environment, *i.e.*, on the external conditions surrounding and acting on it at a given time, and I propose to consider this more especially with regard to animals.

Of environment, as affecting inanimate objects, many admirable examples are afforded us by Physics. For instance, we speak of water as a definite substance having a definite existence, but we know well enough that for the existence of water as such certain definite relations of temperature and pressure are essential. At the normal pressure water will not exist as water at a temperature below 0°C. Similarly, if we raise the temperature beyond a certain limit, 100°C., the existence of water as water again becomes impossible. If we alter the pressure the relations change again, for when the pressure is increased beyond the normal amount we find that the water will remain liquid at a temperature at which it would previously have assumed the solid form of ice.

Let us take another example from Chemistry. Chemical formulae have a very definite appearance, yet every chemist knows well that the reactions they indicate will only occur provided certain conditions of environment be fulfilled, especially those of temperature and pressure; and that a change of no great extent in environment may cause either a completely different reaction or even an actual reversal of the previous action. For instance, Mercury at 300°C. combines with oxygen to form oxide of mercury; now if the temperature is raised to 400°C. this action is reversed.

I have purposely chosen simple and familiar examples in order to illustrate with as little loss of time as possible what we mean by environment. The points to which I wish to direct attention are these: Firstly, that the actual condition of any body may be regarded as the resultant of the several forces acting on it at the time; secondly, that therefore, if we know all the conditions of environment, we can calculate with certainty the actual condition of the body; thirdly, that a change of known amount in one of the elements of the environment will produce a definite and calculable change in the conditions of the body acted on. In other words, that not only has environment a marked and definite action, but that this action is capable of exact measurement.[34]

34　This argument, that the present state of things is wholly dependent on circumstances and can be safely predicted if we know the conditions, is reminiscent of the causal determinism suggested by Laplace a century earlier. AMM's argument is interesting for two reasons: (i) It lacks the appreciation of chaos theory which we would now incorporate into our understanding of causality, i.e., that when more than a very few conditions interact, the ability to predict the outcome is rapidly lost; (ii) AMM uses

Now let me turn to the more immediate subject of my paper, viz., the influence of environment on the structure and habits of animals. It will, of course, be at once conceded that changes in environment not only can but do produce changes of very great magnitude in both the structure and habits of animals—for example, domestic breeds of cattle, &c.; but in most of these cases the conditions are very complicated, and it is impossible to assign to each factor its true value.[35] What I wish to prove to you, if possible, is that change of environment not only produces changes of structure and habit, but also produces definite and calculable changes, so that a definite change in environment is always followed by a definite change of structure, and the result can be predicted with as much certainty as a chemical reaction.[36]

For this purpose all ordinary examples fail us because the environment is too complicated, e.g., pigeon fanciers can produce at pleasure a bird with any number of feathers in the tail they wish, and with almost any variety of plumage, but as to which of the elements of a complicated and artificial environment the result is due we are completely in the dark.[37] For this purpose it is necessary to choose our examples carefully, and the number of available ones that are not of too technical a character for general discussion is very limited.

We have seen that it is very easy to establish a general relation between changes of environment and changes of structure, and that the one series is followed at once by the other. Now this general relation is not sufficient for our proposition, much more rigid proof being necessary before we can accept it. I wish first to ask your attention to one or two cases in which this proof appears to be present, and in which we are able to establish not only a general but a direct causal relationship between environment and structure; as unquestionable indeed as that which holds in the case of oxide of mercury.

Now we know of at least one well authenticated case in which the structure of an animal responds in as precise a manner to simple and definite changes in one of the elements of the environment as certain chemical reactions do to alterations of temperature.

---

examples from the physical sciences to illustrate the point he wishes to make (and will presently develop) about biological organisms. The clear implication is that he wants the audience to view biology as a manifestation of physics and chemistry – no need for a *vital spark* or transcendent life force. He makes no explicit reference to this, nor does he crave the audience's indulgence. Kraft and Alberti (*Studies in the History and Philosophy of Biological & Biomedical Sciences* 2003; **34**: 203–236), in analysing the state of academic biology at the end of the 19th century, note that: "*The tenets of this 'new' biology [as formulated by T.H. Huxley and his allies] included an emphasis on the physico-chemical basis of the living world, microscopic analysis, the study of man as part of the natural world and, by the late 1870s, Darwinian evolution.*"

35   Chaos theory is anticipated, after all!

36   Thus we are required to accept the deterministic argument but (next paragraph) on the basis of examples deliberately chosen to avoid complexity (and the consequent unpredictability.)

37   Breeders make calculated and reliable decisions in order to produce the results they desire, and to that extent the outcome is deterministic. It is not clear how AMM means this to cover unknown *elements in the environment.*

Artemia salina[38] is a small aquatic crustacean about half an inch in length, and somewhat like a small shrimp in appearance; its distribution is peculiar, inasmuch as it is only found in water containing from 4 to 8 per cent. of salt, a much higher proportion than is found in the sea. It consequently can only occur in salt lakes and in such places as the salt pans of Lymington[39] and other similar places where the sea-water is allowed to gradually concentrate by evaporation. Another form, *Artemia Milhausenii*,[40] is known, whose habits are still more peculiar, for it is only found in water containing at least 25 per cent. of salt; its distribution is consequently a very limited one, since it can only occur in certain inland lakes in which, owing to special circumstances, the water attains this very high percentage of saltness.

Artemia Milhausenii differs from *Artemia salina* in so many points of structure that the two forms have always been unhesitatingly ranked as separate species. Of these differences one will suffice for the present purpose. In *Artemia salina* the tail is split at the end into two long pointed lobes, each of which bears a large number of bristles and hairs of a peculiar structure; the tail in the other species, *Artemia Milhausenii*, is abruptly terminated, and presents only a very slight separation into two blunt rounded lobes, which lobes are completely devoid of hairs or bristles. There are besides this many other points of difference between the two forms which need not here be noticed in detail.

Schmankewitsch has recently made the very interesting discovery that the anatomical differences between the two species depend directly and solely on the different percentages of salt in the water they inhabit.[41] He finds if he takes *Artemia salina* living in water containing 4 per cent. of salt, and gradually increases the percentage of salt, that the structure of the *Artemia* becomes gradually modified;[42]

---

38   Brine Shrimp.

39   Lymington on the Solent, on south coast of England, was the largest producer of sea salt during the 18th century until cheaper salt became available from mines in Cheshire. The Lymington salt industry is thought to date back to Roman times.

40   The use of an uppercase first letter for the specific epithet presumably reflects its origin in a proper name.

41   Published as Schmankewitsch WI, Über das Verhältnis der *Artemia salina* Miln. Edw. zur *Artemia salina* mulchauseni Miln. Edw., und dem genus *Branchipus* Schaff. *Zeitschrift fur wissenschaftlich Zoologie* 1875; **25**: 103–116.

42   It is not clear from the wording of this passage and subsequent ones how the observed changes in the shrimps came about. Several possibilities can be imagined: (i) individual shrimps underwent the morphological change (as adults); (ii) the structure of individuals within a breeding population changed gradually over time (i.e. a statistical population change); (iii) there was a selection process such that individuals with particular morphologies survived in certain media whilst others did not. The original Schmankewitsch report indicates that shrimps were cultured in artificial media of varying salt concentrations over a number of generations and that the effects were induced in individuals as they matured (from nauplii, the earliest larval stage), directly by the conditions of the medium. We could describe this as an epigenetic, phenotypic response, not genetic selection, which is just what AMM was attempting to illustrate. Whether the morphological change is reversible in individuals remains unclear from Schmankewitsch's report, as does the time it takes

the lobes of the tail become shorter and shorter, and the hairs diminish in size and number. Ultimately, by increasing the strength of the salt up to 25 per cent, he succeeded in completely transforming *Ariemia salina* into the distinct species *Artemia Milhausenii*. Not only can the transformation of one species into another be effected in the laboratory, but it can be demonstrated to actually occur in Nature. Of this fact Schmankewitsch has recorded the following very striking instance: Two lakes in the neighbourhood of the Black Sea were separated from one another by a dam; of these the upper and larger lake contained water with about 4 per cent. of salt in which *Artemia salina* occurred in great numbers. The water of the lower lake contained about 25 per cent. of salt and had no *Artemia* living in it at all. From some cause or other the dam burst, and the water from the upper lake rushed down into the lower and smaller one, large numbers of *Artemia salina* being carried down by the flood. The dam was repaired, and the water of the lower lake, the strength of which had been reduced by the flood to about 8 per cent, of salt, began slowly and gradually to regain its former strength. Specimens of the *Artemia* were taken out and examined from time to time, and it was found that as the percentage of salt rose the tail-lobes of the *Artemia* gradually got shorter and blunter, the hairs got smaller and fewer, and the whole animal gradually became transformed from *Artemia salina* to *Artemia Milhausenii*. In about three years' time the water had regained its former strength of 25 per cent., and the *Artemia* had become completely transformed into *Artemia Milhausenii*.

Not satisfied with this, Schmankewitsch next tried the reverse experiment. Starting with *Artemia Milhausenii* he gradually decreased the strength of the water in which they lived by adding fresh water, and had the satisfaction of finding that as the percentage of salt diminished so the tail-lobes of his *Artemia* became longer and their hairs more abundant, and when the strength of the water was reduced to about 6 per cent. all the *Artemia* were converted to typical specimens of *Artemia salina*. Having thus satisfied himself of the causal relations between the two species of *Artemia* he determined to push his experiments still further. He therefore went on adding fresh water after he had obtained *Artemia salina,* and by this means converted *Artemia salina* into a well-known fresh-water form that had previously been considered not only a separate species but even a distinct genus—*Branchipus.*[43]

---

for changes to be observable. In later studies, notably that of Gilchrist (*Proceedings of the Zoological Society of London* 1960; **134**: 221–235), shrimps were cultured for several generations with parents maintained in the same media as those in which the young were to be reared. Under these conditions the shrimps reproduced with a generation time of 3–4 weeks. Crucially, Gilchrist studied three subspecies of *Artemia salina* from different parts of the world and found that the response to salinity (in terms of body size and segment proportions, as well as tail morphology and hair number) was modified by both sex and subspecies. This shows that the picture is not as simple as AMM and other authors of the time thought and that there may indeed be a genetic, and thus environmentally selectable, component in the response.

43   The identity of this genus is not clear, for *Branchipus* is not currently recognized as one. Schmankewitsch identifies them on the basis of having nine limbless abdominal segments. Later, Gilchrist (see note 42) understood there to be just the one species, *Artemia salina*, but realized that the different varieties produced under different

From these very striking experiments it would appear that the relation between the structure of the animal and the degree of saltness of the water it inhabits is a perfectly definite one, so much so, that if the percentage of salt be told us, we are able to foretell with perfect confidence and certainty, not only what will be the general character of the animal inhabiting it, but such minutiae as the size and shape of its tail-lobes, and even the number of hairs borne by these lobes.

Here then is a very striking proof of the direct action of environment on structure; a phenomenon as definite as a chemical reaction. Nor must the case be made light of or explained away by saying that it simply shows that zoologists were at fault in describing these forms as distinct genera and species, and that the experiment showed them to be mere varieties; for it must be noticed that the characters by which the three forms differ are characters of the very kind that are invariably employed by zoologists to distinguish the genera and species of Crustacea; while the ultimate criterion of distinct species—fertile breeding—bears the matter out also, for no one of the three forms is capable of living in the medium dwelt in by either of the others.

We also know of several cases among butterflies in which, out of a very complex environment, we are able to put our finger on the one element which is the actual determining cause of the structural changes that we observe. One of the most important of these examples is furnished by two butterflies of widely different appearance, *Vanessa levana* and *Vanessa prorsa*.[44]

Vanessa levana has a red ground colour, with black spots and dashes and a row of blue spots round the margin of the hind wings; *Vanessa prorsa* is deep black, with a broad yellowish-white band across both wings and with no blue spots. These were formerly called distinct species, but have recently been shown to be varieties of one and the same species. The relation between them is a very definite one, and they are what is called seasonally dimorphic: that is, they are double brooded, and have two generations in the course of the year—a winter brood and a summer brood. In the summer or autumn Prorsa is alone met with, and the pupae developed from the autumn brood of eggs of Prorsa hybernate during the winter and hatch in April or May of the following spring, not as Prorsa, but as Levana. This in turn produces a brood of eggs which develop into Prorsa again. This phenomenon was investigated by Weismann,[45] who found it was to be attributed to the direct action of cold and heat. He found that if the pupae of Levana, which would ordinarily have produced Prorsa, were put in a refrigerator for one or two months the butterflies emerged partly as Levana and partly as another form, intermediate in many respects between Levana and Prorsa. Thus the direct action of cold causes pupae, which in the normal state of things would have produced Prorsa, to produce Levana.

The reverse process, however, is not successful, for heat does not change Levana into Prorsa. The explanation of this is, perhaps, not out of place, although it has no

---

degrees of salinity may not always be capable of interbreeding. By modern taxonomy, brine shrimps, including *Artemia*, belong to the Branchiopod class of Arthropods, along with water fleas such as *Daphnia*.

44   The modern generic name is *Araschnia*, of the family *Nymphalidae*.

45   *Studien*, Part I, 1875.

direct bearing on the subject in hand. Levana is the parent[46] form, and in the glacial epoch there was only one single brooded variety and one generation in the year. The lengthening of the summer gave rise to a second generation, and the gradual action of climate caused the Prorsa form to arise. Now, Levana being the parent form, the effect of cold in causing pupae that would normally have become Prorsa to become Levana is simply a case of reversion. The reverse action could clearly not occur. It is a long time since the glacial epoch was over, and therefore probably a long time since the two forms Levana and Prorsa were established; hence the reconversion by cold is only partial.[47]

A similar example is afforded by another butterfly, *Pieris napi*, the green-veined white. This is also double brooded, and occurs as two varieties, each producing the other. The winter form is rather smaller, and has more black at the bases and less at the tips of the wings than the other variety. The green veining on the under surface is also more marked. If the pupae of the summer brood are placed in a refrigerator for three months, and then in a hothouse, they hatch in about three weeks as the winter form.

These illustrations I have cited show the direct action of cold producing a result which can be calculated on beforehand with certainty. By the long-continued application of heat the summer form has been produced, and if the environment is altered and cold substituted for heat, we reverse the action and bring back the summer form to the original winter one.

Now let me refer to an instance in which the evidence is as yet incomplete— an instance somewhat more technical than the previous ones. I choose it purposely because it is of a somewhat abstruse kind, and will therefore serve to show that some of the more intricate problems concerning the exact nature of the relation between environment and structure and habits may be well within our grasp.

Many animals possess the power of changing colours—*e.g.*, the chameleon, frog, many fish, and the cuttle-fish. This power depends on the presence of chromatophores or pigmented bodies found in the skin, which have the power of changing shape. These are of different colours and are arranged in layers, some near the surface, some deeper; the light yellow cells being most superficial, then the red and brown, and the black deepest of all. If the superficial ones contract, the deeper ones will become more apparent, and *vice versâ*. These changes of colour always bear a relation to the surface on which the animal is placed at the time, and are therefore supposed with good reason to be protective. Many attempts have been made to explain this.

Lister believed that the irritation which excites the action of the chromatophores does not act on them directly, but through the intermediation of the optic nerve. It is essential for this action that the eye and optic nerve should be present and healthy,

---

46   This must mean *normal* or *generally occurring*, rather than simple parenthood. In other words, Levana is the original form from which Prorsa is derived.

47   This argument is somewhat thin and circular. It is unclear why the amount of time that has passed since the epoch when the cold form predominated should, by itself, prevent the warm form from being a possibility. The phrase *simply a case of reversion* tells us nothing. AMM lacked the concept of terminal differentiation which, based on our understanding of cellular pluripotency, we might now apply to such a situation: one form is reversibly differentiated and the other is not.

the action ceasing if the eye is destroyed or the nerve cut. In support of this view he pointed out that blind fish are always dark. Pouchet suggested two possible paths of impulse—the spinal cord and the sympathetic system. By dividing the spinal cord the path of impulse was shown to pass along the sympathetic nerves. Again, by dividing some branches of these, he produced zebra-like markings of the skin. Dewar showed that different colours of the spectrum influence the eye differently and cause electric currents of different intensity according to the light employed. These currents are strongest with yellow light, weakest with purple, and nil with black. Semper pointed out that, though electric currents and nerve currents are far from being the same thing, we may not unfairly assume that the nerve current, like the electric current, is greatest with yellow light and nil with black. Thus a black ground, which absorbs nearly all the light, and therefore reflects little or none, will stimulate the eye to a very faint extent, and hence cause a very feeble current, if any at all, and so the chromatophores will remain unstimulated, and as in the case of the blind fish, the fish will remain dark. On the other hand, light from a red or blue object will cause a current of rather greater intensity, and will cause the brown or red chromatophores to expand, and so more or less conceal the black ones. Finally, yellow light will cause the strongest stimulation and all the chromatophores will give out processes; the yellow ones, being most superficial, will be the most conspicuous, and will hide the deeper and darker ones to a greater or less extent.[48]

In the foregoing remarks I have endeavoured to show, first in a general way, and afterwards by the consideration of a few selected examples, that the relation between an animal and its environment is a very definite and direct one, and in many cases a calculable one.[49]

We have seen from certain special cases that an organism responds promptly and with certainty to certain changes in its environment. Can we not extend this and speak of the actual structure of an animal as the resultant of the various forces acting on it, both external and internal? Small changes in environment seem to give rise to results altogether disproportionate with themselves; but then we must remember that the bodies of animals are of very complex chemical composition and in very unstable equilibrium, and in such cases a small force may very easily cause a very considerable rearrangement.[50] Turning to psychology, we see that the tendency of

---

48   The studies by Lister, Pouchet, Dewar and Semper mentioned in this paragraph are unreferenced and difficult to validate. An article by Dr James Weir Jr on *Animal Tinctumutants* appeared in *Popular Science Monthly* in January 1895 and refers, without citation, to studies by Lister published in 1858. It also mentions work by Pouchet and Semper.

49   The contrast between the two sets of examples is striking: The environmentally-induced changes in *Artemia* and the two butterflies are developmental and permanent. The colour changes mediated by chromatophores in the other species are reactive and temporary responses to the animals' background. Although AMM makes no attempt to note the distinction, the predictability of both types of response makes his point for him.

50   This anticipates, surely, the *butterfly wing* argument of chaos theory.

modern research is to show that whatever we do or think is merely the resultant of what we have done or thought or suffered previously.[51]

It is not difficult to show that, even in our own worldly transactions, changes of environment often produce not only direct and immediate changes and readjustments, but also definite and calculable ones.

Thus, as an example, let a be a merchant, and b his purse: the combination ab will at once strike you as a natural and stable one. Indeed there is something about the word merchant that seems to suggest inevitably a well-lined purse. Now let c be a highwayman, and d his pistol: the combination cd is again recognised as a natural and stable one. Now bring the compound ab into the presence of the compound cd, and mark how the stability of the former is shaken; ab, a previously stable compound, becomes at once curiously unstable; the merchant and the purse that we were wont to view as inseparable are split asunder at once, and the several elements become rearranged in a manner that finds perfect expression in the formula

$$ab + cd = a + bcd.$$

Again to take another instance which will remind us of the operation known in chemistry as double decomposition: Let a be a small boy, b a penny in his pocket, c an old woman with an apple stall, d the apples. The combination ab is a remarkably unstable one in the presence of cd, and the affinity between a and d is very great; an interchange takes place as follows:

$$ab + cd = ad + bc.$$

These examples are no doubt somewhat crude, yet they will serve to indicate that if we could define all the conditions we should be able to draw up an equation for a man's life and show how, given the man and the conditions acting on him, there was but one path open to him.

In conclusion let me refer to a statement by Herbert Spencer in which he discusses the influence of environment on animals; starting, however, from a standpoint that I cannot accept, viz., the Lamarckian hypothesis of animals responding by efforts of their own to changes in environment.[52]

The more complete the correspondence between external and internal changes, the more complete the power of response to changes of environment, the higher is

---

51   Modern psychology and neuroscience continue to be perplexed by causation (in thought and deed) and the nature of free will.

52   Despite his acknowledgement of Darwin's work, Spencer is often portrayed as having remained wedded to Lamarckian ideas of the inheritance of acquired characteristics, especially to explain the complexity of higher animals and even the structure of societies. This is the line that AMM seems to take in this paragraph, even extending it (*efforts of their own*) to apparently include intentional change. The Spencer scholar Mark Francis (University of Canterbury, New Zealand) dislikes this simplistic reading of Spencer and the implication that *Lamarckism* was antithetical to *Darwinism*. He writes (personal communication): "*Both Darwin and Lamarck were essentially interested in species change, which was a subject that did not grip Spencer so it is not useful to place him in that debate. ... A broader approach to evolutionary theory would not place Spencer with either Darwinians or Lamarckians. ... Since Spencer was primarily a psychologist and*

the grade of the animal and, as a rule, the longer is its life and the less its fertility. In this respect man stands the highest, for instance, in the power of counterbalancing changes of temperature, and in the supply of food consequent on seasons.[53]

Perfect correspondence to change in environment would be perfect life, eternal existence. Death, whether from natural decay, disease or accident, is simply due to failure of the organism to respond effectively to changes of environment.[54]

---

sociologist who believed that neither the psyche nor the government could direct change, it is difficult to call him Lamarckian in the way one could call a Fabian socialist or Robert Owen Lamarckian."

53   The previous paragraph leads the reader to expect a direct quote from Spencer. The style of *this* paragraph feels sufficiently different from that of the rest of the lecture to indicate that it is not by AMM, yet there is no indication that it comes from elsewhere, nor is there a citation of any kind. It is as if CFM overlooked the need for source attribution or that an error was made when the manuscript was prepared for publication by the original publisher (David Nutt, London). Alternatively, this was an occasion on which CFM found a gap in the lecture manuscript and made an attempt to complete it (see his introduction to the volume). Mark Francis has helpfully directed me to Chapter 6 of Spencer's *The Principles of Biology*, first published in 1864, as the likely source of the material. Chapter 6 of Volume I, Part I, in the revised and enlarged Edition of that work, published in 1898, does not contain directly matching text but has closely related statements, including *"the completeness of the life will be proportionate to the completeness of the correspondence"*; *"the greater correspondence thus established ... must show itself both in greater complexity of life, and greater length of life: a truth which will be fully perceived on remembering the enormous mortality which prevails among lowly-organized creatures, and the gradual increase of longevity and diminution of fertility which we meet with on ascending to creatures with higher and higher developments"*; *"as the life becomes higher the environment itself becomes more complex."* Regarding the meaning of *correspondence*, the previous chapter of *The Principles of Biology* ends with the following: *"... consider the internal relations [of characteristics] as 'definite combinations of simultaneous and successive changes'; the external relations [of sequences and environmental factors] as 'co-existences and sequences'; and the connexion between them as a 'correspondence'."*

54   This paragraph also seems not to be taken from Spencer. Something very similar appeared in *Popular Science Monthly* five years before this lecture was delivered: *"The degree of correspondence between the living thing and its environment is also its degree of life, inasmuch as in effect it connotes an increase in the number and in the mutual dependence of the vital changes which constitute life. If to all changes of the environment there were opposed, as a counterbalance, changes in the living thing, natural death would be no more, nor death by disease or by accident, all of which are signs of a lack of correspondence"* (Dr E Cazelles, *Modern Philosophical Biology*, translated from the French by J Fitzgerald, in *Popular Science Monthly* 1876; **8**: 710–723). *Death* is the subject of Lecture 12.

LECTURE 3

ON EMBRYOLOGY AS AN AID TO ANATOMY

This lecture, given to the Owens College Medical Students' Debating Society, is a masterful presentation of a factually based argument. Dissatisfied with contemporary answers to an intriguing anatomical question, AMM turns to his specialism, embryology, to provide a solution. His answer, which he modestly refers to as a *clue*, comes wholly from an understanding of developmental processes, neatly avoiding any necessity to rationalize body structure in terms of ultimate function or purposeful design.

The second half of the lecture, in which the embryological origin of body components is used to explain the derivation and connection of the cranial nerves, illustrates what an effective teacher AMM must have been. He describes in just a few paragraphs the complex anatomy of the vertebrate body plan. It uses no diagrams, yet after reading it one has a complete and informative picture in one's mind. The argument is factual and convincing but leaves room for knowledge yet to come.

Although there is no recourse to graphics here, AMM certainly made use of visual aids in his teaching and research work. A picture of his Manchester Laboratory taken around 1890 (reproduced in Kraft & Alberti, *Studies in the History and Philosophy of Biological and Biomedical Sciences* 2003; **34**: 203–236), shows a prominently placed table with a neat row of Ziegler Studio wax models of eleven stages in the development of *Amphioxus*. There are also documentary records (Manchester University archives) of him specifically requesting significant sums of College and University money to purchase such items, sometimes in association with the institution's museum collection.

Kraft & Alberti highlight the central importance of anatomy in late 19th-century biology and the lack of a clear distinction between the classroom, the lab and the museum, and between experimental and descriptive research. They suggest that AMM's appointment to the Manchester chair, in competition with an established museum curator with strong but traditional teaching credentials, reflected changing priorities and a move towards professionalization in academic roles. As this lecture demonstrates, AMM was a skilful anatomist but he could exploit his understanding of embryology to move his subject on from description towards investigation.

AMM acknowledges his audience's background more than once in this lecture, apologizing for covering material which may have been familiar to them from their medical studies and even offering sympathy for the tedium associated with mastering it. He certainly knew how to engage his listeners and win them over.

LECTURE 3

ON EMBRYOLOGY AS AN AID TO ANATOMY

T HERE are perhaps few sciences, or branches of science, in which progress has been so rapid and whose importance has been so suddenly and unexpectedly revealed as that of embryology. When we consider the vast number of papers, memoirs, and treatises that are continually being poured forth it is difficult to realise that embryology, as we now understand it, practically commenced in 1837 with the publication of von Baer's treatise on the Development of Animals.[55] The Study of Invertebrate Embryology is still more recent, and in its modern form is only about a dozen years old. No one who compares our text-books of biological science of the present day with those of twelve or even six years ago can fail to be struck with the great change effected by embryology in the methods of teaching this science.

At the outset men hardly knew what to do with this new branch of science; for its study the delicate methods used in histological research were necessary, and hence for a time embryology was included as a branch of physiology and taught by physiologists. But anatomists soon saw that this science belonged by right to them, and that it was absolutely essential for them to master it. A zoologist of the present day relies quite as much on his razor as his scalpel in his endeavour to comprehend the structure of animals.[56]

It is not difficult to see how it is that embryology has so rapidly attained such importance, the reason being that it has afforded an altogether new and very tangible clue to many of the most interesting problems of biology which have been thereby first brought within the comprehension and grasp of man. Thus the zoologist has in embryology obtained a very important clue to the determination of the affinities of animals;[57] in fact we should hesitate to definitely assign a place to some animals whose development was unknown. To the philosophical biologist a clue is thus given to the determination of what, to him, is the problem of problems—viz., phylogeny, or the genetic connection between one animal and another and between their various classes. While to the anatomist embryology offers an explanation of many otherwise completely unintelligible anatomical facts. To the members of a medical society anatomy means, and perhaps I may be permitted to say rather too often means, simply human anatomy. I therefore choose for closer consideration a problem, the main conditions of which are familiar to every one who has dissected a head and

---

55   von Baer, *Über Entwickelungsgeschichte der Thiere: Beobachtung und Reflexion* [*On the Developmental History of Animals; Observations and Reflections*], published in 1828.

56   Presumably the implication here is that a razor cuts slices, suitable for the microscope, while a scalpel dissects.

57   The link between embryology and pedigree is developed throughout the lectures, reaching its most explicit analysis in Lectures 13 and 17.

neck. The problem to which I refer is that of the nerve supply of the muscles by which the various movements of the eyeball are effected.

Whoever has dissected the orbit cannot fail to have been struck with the large number of nerves which not only traverse it but supply parts contained therein; and also with the additional fact that these are either distinct nerves or branches of distinct nerves, and not merely separate branches of one or two cranial nerves. Thus, of the twelve cranial nerves no less than six are distributed wholly or in part to the contents of the orbit. When we consider that of the remaining six, two—the olfactory and the auditory—are nerves of very special function and limited distribution, and that another one—the spinal accessory—has but slight claim to be considered a cranial nerve at all, the fact becomes still more striking. A further and far more significant fact than the above is that of these six nerves four are exclusively distributed to the orbit, and send branches nowhere else. Of these four nerves one is the optic nerve, while the other three supply the muscles by which the movements of the eye are effected. I must ask your indulgence for a moment if I refer to such an exceedingly familiar fact as the arrangement and distribution of these muscles and their nerves. I only do this because it is absolutely necessary to the argument I wish to put before you that this arrangement and distribution should be fresh in your minds.

The movements of the eyeball are effected by six muscles—four recti and two obliqui. Of these, speaking roughly, the superior rectus turns the eye upwards; the inferior rectus turns it downwards; the internal and external turn it inwards and outwards respectively; the superior oblique rotates the eye outwards and downwards, the inferior oblique outwards and upwards. Of these six muscles four are supplied by the third cranial nerve, while each of the remaining two has a separate and distinct cranial nerve to itself, the fourth cranial nerve supplying the superior oblique muscle, and the sixth cranial nerve the external rectus.

Now it is a very remarkable fact that this small group of muscles should have three cranial nerves to supply them, and still more remarkable that these nerves should do little or nothing else. The third nerve, besides supplying the muscles mentioned, also supplies the elevator muscle of the upper eyelid and sends branches which penetrate the eye and supply the ciliary muscle *and* iris. The fourth and sixth nerves, however, do nothing else, and form unique instances of nerves with separate origins from nerve centres distributed exclusively to single muscles, and it is very curious that these two muscles should both be connected with the eye. In order to complete the conditions of the problem it is only necessary to add that the distribution of the nerves and muscles I have just noticed is not peculiar to man but occurs throughout the whole vertebrate series almost without exception. Wherever we meet with a vertebrate having well-developed eyes, there we find this same arrangement of muscles and nerves. The exceptions to this rule that have been quoted are probably only apparent. The few exceptions that are quoted are worth noticing. In *Amphioxus*[58] the eye is a mere spot of pigment and has no muscles at all; in *Myxinoids*[59] the eyes are small and the eye-muscles imperfect, but all three nerves, though very small, are

---

58   A marine, sand-dwelling invertebrate chordate, sometimes known from its shape as a lancelet.

59   Presumably the *Myxinidae*, or Hagfish: eel-shaped marine fish with a skull but no vertebrae.

found to be present and with their usual distribution; in *Lepidosiren*[60] there are no oblique muscles, and the third, fourth, and sixth nerves are said to be absent; the observations on this point are, however, imperfect. In some of the lower *Amphibia*, the newt for instance, all three nerves are present, at any rate proximally, although distally they may become closely bound up with branches of the fifth nerve. With these exceptions, in all fish, amphibia, reptiles, birds, and mammals these six muscles are all present and all have their typical nerve supply. Let me again remind you that these nerves supply nothing else; the third nerve supplies no other parts than those I have mentioned; the fourth nerve never supplies anything besides the superior oblique muscle; the sixth nerve invariably supplies the external rectus, but may also supply two other muscles when these are present—viz., the retractor muscle of the eye, which is found in some newts, many reptiles and most mammals, except the monkeys and ourselves, and the special muscle of the nictitating membrane,[61] which is met with in birds and reptiles.

Such a state of things so widely spread must have a reason for its existence, but hitherto no satisfactory or even intelligible answer has been given to the problem, and anatomy still leaves us in total ignorance as to why the external rectus muscle has a special nerve all to itself. Explanations have of course been attempted, for the problem cannot fail to strike any one who is acquainted with the anatomy of the parts in question. A common attempt is to say that the movements of the eye are so complex and have to be so extremely nicely adjusted that a complex nerve supply is necessary. This, I would submit, is no explanation at all; it does not tell us in the least why a small group of six muscles should have three cranial nerves to supply them. For we find the same nerve supply in the lowest vertebrates, in which we cannot suppose these movements to be very complex. Nor does it suffice to say that the nerve supply is due to the two eyes having to work together, and the internal rectus of one eye having to work habitually with the external rectus of the other, for in these lowest vertebrates[62] the eyes are situated at the sides of the head, so that they have totally distinct fields of vision and hence do not work together at all. Sir Charles Bell, the illustrious

---

60   *Lepidosiren paradoxa*, the Lungfish of South America, which has gills as a juvenile but lungs in the adult form.

61   A kind of membranous third eyelid which passes across and lubricates the eyeball.

62   This way of referring to non-mammalian vertebrates is common throughout the lectures. It remained in frequent use even into the second half of the 20th century. Its meaning, now as then, is generally implied by usage rather than by consistent definition. Using *high* and *low* to segregate animal groups, or plant groups for that matter, must, in the final analysis, reflect human arrogance and the desire to assign everything to its appropriate rank. This was encouraged by the Bible (Man's dominion over the rest of the biosphere, as provided for in Genesis), and Herbert Spencer (*Principles of Biology*, Williams & Norgate, 1898) uses the terminology freely. The difficulty of interpretation comes when we try to establish congruence between rank, complexity and importance. The only successful way of resolving that conundrum is to consider man as the pinnacle of evolution (as Spencer appears to do) and to invoke purpose as the reason for him being so (see the last paragraphs of Lecture 2 on that point). Such views are alien to modern biology, even if we are content to acknowledge different degrees of complexity amongst contemporary organism. Richard Dawkins

anatomist, who was keenly alive to the interest of the problem before us, attempted an explanation as follows: Observing that during sleep and also during certain involuntary acts, such as sneezing, the eyeball is turned upwards beneath the upper eyelid, and finding by experiment that the recti muscles are voluntary, he was led to regard the oblique muscles as involuntary, especially the superior oblique, whose separate nerve supply was supposed to be thus accounted for. However, as Bell himself points out, this theory involves an assumption which more recent research has failed to verify. He showed that division of the superior oblique muscle causes the upward rolling of the eyeball to be increased, and that therefore if this movement is due to an impulse transmitted along the fourth nerve this impulse must be of such a nature as to cause not contraction but relaxation of the muscle—i.e., that stimulation of the fourth nerve causes relaxation of the superior oblique muscle and not contraction.

I have referred to the views of Sir Charles Bell for two very sufficient reasons: firstly, because this is, so far as I am aware, the only rational attempt that has been made to grapple with the problem before us; secondly, because any account of such a question without reference to the opinion of the great anatomist who did so much to render it possible for us to understand problems connected with the distribution of the nerves would be a gross injustice.

So far we have failed utterly to master our problem; anatomy and physiology have alike failed to give us the clue we require, and for my part I see no reason why, if we had nothing else to help us, this should not always be so, and the problem be ranked as insoluble. It is under circumstances such as these that the anatomist turns to embryology—that sheet-anchor of philosophical anatomy—to which, when in difficulty, he has of late years turned so often and so rarely in vain; and I wish now to ask your attention for a short time while I endeavour to make clear to you the evidence which embryology offers on this point. I must again ask your indulgence if I have to refer to matters many of which are perhaps painfully familiar to you, while others may appear at first sight to have nothing whatever to do with the matter in hand. This evidence applies only at present to Elasmobranch fishes,[63] but the conditions of the problem are absolutely the same as in man.

You all know that there is in the trunk a space between the body walls and the alimentary canal called the pleuro-peritoneal cavity or body cavity, or, still better, the coelom, this space or cavity being absent in the head and neck. That is, if you were to push a sharp instrument through the walls of the abdomen into the intestine the instrument would not pass through solid tissue along its whole course but would traverse a cavity—the coelom—before reaching the intestine. While if you performed a similar operation in the neck the instrument would pass through no cavity but would penetrate solid tissue all the way until the oesophagus was reached. Now the whole body is formed from three cellular layers or strata—the epiblast, mesoblast, and hypoblast[64]—and if we make a diagrammatic section through the body the epiblast will form the outermost layer or skin; the hypoblast will form the lining

---

('The Information Challenge' in *A Devil's Chaplain* (London: Weidenfeld & Nicolson, 2003)) discusses this at length, neatly defining the complexity of an organism by the number of words which would be required to provide a complete description of it.

63   Essentially, the cartilaginous fish: sharks, rays and their allies.

64   Also (and more commonly today) called ectoderm, mesoderm and endoderm.

of the alimentary canal; all the rest will be mesoblast. The body cavity or coelom is entirely within the substance of the mesoblast. In the earliest stage of development the mesoblast is solid and there is no coelom. In the trunk the mesoblast very early splits into two layers which become separated from each other and so give rise to the coelom. It was formerly assumed to be a sharp distinction between the trunk on the one hand and the head and neck on the other, that in the former the mesoblast splits in this manner and that in the latter it does not do so. This distinction is now known not to hold good, for the splitting of the mesoblast has been shown to extend to the head, but the cavities on each side do not meet in the median ventral line. Again, in the trunk the coelomic cavity becomes divided on either side into an upper or dorsal part, and a lower or ventral part. Each of the dorsal portions becomes cut up transversely into a number of segments arranged one behind the other in series,[65] one such segment occurring in each primary body segment or protovertebra. In the head the coelomic cavity is first cut up transversely by the visceral clefts into a series of cavities, one in each visceral arch, and then each cavity divides again into dorsal and ventral portions. The ultimate result is the same as in the trunk, but the order of division is different.

Of these divisions of the coelom in the head, or head cavities as they are called, we are only concerned with the three most anterior. The first head cavity— the premandibular—is in front of the mandible and immediately behind the eye; the second—the mandibular is situated in the mandibular arch; the third—the hyoidean—is in the hyoid arch. Now in the trunk the walls of both the dorsal and ventral divisions of the coelom become converted into muscles, and the cavities of the dorsal division become obliterated owing to the great increase in the thickness of their walls. In the head cavities the walls of both dorsal and ventral divisions become converted into muscles in the same way, and the only difference is that both dorsal and ventral cavities become obliterated instead of the dorsal only. If we bear in mind that the transverse divisions of the dorsal portion of the coelom into segments is obviously part of the general segmentation of the body, and that the similar division of the head cavity is manifestly of the same nature, we see that we may speak of these first three head cavities as indicating three distinct and successive segments of the head comparable to three distinct and successive body segments.

We have now accomplished by far the most difficult and tedious portion of our task, but let me direct your attention to another set of organs. The central nervous system consists at an early period in all Vertebrates of a tube closed at both ends stretching along the back of the animal from head to tail. This tube is not of uniform calibre, but its anterior part—the future brain—is from the first wider than the hinder part—the future spinal cord. The anterior part is bent in the shape of a hook, and presents along its whole length a series of alternate dilatations and constrictions which are much larger and more conspicuous in the brain than elsewhere, but occur along the whole length of the cord. The first dilatation is the fore brain, and from it the cerebral hemispheres project; the second is the mid brain; this is followed by a series of vesicles, rapidly decreasing in size, and called collectively the hind brain; behind this we have the spinal cord. In the spinal cord the dilatations, though only feebly marked, manifestly correspond to the segments, and from each one a pair of spinal

---

65   Now called somites.

nerves arises. Similarly in the brain from each well-marked vesicle a pair of nerves arises. Thus, from the mid brain the third pair of nerves arises and runs backwards to the interval between the first and second head cavities. From the first vesicle of the hind brain arises the fifth nerve, and this passes down between the second and third head cavities. From the second vesicle of the hind brain the seventh or facial nerve springs and runs down behind the third head cavity. The division of the head into segments is thus very clearly and satisfactorily shown, for all the different elements that could afford evidence—viz., brain vesicles, nerves, head cavities, and visceral clefts and arches—all point to the same divisions and agree among themselves.

We have now got all the links of our evidence nearly complete; let us fit them together and give them the finishing strokes. We have seen that the walls of the head cavities like their homologues in the body develop into muscles. Now the first head cavity lies immediately behind the eye and extends round it so as to invest it like a cup. From its walls are developed the rectus superior, rectus inferior, rectus internus and obliquus inferior muscles—i.e., all the muscles supplied by the third nerve. At last we see why the third nerve supplies these and no other muscles; it does so because it is the nerve belonging to the segment in which the first head cavity lies; and therefore supplies the muscles that are formed out of the walls of that cavity. This reason is a complete and sufficient one.

Concerning the obliquus superior we are still in the dark; it does not appear to have any connection with the first head cavity, and this is sufficient reason why it should not be supplied by the third nerve. The development of the nerve that does supply it—the fourth—is at present absolutely unknown. Concerning the rectus externus muscle we are in a better position. The sixth nerve which supplies it bears the same relation to the seventh nerve that the anterior root of a spinal nerve does to the posterior root. The rectus externus has no connection with the first head cavity, but lies altogether superficially to it, and for some time behind it; it appears to be developed partly from the third head cavity and possibly in part from the second. We see now clearly why it is not supplied by the third nerve, the reason being that it does not belong to the segment which is supplied by that nerve, but to one further back.

Thus, we have not quite a complete, but quite a novel and, I think, an intelligible clue to the solution of our problem.

LECTURE 4

THE THEORY OF CHANGE OF FUNCTION

This lecture, the shortest in the collection, shares much of the cogent directness which characterizes Lecture 3 and in many ways the two lectures form a complementary pair. Here the emphasis is on resolving an issue embedded in Darwin's *On the Origin of Species*, using a range of anatomical examples.

AMM follows Darwin in using the word *perfection* when referring to the development of a complex organ by natural selection. We may tend to read this word as signifying the attainment of a design goal, with all the connotations of purpose which that implies. But this is surely not what either author meant: such an organ (an eye, a hand, etc.) is perfect only because we marvel at its construction and its ability to carry out the function we have come to expect of it.

Indeed, this could be said to be the philosophical burden of the lecture: to identify an organ as having changed its function presupposes that the original function was the one for which it was first intended. The safest defence against accusations of teleological interpretation is to recall, as Darwin undoubtedly understood, that evolution is opportunistic: organisms survive by exploiting whatever features they possess, and those features persist in the population because they are useful. *Improvements*, as AMM explains, are simply changes that happen to confer advantage.

AMM returns to the embryological role of gill clefts in the penultimate paragraph, for they provide a particularly persuasive example of change of use. The context he provides for this—olfaction—illustrates the distinction between what we would now refer to as divergent evolution (represented by synapomorphic features, shared by all members of a clade) and convergent evolution (represented by independently emerging, homoplastic or analogous features).

LECTURE 4

THE THEORY OF CHANGE OF FUNCTION

WHEN asked to read the opening paper of the second session of the Owens College Biological Society, I felt that the subject most suitable for the occasion would be one gathered from the works of that great Englishman whose name it was at one time proposed that this Society should bear; that prince of biologists who has taught us not to be satisfied with a knowledge of facts but to seek earnestly for the reason of these facts; who has taught us both what to seek and how to seek, and who has thus rendered philosophical biology not only a possibility but an actuality.

In selecting the particular aspect of Mr. Darwin's wonderful theory, to which I should draw your attention, I have been influenced mainly by two circumstances. In the first place it appeared more profitable, and therefore more likely to prove acceptable to the Society, to select some portion to which exception has been taken rather than one which has met with general acceptance; to consider one of the numerous difficulties in the way of accepting the theory of Natural Selection, difficulties pointed out by none so clearly as Mr. Darwin himself.

The second circumstance to which I have alluded as influencing my choice was the accidental fact that some work on which I happened to be engaged some little time ago brought me into very violent, and for the time very embarrasing contact with one of these said difficulties, and thus compelled me to take its bearings and measurements very accurately in order to discover in what way it might most conveniently be circumvented or surmounted, or, if possible, removed altogether.

I purpose then asking your attention to one of the more serious of the many alleged difficulties in accepting the doctrine of Natural Selection and to a brief consideration of the means proposed for removing that difficulty. Let me first attempt to define clearly the nature of this obstacle. In the higher animals we meet with a great number of very complex organs, each with very definite functions, such as the eye, the ear, and the hand. Now, according to the Darwinian theory, the mode in which such organs have attained their present complexity and perfection is as follows:

Once upon a time the ancestors of these animals possessed eyes less perfect and less complex than those they have at present. Now no two animals of the same species have eyes absolutely identical with one another; slight differences always occur, and of these slight differences it must happen that certain ones are improvements, are changes for the better, or what we call useful modifications. All the animals possessing these useful modifications will be slightly better off and will have a slight advantage over their companions who have not got them. Now more animals of every species are born than can possibly live. Hence the greater proportion of animals that are born die, and die young before they have produced any offspring.[66] It is clear that those animals which have this slight advantage over their brethren will by virtue of this fact have a better chance of living. Now we know as a fact that such variations tend to be inherited—i.e., to be handed down from the parents in whom they first appeared to their offspring, and not only so but also that they tend to appear in the offspring in a rather more strongly developed form.[67]

Hence we get this state of affairs; certain forms of a particular species, say of deer, tend to survive because they have some slight accidental modification of their eyes which gives them a slight but distinct advantage over their brethren; they transmit this modification and with it this advantage to their offspring; certain of their offspring present the modification in a greater degree not only than their brethren but also than their parents. These will get a slight additional advantage and will tend to survive, and so on for successive generations. The gradual accumulation of these slight modifications will in time cause a perceptible change in the structure of the eye, and as each successive modification is preserved only because it is useful it is clear that the eye will as a whole have improved and have become more perfect. Similarly with the hand or with any other of the organs of the body; formerly less complex, it has acquired its present complexity and perfection as the result of accumulations of a long series of

---

66 *Hence* suggests that a greater than even chance of mortality is a necessary consequence of too many births. This may well be the case with animals (and other organisms) whose success relies on prolificacy with minimal parental investment (so called "r" species) but it is not inevitably so for those with few offspring and enduring parental care ("K" species; including most large mammals). Lecture 6 explains AMM's standpoint more fully.

67 The meaning of the *developmental strength* of a character in this statement is unclear. Nor is it clear why advantageous characters should tend to appear (immediately) in stronger form. The same implication appears in the first sentence of the next paragraph, but not in the rest of that paragraph or the remainder of the lecture. Darwin (*Origin of Species*, Chapter 4) clearly understood that *slight but useful* differences accumulate over many generations: "*Variation itself is apparently always a very slow process*" and "*We see nothing of these slow changes in progress, until the hand of time has marked the long lapses of ages*"; also, "*Natural selection acts solely through the preservation of variations in some way advantageous, which consequently endure.*" In Chapter 6, Darwin says: "*In living bodies, variation will cause the slight alterations, generation will multiply them almost infinitely, and natural selection will pick out with unerring skill each improvement.*" Darwin's closest statement to AMM's seems to be "*The laws of correlation of growth, the importance of which should never be overlooked, will ensure some differences; but, as a general rule, I cannot doubt that the continued selection of slight variations, either in the leaves, the flowers, or the fruit, will produce races differing from each other chiefly in these characters*" (*Origin of Species*, Chapter 1) but this hardly supports AMM's point as written. Alternatively, AMM meant "*But if variations useful to any organic being do occur, assuredly individuals thus characterized will have the best chance of being preserved in the struggle for life; and from the strong principle of inheritance they will tend to produce offspring similarly characterised. This principle of preservation ...*" (*Origin of Species*, Chapter 4).

small but useful modifications. Each of these modifications must give some distinct advantage to its possessor or else it would not be preserved or perpetuated.

It has been objected that this explanation is not sufficient; that although it will account for the gradual perfection and increasing complexity of an organ after it has attained a certain size and after it has assumed its definite function, yet that it fails completely to explain the first origin of such organs; for in their earliest origin they must have been very minute and absolutely incapable of fulfilling or even aiding in the function which they are afterwards destined to perform.

This objection, which is the one I wish to consider to-night, must be confessed at once to be of very great weight, and is indeed freely acknowledged to be so by Darwin himself.[68] By some writers, such as Mivart (almost the only English naturalist of any repute who at the present day rejects the doctrine of Natural Selection), it is indeed considered as absolutely fatal to the whole theory. We shall perhaps get a clearer idea of the nature and force of the objection by considering one or two examples which bring it forward prominently.

The limb of a Vertebrate in its earliest stage is a small bud arising from the surface of the body and is absolutely useless for any of the purposes to which we put our arms. If this represents the primitive condition, why should such an organ have been preserved at all? Again, a more striking example and one yet not thoroughly explained is found in the wing of the bat. This is a great fold of skin connecting together the much elongated fingers, and also connecting these with the sides of the body extending downwards so as to involve the legs as well. This constitutes as we know a very efficient flying apparatus. The theory of evolution undoubtedly requires that bats should be descended from, should have had as ancestors, mammals possessing fingers of normal length and devoid of the lateral expansion of the skin.

---

68   Darwin faces up to this criticism in *Origin of Species* Chapter 6, *Difficulties on Theory*: "*... is it possible that an animal having, for instance, the structure and habits of a bat, could have been formed by the modification of some animal with wholly different habits? Can we believe that natural selection could produce, on the one hand, organs of trifling importance, such as the tail of a giraffe, which serves as a fly-flapper, and, on the other hand, organs of such wonderful structure, as the eye, of which we hardly as yet fully understand the inimitable perfection?*" The eye is his primary example of an organ "*of extreme perfection and complication*" developing by small increments in complexity, under the pressure of natural selection. Tails in general provide the context for extending the argument to change of use: "*Organs now of trifling importance have probably in some cases been of high importance to an early progenitor, and, after having been slowly perfected at a former period, have been transmitted in nearly the same state, although now become of very slight use; and any actually injurious deviations in their structure will always have been checked by natural selection. Seeing how important an organ of locomotion the tail is in most aquatic animals, its general presence and use for many purposes in so many land animals, which in their lungs or modified swim-blad-ders betray their aquatic origin, may perhaps be thus accounted for. A well-developed tail having been formed in an aquatic animal, it might subsequently come to be worked in for all sorts of purposes, as a fly-flapper, an organ of prehension, or as an aid in turning, as with the dog, though the aid must be slight, for the hare, with hardly any tail, can double quickly enough.*"

The theory of Natural Selection demands further that the various steps in the transformations of this ancestral form to the existing bat should have been slow and gradual. Now it is obvious that a very slight lengthening of the fingers, coupled with a very slight increase in the extent to which they were webbed, would in no way enable the ancestral mammal to fly; and hence that such accidental variations could never have been preserved and perpetuated because of their utility as flying organs.

Cases such as these have long been felt to offer very serious difficulty and to throw serious doubt, not on the reality of Natural Selection, but on its sufficiency. Darwin himself suggested a possible solution of the difficulty, at any rate in certain cases, by pointing out that an organ which had already attained considerable size and complexity might, owing to some change in the habits or surroundings of the animal possessing it, have not its structure but its function changed—i.e., that an organ already in use for one purpose might become employed for some other purpose, might gradually lose its original function and undergo further modifications fitting it better for its adopted function.

This idea, which clearly affords us a possible mode of getting over the difficulty, is only briefly alluded to by Darwin, but was taken up and developed by Dr. Dohrn who, in a pamphlet published in 1875,[69] discussed it at considerable length.

A very good example of the principle in question is afforded by the swimming bladder of fishes. This, in most fish, is a closed sac lying just underneath the vertebral column. In many fish it acquires a connection by a duct with some part of the alimentary canal. It then becomes an accessory breathing organ, especially in those fish which are capable of living out of water for a time—e.g., the Protopterus of Africa. An interesting series of modifications exists connecting the air bladder with the lung of the higher Vertebrates, which is undoubtedly the same organ. The air bladder of fishes is in fact a very good example of change of function, inasmuch as it is used originally for purposes of flotation and afterwards it is preserved as a lung. This example is instructive also because the evidence of embryology, on which we are accustomed to rely, fails utterly here. The lung develops as a pit-like depression in the floor of the oesophagus. Now this could not have been the earliest origin of the lung, for it would be utterly useless as such for the simple reason that food would always be falling into it.[70]

In developing the theory of Change of Function Dr. Dohrn points out that it is a very common thing, if not the general rule, for an organ to have not one function alone but a number of different functions to perform, of which functions one at any

---

69   *Der Ursprung der Wirbelthiere und das Princip des Functionswechsels: Genealogische Skizzen* which proposes the *change of function* theory of the origin of vertebrates.

70   This confusion, about the origins of swim bladders and lungs, goes to the heart of the understanding of convergent and divergent evolution. AMM assumes that they are, phylogenetically speaking, the same organ but can find no embryological evidence for this. Whilst his objection to the idea seems trivial, he seems not to entertain the possibility that they evolved independently and that their similarity (they contain air) is coincidental. The current view is that both organs are in fact embryologically derived from the upper part of the gut, involving the expression of similar (*Wnt*) genes, and are to that extent homologous. Whether there was a single common ancestor with such a structure remains unknown.

given time is predominant and the rest subordinate; but that it is quite conceivable that some slight change of circumstance might cause the relative importance to be changed, and one of the subordinate functions to become the primary one, or to put it in his own words: *"Each function is a resultant of several components of which one is the principal or primary function, the other secondary. Diminution of the primary function and increase of a secondary function alters the total function; the secondary gradually becomes the primary, the total function is changed and the issue of the whole process is the transformation of the organ."*

For instance, the primary function of the stomach is undoubtedly the secretion of gastric juice; a secondary function is the movement of its muscular walls aiding the action of the gastric juice by bringing the contents into closer contact with it. Now in no animal are the glands absolutely uniformly dispersed over the stomach walls, and it is readily conceivable that both glands and muscular wall might be better developed in one half of the stomach than the other, and indeed this condition actually occurs in the stomach of the rat. A continuance of this modification brings us to the condition met with in the ruminant stomach. Here the first compartment is a mere receptacle for the storing of food, and in this no digestion takes place at all. Again, from the same starting-point, we have another series. Suppose the glands to aggregate at one end of the stomach, and the muscular coat to be thickened at the other; if we carry this far enough we get to the condition of the stomach of the bird where peptic glands are confined to one part, muscles to the other. Here the second portion has completely lost its original primary function, this having become replaced by a secondary function which has now become primary.[71]

The modifications of the limbs of Arthropods afford numerous and admirable illustrations of the case before us. Here we have organs whose primary function is undoubtedly locomotion, but of which a certain number, greater or less, in different groups, have become modified so as to aid in the mastication of food. Naturally, those nearest the mouth are the ones so modified, and of these, those at the actual sides of the mouth are most likely to undergo the greatest modifications. Such modifications, if extreme, would unfit them for their primary locomotor function, hence we find the first post oral pair of appendages, the mandibles, in most Arthropods completely and permanently altered, both structurally and functionally, the hinder ones being altered to less extent.

Of the two divisions of the limb the inner or endopodite is the nearer to the mouth, and consequently most likely to be useful for purposes of feeding. On the other hand, for swimming, the exopodite is of equal, if not greater, importance; for walking the endopodite is of most use as being more directly beneath the body. Another good example of change of function is found in the hyomandibular cleft.

---

71   The idea of a single stomach, even if divided into functionally different sections, does not quite fit with current interpretation. In birds, the proventriculus (which is peptically secretory and absorptive) is distinguished from the muscular (and non-secretory) gizzard; the former is a stomach in the mammalian sense, whilst the latter is best viewed as a separate modification of the small intestine. In ruminants, the reticulum and rumen are elaborations of the oesophagus, the abomasum is absorptive and the omasum is the secretory stomach. Whether functions have thus been lost or gained, in the sense that AMM means it, is a matter of interpretation.

This, like certain other clefts, has become saved from destruction by becoming modified into an accessory organ of hearing.

The olfactory organ furnishes another illustration of the theory, and with regard to this let us first consider what appear to be the difficulties of the case. All Vertebrates possess olfactory organs with the solitary exception of Amphioxus, that curiously exceptional Vertebrate whose anatomy seems to be made up of contradictions. These olfactory organs have in all cases the same essential structure. Certain differences occur, but these are so slight as to leave no doubt whatever that the vertebrate olfactory organs are, wherever they occur, the same organs. As soon, however, as we get beyond Vertebrates we meet with a difficulty. There is no doubt that Vertebrates came into existence later than Invertebrates, therefore Vertebrates must either have inherited their olfactory organs from their invertebrate ancestors, or must have acquired them for themselves. An olfactory organ is found in Invertebrates; some insects almost certainly have them, but these could not have been the ancestors of the vertebrate olfactory organ. Now if the Vertebrates did not inherit them as such they must have got them some other way, either acquired *de novo* or by change of function. Now in its time of appearance, its mode of development, the histology of its epithelium, there is a very close resemblance between the olfactory organ and a gill. Hence there are strong reasons for regarding the olfactory organ of Vertebrates as a modified gill. It has been stated that the mode of development of the olfactory nerve as an outgrowth from the cerebral hemisphere constitutes a serious objection to this view as differing from the other cranial nerves. This objection does not now hold good, since it has been shown that the olfactory nerve develops in the same way as other cranial nerves.

In conclusion, we seem to have in this principle of change of function a real and practicable solution to the chief difficulty in the way of accepting the doctrine of Natural Selection, a solution that has long been overlooked, and whose real importance has, I believe, yet to be appreciated.

LECTURE 5

BUTTERFLIES

CFM was unsure of the circumstances of this lecture but the AMM papers in the Manchester Archives record that a lecture with this title was given to the Ancoats Society on 12 January 1890 as part of the Eighth Sunday Lecture Series to Men and Women, New Islington Hall, Ancoats, Manchester. (See also Lecture 10 and **Lectures on the Darwinian Theory: Context**.)

The title is clearly aimed at a general audience. The content is designed to appeal to non-expert interest rather than addressing detailed biological problems in the manner of the previous two lectures. Nevertheless, the style is inquisitorial, stemming from curiosity about the beauty of nature and intriguing observations of animal habits. The lecture makes no particular point nor reaches any distinct conclusion, but the audience would have been left with an impression of there being more to know, much more to find out.

AMM takes pains, in the gentlest way, to avoid anthropomorphism whilst at the same time not rejecting the possibility of conscious decision-making in animals. The balance is a fine one, negotiated with skill. His examples, not all of which are lepidopteran, come from the current work of naturalists and would have served to open the minds of the audience. The final description of pupae and the various means by which larvae make their escape demonstrates AMM's masterful ability to create pictures with words.

From an academic perspective this lecture may suggest that AMM was somewhat free with the work of others as regards attribution, although there is no evidence that he tried to mislead. He evidently used Wallace to furnish examples, repeating them without necessarily examining original sources, although one cannot know exactly how he portrayed them to the audience.

The broader topic of coloration and mimicry is dealt with more extensively in Lecture 18, given five years later, also to Ancoats.

LECTURE 5

BUTTERFLIES

THE beauty of butterflies is partly due to their shape and proportions and to their very graceful outlines but it is in their colouring that their chief beauty lies; and concerning this colouring I wish to say a few words, even at the risk of repeating what is already familiar.

The colour of a butterfly is due to the scales which cover both surfaces of the wings and which are easily rubbed off by our fingers. These scales are of various shapes and overlap each other like the tiles on a roof, each having a short stalk for insertion into a small depression in the wing. Their variety of colour is extraordinarily great and is due partly to actual pigments; the brightest tints however, and more especially the metallic lines seen in some foreign butterflies, and perhaps best of all in humming-birds, are due to fine lines on the surface of the wings producing what are known to physicists as interference colours.

Let us consider the question. Why should butterflies be thus gaily clothed? The older naturalists thought it sufficient to say that butterflies were beautiful because their beauty gave us pleasure, and that it was for our enjoyment that these delicate tints and gorgeous hues existed. But this is clearly incorrect, for the most magnificent butterflies are found in parts where man but seldom visits, for instance, tropical America and the East Indies. The true explanations are varied and not the same in all cases.

Sexual Colours.—With butterflies as with birds there is often a striking difference between the sexes, the male as a rule being the more gaily coloured. Among birds, for instance, the cock bird is often most gaily coloured; the hen bird being more soberly coloured, and indeed often plain. The peacock, drake, and bird of paradise will at once occur to our minds as illustrations. Mr. Wallace pointed out that the sober colouring of the female was explained as protective so as to escape detection when sitting on eggs on an open nest. The male usually taking no share in incubation had no special need for protective colouring. Mr. Darwin explains the specially brilliant colours of males as due to what he calls sexual selection, that is, to the fact of the female preferring a smart male to an untidy one. So with butterflies and moths; and if we watch butterflies or birds coquetting we soon convince ourselves that the males know they are beautiful, and mean to show off their beauty to its best advantage in the hopes of gaining the affection of the female.

It has been objected that this is going too far; that the proper appreciation of the colours and tints of butterflies, birds, or flowers requires considerable training and a distinct intellectual effort on our part, and that it is absurd to suppose that what we ourselves only do imperfectly and with difficulty a butterfly can do just as well and without effort. But I do not think we ought to reject an explanation just because it happens not to be particularly flattering to ourselves. In the case of flowers it has been conclusively shown that their colours, markings and odours are developed to attract insects bees, beetles, and butterflies—to visit them and so fertilise them; in other words, to advertise themselves.

The colours of flowers are as beautiful and as varied as those of butterflies. And if, as it is acknowledged, butterflies and birds understand and appreciate the differences between the colours of gay flowers, and if those colours are developed simply to attract them, why deny them the power of appreciating similar colours in themselves and in one another? Again, the appreciation of colour in ourselves has greatly developed; for if we study the history of painting and compare the oldest pictures, whether oil or water colour, with more recent ones we note a distinct development in the power of appreciating the full effect of colour, or what is known as the colour sense.

Protective Colours.—So far we have considered the colours of the upper surface only. With regard to the colours of the under surface the case is different; here the colours are protective, and butterflies have the habit of turning up their wings when resting, so as to expose the under surface only. The under surface is coloured like the leaves, twigs, and especially the flowers on which they most love to perch, for it is then that they are most exposed to the attacks of enemies and have the most need for protection. To appreciate this it is necessary to see them in their haunts, and to watch them at home. The best marked instance is *Kallima*,[72] which is met with in India, the Malay Archipelago and Sumatra. It is very common in these places, and is a large showy butterfly with orange and purple colouring on the upper surface, and is a rapid flier frequenting dry forests. It always settles where there is some dead and decaying foliage, for the colouring on the under surface of the wings bears a remarkable resemblance to that of a dead leaf, and when the wings are turned up and the head and body hidden between them, it is often very difficult to distinguish them from dead leaves, the resemblance being rendered even more close by the short tail which looks like the stalk of a leaf, and by the markings on the under surface which closely imitate the mid-rib and veins of a leaf. Speaking of this insect Mr. Wallace says:[73] *"The colour is very remarkable for its extreme amount of variability, from deep reddish-brown to olive or pale yellow, hardly two specimens being exactly alike, but all coming within the range of leaves in various stages of decay. Still more curious is the fact that the paler wings, which imitate leaves most decayed, are usually covered with small black dots, often gathered into circular groups, and so exactly resembling the minute fungi on decaying leaves that it is hard at first to believe that the insects themselves are not attacked by some such fungus. The concealment produced by this wonderful imitation is most complete, and in Sumatra I have often seen one enter a bush and then disappear like magic. Once I was so fortunate as to see the exact spot on which the insect settled, but even then I lost sight of it for some time, and only after a persistent search discovered that it was close before my eyes."*

This example serves as an extreme case of what is really a general law among butterflies.

The mode in which this protective colouring is acquired is explained by Natural Selection. At first there is a more or less accidental resemblance. Now large numbers of butterflies are killed by birds, lizards, and other animals, and any whose markings and habits of perching rendered them less easy to see would have a better chance of

---

72   The Oakleaf butterflies of the family *Nymphalidae*.

73   Wallace, A.R., *Darwinism: An Exposition of the Theory of Natural Selection with Some of its Applications* (London & New York: Macmillan, 1889), Chapter VIII: "The Origin and Uses of Colour in Animals", p. 207.

escaping their enemies; these, therefore, will survive and transmit their peculiarities to their offspring; the survivors of these in turn transmitting the peculiarities in a more strongly marked form, and so the protective colouring becomes more marked from generation to generation and the unprotected ones perish.

Warning Colours. —These are found in specially protected individuals which are usually uneatable, or at any rate unpalatable, and whose nastiness it is desirable to advertise in order that they may not be eaten, or at any rate killed, by mistake. Their object, therefore, is not to escape notice, but to be readily seen and recognised. The best examples of these are found in three great families of butterflies—the *Heliconidae*, found in South America, the *Danaidae*, found in Asia and tropical regions generally, and the *Acraeidae* of Africa.[74] These have large but rather weak wings and fly slowly. They are always very abundant, and all have conspicuous colours or markings and often a peculiar form of flight; characters by which they can be recognised at a glance. The colours are nearly always the same on both upper and under surfaces of the wings, and they never try to conceal themselves, but rest on the upper surfaces of leaves and flowers. Moreover, they all have juices which exhale a powerful scent; so that if they are killed by pinching the body a liquid exudes which stains the fingers yellow, and leaves an odour which can only be removed by repeated washing. This odour is not very offensive to man, but has been shown by experiment to be so to birds and other insect-eating animals.

Another example is furnished by the skunk, which, although not included in the immediate subject of this lecture, may be mentioned as an extreme case illustrating the point we are considering.

Concerning this animal I cannot do better than quote Mr. Wallace again:[75] *"This animal possesses, as is well known, a most offensive secretion which it has the power of ejecting over its enemies, and which effectually protects it from attack. The odour of this substance is so penetrating that it taints and renders useless everything it touches or in its vicinity. Provisions near it become uneatable, and clothes saturated with it will retain the smell for several weeks even though repeatedly washed and dried. A drop of the liquid in the eyes will cause blindness, and Indians are said sometimes to lose their sight from this cause. Owing to this remarkable power of offence the skunk is rarely attacked by other animals, and its black and white fur and the bushy white tail carried erect when disturbed, form the danger signals by which it is easily distinguished in the twilight or moonlight from unprotected animals. Its consciousness that it needs only to be seen to be avoided gives it that slowness of motion and fearlessness of aspect which are, as we shall see, characteristic of most creatures so protected."*

Recognition Colours.—A good instance of this class of colouring is seen in the upturned white tail of the rabbit which, although making it conspicuous to its enemies as well as friends, is probably a signal of danger to other rabbits; and when feeding together, in accordance with their social habits, soon after sunset or on moonlight nights, the upturned tails of those in front serve as guides to those behind to run home on the appearance of an enemy. Many birds, antelopes, and other animals have markings believed to serve a similar purpose, and probably the principle of distinctive

---

74   The Heloiconidae, Danaidae (Milkweed butterflies) and Acraeidae (Brush-footed butterflies) are now classified as subfamilies of the family *Nymphalidae*.

75   Wallace (see footnote 73), Chapter IX: "Warning Coloration and Mimicry", p. 233.

colouring for recognition has something to do with the great diversity of colour met with in butterflies.

Mimicry.—Many butterflies escape destruction through mimicking the appearance of uneatable or venomous forms; for instance, *Leptalis*,[76] a form allied to the common garden white, mimics the *Heliconidae*, a widely distinct family, in the shape of its body and wings, in its colour, and even in its habits and mode of flight, so much so that they are difficult to distinguish. They are much rarer than the forms they mimic, and may aptly be compared to the Ass in the Lion's Skin. It is often only the female which mimics, for this has greater need for protection. The deception is indeed often so good that it may deceive not merely an expert naturalist but even the insects themselves, and Fritz Müller[77] says *"I have repeatedly seen the male pursuing the mimicked species till after closely approaching and becoming aware of his error he suddenly returned."* Other examples of mimicry are found in the beehawk moth, which mimics the humble bee; in the clearwing moths, which receive names such as apiforme, vespiforme, from the insects they mimic; and, among other animals, the harmless snakes mimic the venomous ones. Again, the "devil's coach horse",[78] the beetle with the habit of turning its tail over its back and pretending to have a sting, certainly deceives children and perhaps grown-up people as well. Many other instances of mimicry could be cited; a certain spider simulates the droppings of birds for instance,[79] and a blue butterfly was actually seen by Wallace[80] resting on what was apparently dung but in reality a spider. Nor is the advantage of the power of mimicry confined to animals, for among ourselves the power of pretending to be other than what we really are is often of the greatest possible service to us. For example, the success of a detective largely depends on his being able to pass himself off as something else and to entirely conceal his real calling.

---

76   *Dismorphia (Leptalis) theucharila*, the clear-winged mimic white butterfly of the Pieridae family.

77   J.F.T. Müller (Itajahy, Sᵃ Catharina, Brazil) in a letter to Charles Darwin of 14 June 1871. See *Darwin Correspondence Project*, letter DCP-LETT-7820, www.darwinproject.ac.uk/letter/DCP-LETT-7820.xml

78   *Ocypus olens*, also called the Cocktail Beetle.

79   Presumably the Bird Dropping Spider, *Celaenia excavate*.

80   This was not actually seen by Wallace. In *Darwinism*, Wallace wrote (footnote 73, p. 211): "Mr. H. O. Forbes has described a most interesting example of this kind of simulation in Java. While pursuing a large butterfly through the jungle, he was stopped by a dense bush, on a leaf of which he observed one of the skipper butterflies sitting on a bird's dropping. "I had often," he says, "observed small Blues at rest on similar spots on the ground, and have wondered what such a refined and beautiful family as the Lycænidæ could find to enjoy, in food apparently so incongruous for a butterfly. I approached with gentle steps, but ready net, to see if possible how the present species was engaged. It permitted me to get quite close, and even to seize it between my fingers; to my surprise, however, part of the body remained behind, adhering as I thought to the excreta. I looked closely, and finally touched with my finger the excreta to find if it were glutinous. To my delighted astonishment I found that my eyes had been most perfectly deceived, and that what seemed to be the excreta was a most artfully coloured spider, lying on its back with its feet crossed over and closely adpressed to the body."

There is an old view which holds that colours are due to the direct action of the sun, but this will not explain the constancy of colouring in most species, or the bright colours of butterflies in climates such as our own. There are some cases, however, in which differences in temperature, at any rate in the season of the year, appear to have had a distinct influence on colour, the best known examples being the common continental butterflies *Vanessa levana* and *Vanessa prorsa* and the English butterfly *Pieris napi*.[81]

Colour is perhaps in part a matter of indifference, and a part may be red because there is no reason why it should be otherwise. The red colour of blood, the red colour of many deep sea animals, the blue colour of the sky, and perhaps the green colour of the grass and vegetation generally, may, for all we know, be indifferent.[82]

The Senses of Butterflies.—Our knowledge of this very interesting subject is at present very imperfect. We know that the sense of smell must, at any rate in some forms which have a liking for dead and decaying animal matter, be very acute. The eyes are compound and their facets extraordinarily numerous, as many as 12,000 being present in the eye of the death's-head moth, and 17,000 in the swallow-tailed butterfly. The range of vision in butterflies is unknown.[83] Of the sense of hearing in butterflies we know very little, but in moths and other insects the antennae are often fringed; and some most interesting experiments have been made with the antennae of the male mosquito, special hairs of which were found to vibrate to particular notes, those hairs being most affected which were at right angles to the direction from which the sound came. The sound is thus most intense if directly in front of the head; if one antenna is affected more than the other the head is turned till both are equally affected, and so the mosquito is enabled to direct its flight directly towards the point from which the sound originates; this power being specially used to aid it in finding the female.

The Life-History of Butterflies.—A butterfly is not always a butterfly, but goes through several phases—viz., the egg, the larva or caterpillar, the pupa or chrysalis, and the adult form or imago. The caterpillar phase is the nutritive one; the adult phase is reproductive and usually short-lived, often not more than a few days, and sometimes limited to a few hours. The *eggs* are usually protectively coloured, and often show very beautiful markings under the microscope. It has been shown that while at least nine-tenths of the eggs and larvae of North American butterflies are destroyed by parasites or disease, such parasites never attack the eggs of the

---

81   {M} For a full account of these see paper on "Environment", Lecture 2.

82   Presumably, by *indifferent*, AMM means *of no functional or adaptive significance*. We can easily explain each of his examples using physics and chemistry (by the reflectance and absorbance of different wavelengths) and, except for the colour of the sky, also suggest that the pigments responsible evolved under selective pressure. In the case of haemoglobin and chlorophyll (both of whose functions were poorly understood prior to the 20th century) that pressure probably had to do with biochemical efficiency rather the selection of colour *per se*. The red coloration of deep-sea animals may well have been directly selected given that colours at the long-wavelength end of the visible spectrum are absorbed by water less than those of shorter wavelengths.

83   It is now known: in several species, vision across the spectrum and into the ultraviolet region is effected through a complex pallet of visual pigments.

Danaidae, and it is possible that these are distasteful to their foes at all stages of their existence. The eggs of butterflies are always laid in safe places and as near as possible to the proper food for the young; the butterfly never sees her young and can have no clear idea as to the respective merits of cabbages, carrots, and oak-leaves, yet she makes no mistake.

The *larva*, or caterpillar, is soft and fleshy and is very easily injured. It is furnished with a biting mouth, soft fleshy fore-legs, and a terminal sucker. It has three pairs of harder jointed legs corresponding to the legs of a butterfly and which are sometimes very long, as in the lobster moth. The great and only work of cater-pillars is to eat, and this they do all night, many of them all day as well. They feed on the leaves of plants by means of their powerful jaws, and have no need even to stop to take breath, for their breathing is carried on by means of spiracles or pores along the sides of the body which lead to the tracheal tubes by which the respiration of insects is effected. Their rapid growth soon renders their skin too tight and the outer layer, or cuticle, is thrown off like that of a crab or lobster; after this they eat more ravenously than ever to make up for lost time, often commencing with their cast-off clothes. As a rule there are several of these castings before the caterpillar attains its full size. Their gain in weight is prodigious, the caterpillar of one of the hawk moths for instance, *Acherontia*,[84] weighing about one-eighth of a grain[85] on leaving the egg, in thirty days increases its weight 10,000 times. Even this is not the limit, for some caterpillars live three years; for instance, the goat moth, which grows to 72,000 times its weight on hatching.

The shapes and colours of caterpillars follow the same laws as those concerning the adult insect, the colour being generally protective because the caterpillar is soft and fleshy and good eating for birds. The colour for this reason is usually green, and those feeding on grass are striped longitudinally, those on larger leaves obliquely, this forming a very effective protection. Others are brown, and so like twigs in shape that even the most experienced may be deceived. Mr. Jenner Weir writes:[86] *"After being thirty years an entomologist I was deceived myself. I took out my pruning knife to cut from a plum tree a spur which, I thought I had overlooked. This turned out to be a larva of a geometer[87] two inches long. I showed it to several members of my family and defined a space of four inches in which it was to be seen, but none of them could perceive that it was a caterpillar."*

Warning colours are well seen in caterpillars. All green and brown ones are readily and greedily eaten by birds, lizards, and frogs, but many are conspicuously coloured and do not attempt to conceal themselves, and these are usually nauseous to the taste. For instance, the gooseberry moth caterpillar was given to frogs to eat, whereupon they *"sprang forward and licked them eagerly into their mouths; no sooner had they done so than they seemed to become aware of the mistake they had made, and*

---

84   Death's-head Hawkmoth.

85   A fraction over 8 mg.

86   *The Difficulties of Natural Selection*, a letter to *Nature* (29 December 1870, p. 166); cited in Wallace (footnote 73), Chapter VIII, p. 208. The original paragraph containing the quoted statement concludes with: *"Surely this was a case of protective mimicry."*

87   A moth belonging to the family *Geometridae*.

sat with gaping mouths rolling their tongues about until they had got quit of the nauseous caterpillars, which seemed perfectly uninjured and walked off as briskly as ever."[88]

Terrifying colours are also met with; in the caterpillars of the puss moth and hawk moth the eye spots are of this nature, and the caterpillar of *Bombix regia*, the "hickory horned devil" of the Southern States of North America, is ornamented with an immense crown of orange-red tubercles which if disturbed it erects and shakes from side to side in a very alarming manner, the negroes believing it to be as deadly as the rattlesnake, whereas it is really perfectly harmless.

The *pupa*, or chrysalis, is the last stage of existence of the caterpillar. The change from caterpillar to butterfly is an enormous on; not merely as regards the possession of wings, but as regards the whole organisation; the mouth, the digestive system, the nerves, the eye and all parts being most profoundly modified. The chrysalis period is the stage of rest, and constitutes a protected condition in which these changes can be effected. This stage sometimes lasts a few weeks or even days only, sometimes months or during a whole winter. In the case of the small eggar moth, insects of the same brood appear a few at a time each year up to fourteen or fifteen years. Now the time of appearance of the moth being February, it is obvious that in a severe winter the whole brood might be killed off if they all appeared at the same time, hence the advantage of appearing a few at a time in successive years.

Pupae are of various forms: (I) *Underground pupae*, which are found in crevices in walls and the roots of trees, the holes in which they lie being often carefully lined. These pupae have some power of motion, and work their way to the surface before the imago emerges from the pupa. (2) *Suspended pupae*. A good example of these is found in the pupa of the common tortoiseshell butterfly. The larva having chosen a suitable place, by means of the glands which open on the upper lip, spins a little button of silk strong enough to support its weight; it then thrusts its tail into the button and swings head downwards. It has now to get out of its skin without letting go its hold, and this it does by gradually working out of its skin, which has previously been split along the back, and pushing it towards its tail. Before completely escaping, it gets the tip of its tail free while the part in front of the tail is still fixed, the lining of the trachea and intestine helping to suspend it; it then stretches its tail up to the button of silk, fixes itself and spins round several times to secure a firm hold, finally casting away its larval skin. Other larvae sling themselves up by a girth of silk, and some are flexible enough to attach the thread on one side, carry it over and fix it on the other side, repeating this process several times. Others again spin the girth of silk first and then slip their heads under it; some, such as the swallow-tail, hold the silk in their claws till it is strong enough, and then slip it over their heads.

Many other interesting points could be mentioned in connection with butterflies, but those I have briefly touched upon will serve to indicate what interesting questions arise from the study of butterflies and other insects, and what good examples they afford us of many of the problems of Natural History.

---

88   This quote is from A.G. Butler (friend of Mr Jenner Weir): "Remarks upon Certain Caterpillars, etc., which Are Unpalatable to their Enemies", *Transactions of the Entomological Society of London*, 1861; Part I: 27–29. It is cited in Wallace (footnote 73), Chapter VIII, p. 237.

LECTURE 6

FRESH-WATER ANIMALS

A MM became President of the Manchester Microscopical Society in 1887 and remained so until 1893. The Manchester Archive papers record that this presidential lecture was given on Saturday, 29th January 1887 in the lecture hall of the Athenaeum and was accompanied by demonstrations using "a large collection of microscopes". Lectures 7, 8, 9, 11 and 12 were also given to this Society. The Society's records show that in 1887 it had 214 members.

In some ways, this is an unsatisfactory lecture: the evidence presented largely fits the arguments being made but feels selective, and the conclusions reached (antepenultimate paragraph) are little more than restatements of those original arguments. A very limited range of reasons (for occupancy of fresh or sea water habitats) is given and the reader comes finally to feel that, given more time and more information, counter-evidence could be produced in all cases. The exhortation to members of the Society for continued research (final two paragraphs) indicates that AMM wishes, after all, for his ideas to be treated as hypotheses available for testing.

CFM seems to have taken a particular interest in this Lecture, judging by the footnotes in which he updates the information with recent discoveries. He was himself a student at Owens College (see the Biography).

LECTURE 6

FRESH-WATER ANIMALS

I PROPOSE to speak this evening about the fauna of fresh-water ponds and streams, a subject of very special interest to the members of the Society, many of whom have a far more intimate and accurate acquaintance with special groups than I can either claim or reasonably hope to possess. Still, there are some points of general interest which may have escaped the observation of those who have chiefly concerned themselves with one or two particular groups; and it is with regard to these more general conclusions that I propose to speak to-night.

As regards geographical distribution the entire animal kingdom may conveniently be divided into terrestrial and aquatic forms, and the latter again subdivided into those that live in the sea, and those inhabiting fresh water. The marine fauna is infinitely more abundant, and includes a far greater number of species than that found in fresh water; and we can hardly be surprised at this. A glance at a map of the world will show that the area covered by the ocean is much greater than that occupied by land—the proportion being about three to one. Moreover, only a very small, portion of the land is taken up by rivers and ponds; so that the area over which fresh-water forms can exist is necessarily much more restricted than that inhabited by marine animals.

Again, geologists show us that the land is subject to constant change in level and in extent, owing to upheaval or depression; and these oscillations, especially those of depression, must cause great disturbance, or even local annihilation, of the fresh-water fauna. By upheaval on the other hand, shallow water marine areas may be cut off from the sea, and converted into brackish marshes, and ultimately into fresh-water ponds; the animals inhabiting them being either killed off, or else adapting themselves to the altered environment, and becoming converted into fresh-water forms.

From these and other similar considerations it may be concluded that fresh-water animals must, with few exceptions, have been derived from marine forms, and my chief object this evening is to consider the various modifications, either in structure or life-texture, which marine animals undergo in consequence of becoming adapted to fresh-water life.

Why should marine animals strive to work their way up rivers? Why should they not stop in the sea? The reason is not far to seek, and is to be found in that struggle for existence we hear so much about, but the full extent of which we too often fail to realise. Very large numbers of every species of animal perish before reaching maturity. If we reflect for a moment on the really enormous number of eggs which fish and other marine animals lay, it is clear that if all these eggs, in the case of any single species, were to come to maturity, there would very soon be little room in the sea for anything else. A mackerel of a pound weight, for instance, will lay eighty thousand or ninety thousand eggs, a cod may produce five millions, and a conger-eel no less than fifteen millions. However, the vast majority in every species never come to maturity, but are devoured while young by other animals, as food; while a very small minority, favoured by the possession of some slight accidental advantage, escape destruction, survive, and perpetuate the species. So keen is the struggle for existence in the sea, especially in the shallow waters round the coasts, where life is far more abundant and competition more severe than at greater depth; that many forms, belonging to different groups, have, to escape it, worked their way up the rivers, and adapted themselves to fresh-water life.

Let us now see what these fresh-water animals are like, to what groups they belong, and what are their most important characteristics. The first of the eight large groups into which the animal kingdom is usually divided is that of the Protozoa. These include the simplest unicellular forms of animal life, and occur very abundantly in fresh water, though certain important subdivisions—e.g., the Radiolaria, are almost exclusively marine. Of the next group, or Sponges, some thirty-nine or forty families are recognised, of which one only occurs in fresh water, the rest being exclusively marine. Among the Coelenterata, to which the polypes,[89] sea anemones and corals belong, much the same state of things occurs, for of about seventy families only three have fresh-water representatives; whilst of the next group, the Echinodermata or starfish, sea-urchins, etc., not a single species inhabits fresh water. In the remaining groups we find the freshwater forms rather more abundant. The Vermes, or "worms," using the term in its widest sense, have a large number of fresh-water representatives, such as the river worms, Rotifera; but here also the greater number of families of worms have no fresh-water members.[90] Among the Mollusca we find that several of the best marked groups never get into fresh water, while other groups, as the snails and bivalves, have a large number of fresh-water representatives. The seventh group, or Arthropoda, tells much the same tale; for though the lower forms, such as the Entomostraca,[91] occur abundantly in fresh water, the higher families are very poorly represented. Turning to the last group, that of the Vertebrates, we find that there are a very large number of fresh-water fish; yet even here, out of 137 families of fish recognised by Dr. Gunther and other leading authorities, there are only thirty-five that get into fresh water, and many of these are only represented by single species. On the other hand the Amphibians, such as newts and frogs, form perhaps the most purely freshwater group of animals known. They are, with very few exceptions, aquatic, at any rate in their earlier or tadpole stages and yet no single Amphibian is found in salt water.

The general conclusion we arrive at is that only a very small proportion of the families of marine animals have made their way into fresh water. Of the eight large groups into which the animal kingdom is divided, one is exclusively marine, two others have an exceedingly small number of freshwater representatives, whilst in the remainder, fresh-water forms, though more abundant, are confined to certain families.

It becomes now of interest to enquire why it is that the bulk of marine animals do not make their way up the rivers, and thereby escape from the enemies that devour

---

89   The more usual spelling now is polyps.

90   This scarcely does justice to *worms* in the sense that we use the term today. Rotifers may have a worm-like shape but comprise a completely separate group of animals from polychaetes (round and ringed worms) and platyhelminthes (flat worms), neither of which are mentioned.

91   An obsolete term for the Branchiopoda, Cephalocarida, Ostracoda and Maxillopoda.

them in such enormous numbers. For a long time it was supposed that the real and sufficient reason was that marine animals are unable to live in fresh water; but it is now known that though this may apply to some cases, it certainly will not to all. A series of remarkable experiments were made some years ago, at Marseilles, by Beudant.[92] He took a large number of specimens of different species of marine snails and bivalves, and by gradually adding fresh water to the salt water in which they were originally living he succeeded in gradually converting them into fresh-water animals. During the experiments only 37 per cent. of the animals died, 63 per cent. surviving the change from sea water to fresh water. This result is rendered still more remarkable by a check experiment performed at the same time, in which an equal number of individuals of the same species as before were taken, and kept in salt-water tanks, when it was found that 34 per cent. died; so that the difference in mortality between those living in their natural medium, and those constrained to change from marine to freshwater habitat, was but 3 per cent.

A still more remarkable case is that of the two Entomostracan genera, Artemia and Branchipus, of which the former is marine, the latter freshwater.[93] Of Artemia two species are known, the differences between which are very well marked. *Artemia salina* lives in water containing from 4 to 6 per cent, of salt, and is found in the brine pans of salt works, and in other similar places. The other species, *Artemia Milhausenii*, requires water containing not less than 25 per cent. of salt. By direct experiment it has been shown that the differences between the two species depend simply on the percentage of salt in the water in which they live; and that by gradually adding salt, or adding water, *Artemia salina* may be converted into *Artemia Milhausenii* or *vice versa*. This experiment has been performed not merely in the laboratory, but also by Nature herself.[94]

The reverse experiment also succeeded, and by gradually adding fresh water, and so reducing the strength of the solution, *Artemia Milhausenii* has been converted into *Artemia salina*. The experiments have been pushed still further: and by addition of fresh water *Artemia salina* has been converted into the fresh-water genus *Branchipus*, which is of larger size, and in many respects very different.

A number of other cases could readily be quoted illustrating the same point, namely, that many marine animals which never do make their way into rivers, are quite capable of living in fresh water, if the change is made sufficiently gradually. If therefore it is not the difference between sea water and fresh water that prevents marine animals from entering rivers, there must be some other cause or causes at work. One of these is, undoubtedly, the severity of the climatic conditions to which fresh-water animals are exposed, as compared with the practically uniform temperature of the sea. Further, the variation in the amount of water in rivers at times of

---

92   Several authors mention work by Beudant in 1816 on the saline sensitivity of ormer (abalone; *Haliotidae*), but by anecdotal reference rather than direct citation. There seems to be no available source for the work which AMM refers to here. The identity of Beudant is unclear but it is probably Françoise Sulpice Beaudant (1787–1850), a French mineralogist who studied in Marseilles. Whether he carried out experiments on molluscs is not known.

93   See Lecture 2 for a more detailed account of this topic.

94   {M} *Vide* Lecture on "Environment," Lecture 2.

drought and flood respectively, the changes in the strength of the current and modifications in the character of the water from sewage and other contamination, all conspire to render the environment a very shifting one and at times a very harmful one, as compared with the much greater uniformity of external conditions under which the marine fauna exists.

But even these causes, potent as they undoubtedly are, will not suffice to account for so few marine families getting into fresh water. Some further explanation is required, and this is, I believe, to be found in a suggestion made originally by Professor Sollas, of Dublin, who has pointed out that though the adult animal might be thoroughly well adapted to fresh-water life, yet it would by no means follow that the early stages of development would be equally well suited to the change. Professor Sollas also shows that the early larval stages of most of the marine animals are forms peculiarly ill-adapted to living in fresh-water streams, forms indeed which could not hold their own under such conditions. Thus, in the *Echinodermata*, which we have already seen are exclusively marine, the young hatch as very minute larvae swimming freely in the water by means of cilia; and larvae similarly occur in almost all the other groups of marine invertebrates. Now such small ciliated larvae are altogether unsuited to fresh-water streams, and could never hold their own in them. They are quite incapable of swimming against even weak currents, and the inevitable consequence is that each succeeding generation would be carried further and further down stream, and ultimately the whole species carried out to sea. This is a very important point, and as it is one that has not yet received general attention I think it will be well that we should enquire into it more fully, taking the several groups one by one and seeing how far it will serve to explain the special characters in the distribution of fresh-water animals.

Before doing so, there is one further point of a preliminary nature that will require attention. It is a very familiar fact that the eggs of different animals vary greatly in size. Thus, the egg of a herring is about the size of a pin's head, that of a salmon is as big as a pea, while that of a dog-fish, or of a hen is very much larger still. Every egg contains two chief kinds of matter, germ-yolk and food-yolk, the former of which develops directly into the embryo, while the latter is simply a store of food material generally in the form of minute granules dispersed through the germ-yolk, which can be drawn upon as required and at the expense of which the embryo is able to develop. The difference in size between one egg and another concerns almost exclusively the amount of food-yolk, which is very abundant in large eggs, but comparatively scanty in small ones.[95] Until the food-yolk is used up there is no need for the embryo to hatch: consequently embryos developed from large eggs will be of much larger size and greater strength at the time of hatching than those developed from smaller eggs. This is well illustrated in the development of frogs. In the ordinary frog the egg has but a comparatively small amount of food-yolk and the embryo hatches as a tadpole, an animal of small size and of much simpler organisation than the frog. In the little West Indian frog Hylodes however, the eggs are of larger size—*i.e.*, they contain more

---

95   A further difference is that in the microlecithal eggs of amphibians the yolk divides along with the embryo during the pre-hatching stage, whereas in the macrolecithal eggs of birds the large yolk remains undivided as the embryo develops. (For further discussion see Lecture 10.)

food-yolk—and the embryo consequently hatches, not as a tadpole, but as a fully-formed frog.

As small free-swimming larvae are as a rule unable to hold their own in fresh water it follows that any increase in the size of egg—*i.e.*, in the amount of food-yolk—will be a direct advantage to a fresh-water animal, enabling it to hatch of greater size and strength and consequently better able to resist the currents of the river. Hence we find that in fresh-water animals the eggs are usually larger than in their marine allies. A crayfish, for example, though only a third the length of a lobster lays actually bigger eggs. Having thus seen what are the special conditions under which freshwater animals exist, we may now proceed with our inquiry and consider in what way the fresh-water representatives of the several groups of animals meet these conditions.

Concerning the Protozoa I am not aware of any points, either in structure or life-history, that would distinguish the fresh-water from the marine forms. The mode of life is apparently the same in the two cases; though it is very possible that closer examination would show that differences do exist, at any rate in certain cases. From this point of view a careful study of the fresh-water Protozoa might very possibly yield results of interest and importance. Among Sponges Spongilla is the most familiar one occurring in fresh water. Spongilla reproduces sexually, like the marine sponges; but, unlike the latter, it can also reproduce by means of special buds or "gemmules."[96]   Each gemmule consists of a little spherical group of cells, formed in the deeper part of the sponge, of which the superficially placed cells become specially modified so as to form a thick projecting capsule strengthened by peculiarly formed silicious spicules. Such gemmules are formed usually in the autumn. Owing to their protective capsules they are enabled to survive the cold, and other adverse conditions of the winter months, which are often fatal to the parent sponge; and in the spring the capsule ruptures, and the contained cells crawling out commence to develop into little sponges. Such gemmules, or clusters of specially protected cells, are unknown among marine sponges, and there can be no doubt that their occurrence in Spongilla is to be regarded as a special adaptation to the freshwater habits of this genus.

Among Coelenterates the production of small free swimming ciliated embryos is almost universal, and this is almost certainly the reason why so very few fresh-water members of the group are found. Of those that do occur, the best known is the fresh-water Hydra, whose marvellous powers of recovery from injury have been so admirably investigated and described by Mr. Dunkerley.[97,98] As regards its life-history Hydra exhibits several peculiarities, which I believe, are to be associated with its fresh-water habits. In the first place, unlike its marine allies, it not only does not give rise to free-swimming reproductive zooids or jelly-fish, but does not even show

---

96   This word appears in the Lectures with two meanings: as here, referring to the asexual reproductive cells of Spongilla, and elsewhere in the context of Darwin in his (unformed, provisional) explanation of the apparent particulate mechanism of inheritance (see Lectures 7 and 11).

97   {M} "Hydra: its Anatomy and Development," by J. W. Dunkerley, F.R.M.S.

98   Dunkerley JW, Hydra: its anatomy and development. *Manchester Microscopical Society Reports*, 1883–84; 19–33. No further information is available about Dunkerley although he may have been a dental surgeon. FRMS = Fellow of the Royal Microscopical Society. Presumably he was a member of the Manchester Society when this lecture was given.

the slightest tendency to form such zooids. Secondly, only a single ovum, and this of very large size, is found in the ovary, a point in which Hydra is unique among Coelenterates, and a point the special advantage of which, in fresh-water forms, has already been fully discussed. Furthermore, this single egg has a special protective capsule formed around it, and the young at the time of hatching has the form of the parent and adopts at once its mode of life.

Although jelly-fish, which are weak swimmers, are apparently altogether unsuited to fresh-water life, yet it is of interest to note that one truly fresh-water form, Limnocodium,[99] is known, while several others occur in brackish water.[100] Limnocodium has however only been found as yet in ponds, not in streams, and the mode of its development is not known. It was discovered a few years ago in large numbers in an artificially heated pond of the Regent's Park Botanical Gardens, and may possibly have been introduced with the Victoria regia or other plants living in the pond.

Concerning "worms," few worms are more familiar to microscopists than the Rotifers; active little animals, sometimes swimming freely, some-times attached and living in tubes, and sometimes massed together into colonies. Nothing is more remarkable concerning them than their extraordinary power of resisting desiccation. Mr. Sykes[101] has recently sent me some dust he collected more than a year ago from a gutter on a house-top, and containing some of these dried-up Rotifers. On placing a little of the dust in warm water for about half an hour, a number of the Rotifers came out and swam about actively. Rotifers have been known to exist in this dried-up condition for many years, and there is no doubt that their geographical distribution, and their persistence as fresh-water forms is largely dependent on this power, for while in the desiccated condition they can withstand the action of cold or drought, and can also be blown by the wind for very considerable distances from pond to pond, or stream to stream.

Polyzoa[102] again, of which some of the most beautiful and interesting forms are fresh-water, exhibit special modifications in accordance with their habitat. Almost all the fresh-water Polyzoa give rise to specially protected buds or "statoblasts," which are collections of cells very similar to the gemmules of Spongilla,[103] and like these, enclosed in hard protective capsules. These statoblasts survive the winter, which is usually fatal to the parent, and in the spring a young fully formed Polyzoon emerges from each, and at once adopts the mode of life of the parent.

Among the Entomostraca, such as Daphnia and Cyclops, we meet with what seem at first sight to be striking exceptions to the rules we have laid down concerning fresh-water animals; for while some, such as Daphnia, produce large and specially protected eggs which hatch in the full form of the parent, yet a very large number of others, such as Cyclops and Cypris, produce very small eggs, from which small

---

99   Now called *Craspedacusta* (sp. *sowerbii*).

100   {M} A second genus of fresh-water Medusae (*Limnocnida*) has since been found in Lake Tanganyika.—ED.

101   No further information has been found. Presumably he was a member of the Manchester Microscopical Society.

102   Now called Bryozoa, although that term was evidently in use as early as 1831.

103   {M} These have since been shown (*Ephydatia Mülleri*) to retain the power of giving rise to free embryos after two years desiccation.—ED.

free-swimming embryos arise. We may note however that in these cases the small free-swimming larva, or nauplius, though very unlike the parent, is provided with very powerful swimming appendages, and is therefore able to hold its own against streams which would be fatal to those small larvae which depend for locomotion on the action of cilia. In fact in these Entomostraca the mode of life of larva and adult is the same, and hence if the adult can hold its own, there is little reason why the larva should not do so also. Still it is a point that requires further investigation, how it comes about that Daphnia should lay large eggs, while the allied genera, Cyclops and Cypris, living under the same conditions, and in the same ponds and streams, produce very small eggs.

Concerning these fresh-water Entomostraca there is another point of much interest, that would well repay further examination. In the sea there are marked differences between the shore and shallow-water animals living near to the land, and on the other hand the oceanic or pelagic forms that are met with in the open sea hundreds of miles from land. The same distribution applies to the large fresh-water lakes, in which shallow-water and pelagic fauna may also be recognised.

The pelagic forms are very generally characterised by the possession of eyes of unusual and often of gigantic size, and by their habit of remaining down at some depth during the daytime and only coming to the surface at night. Professor Weismann has suggested that these two peculiarities may be associated together, in this way. Ordinary daylight only penetrates to a limited depth in water, and it has been shown that below 25 fathoms[104] photographic paper is not acted on by daylight. It is also found that 25 fathoms is about the depth to which the pelagic Entomostraca descend during bright sunshine. The object of so descending is apparently to utilise the light, so as to range during the four and twenty hours over their whole hunting ground for food. Were they to remain at the surface in the daytime they would lose their sole chance of obtaining food from the greater depths. It is for this reason again that the eyes are of such great size, for it is clear that forms with larger and more perfect eyes would have a better chance in the pursuit of food than those less well equipped; and hence by the action of natural selection, the large-eyed forms would survive, and any further improvement of the eyes would be preserved and transmitted to their descendants.

Turning now to the Mollusca, one of the best known and interesting fresh-water forms is the common river mussel, *Anodonta cygnea*.[105] Like bivalves in general, Anodonta produces eggs of small size; however, these are not immediately passed from the body of the parent, but are transferred to the outer gills where they remain for some time, and where they pass through all the early stages of their development.

The gill consists of a couple of laminae fused together along their ventral edges; each lamina is not a continuous membrane, but a trellis-work composed of very numerous vertical bars, crossed and connected together by a smaller number of horizontal bars. The two laminae are further bolted together at intervals by cross bars. The eggs, which are exceedingly numerous, form a bulky mass lying between the two laminae. Owing to the cilia which clothe the outer surface of the bars, streams of water are constantly passing through the meshes of the trellis-work; and in this

---

104   A little under 46 metres.
105   The Swan Mussel.

way the embryos not merely obtain perfect protection during the early stages of development, but are also abundantly supplied both with oxygen for respiration, and with food in the shape of minute particles of vegetable or animal matter suspended in the water. The outgoing stream will also serve to carry away any excretory or faecal matter passed out by the embryos.

At the time of leaving the parent the young Anodonta is a fully formed bivalve, but is still very unlike the parent, and of too small size, and too weak to hold its own against the currents of the streams in which the adult lives. The young at this period have bivalved shells, the two halves of which are triangular, with the apices incurved so as to form sharp serrated projections, which, when the valves are closed, form a very efficient pair of pincers; they have a very rudimentary foot and gills, and in other respects differ markedly from the adult.

On passing out from the gills of the parent into the water, they swim by snapping movements of the valves, like a Pecten,[106] but very speedily attach themselves by the pincer-like processes of their shells to fish, such as the stickleback, or else to the legs of water birds. The skin either of the fish or bird, irritated by the pincers, swells up and forms a capsule within which the young Anodonta completes its development. When it has attained the adult form it leaves its host, drops down to the bottom of the stream, and henceforth leads the life of the adult, half buried in the mud at the bottom of the stream, along which it slowly ploughs its way by means of its powerful muscular foot.

No better illustration could possibly be given of the special modifications acquired by fresh-water animals as regards their life-history. Marine bivalves, such as the oyster or cockle, all give rise to small ciliated embryos, adapted to a free swimming existence. So also does Anodonta, but the embryos are not set free in the streams, where as we have seen they would be entirely unable to hold their own, but are retained within the gills of the parent until they have attained sufficient size and strength to look after themselves. The second phase in their life-history, during which they are attached to fish or birds, has probably been acquired rather as a means of ensuring wider distribution than as a precaution against the strength of the currents. The adult Anodonta leads a very sedentary life, while through the stickleback or bird, new colonies may be started miles further up stream, or even in other streams some distance off.

Among fish very numerous examples are met with of marine animals that have taken partly or completely to fresh-water life. In many cases, as in the salmon, the fish live normally in the sea, and run up rivers merely for the sake of laying eggs in places where they are exposed to less danger than in the sea. It is the species and not the individual that here benefits,[107] for while the salmon on running up the rivers in the winter months is fat and in good condition, on its return journey, after laying eggs, it is in wretched condition, and very many die on their way back to the sea. Many other fish, such as lampreys and sturgeons, have similar habits, ascending rivers during

---

106   Scallop.

107   One can understand what AMM means by this although with a modern perspective it ignites the debate over individual vs. group selection. One resolution, of course, is to view individuals simply as vehicles for the transmission and spread of genes. See also footnote 324, Lecture 12.

the spawning season. On the other hand many characteristically fresh-water fish, as the stickleback, frequently descend the rivers to the sea, in which they are perfectly at home.

The large size of the eggs of fresh-water fish, as the trout, or such fish as breed in fresh water, as the salmon, is worthy of notice. In consequence of the larger supply of food contained in the eggs, as compared with those of marine fish, the young hatch of larger size and greater strength, and are therefore better able to cope from the time of their birth with the downward currents of the streams and rivers.

The Amphibians may be briefly alluded to in conclusion. They are a characteristically fresh-water group of animals, the chief interest about which is that during their development, at least in the higher forms, such as the frogs, they pass from a gill-breathing aquatic fish-like stage, to a lung-breathing terrestrial condition, and show us very clearly steps by which air breathing vertebrates have been evolved from the more primitive water-breathing forms.

Summarising the results arrived at, we may say that fresh-water animals have almost certainly descended from marine animals which have become habituated to fresh-water life: that very little modification, if any, is required in the adult structure to fit an animal for the change in habitat; and that many purely marine animals will live readily in fresh-water, if the change from one to another is sufficiently gradual. We have further seen that the reasons why so few groups of marine animals have given rise to fresh-water representatives, concern not so much the adult condition as the earlier larval stages; and that it is the inability of small free-swimming ciliated larvae to hold their own against the currents in the streams and rivers that is probably the main cause of this paucity of fresh-water species.

An examination of the life-history of those forms that have established themselves in fresh-water, has shown that in many cases there are very special devices of a curiously interesting kind to enable the animals to get over these difficulties. We have only had time to notice a few of the more striking instances, and very much yet remains to be done before we can explain fully all the peculiarities of the fresh-water fauna.

The aspect of the question which I have touched on this evening, has only very recently attracted notice; the subject is one of great importance and interest, and I would venture to commend it very earnestly to the attention of members of the Society, as one the further and more systematic investigation of which would, beyond doubt, yield results of high scientific values.

LECTURE 7

INHERITANCE

This lecture, also given to the Manchester Microscopical Society as the 1888 Presidential Address, starts with a eulogy to microscopy and the benefits of its application in biology and other pursuits. Following this lengthy introduction, AMM formulates his main question in a typically direct and accessible way: *Why is a child like its father?* After some contextual observations on inheritance, he sets out with equal clarity the two leading theories of the day, by Darwin (almost entirely in its originator's own words) and by August Weismann. He then proceeds to point out the attractions and inadequacies of each: Darwin's gemmules provide a mechanism but defy belief on quantitative grounds, whilst Weismann's germ-plasma theory (a working hypothesis involving a separation of inherited matter—flowing in an "immortal river"—from the rest of the body, which he first set out in 1885 and defended with modifications throughout the rest of his life) provides the necessary continuity without offering a mechanism.

As with our other readings of these lectures, our current response to AMM's question is inevitably coloured by our modern understanding of genetics, a set of bio-chemical and cellular processes of which he and his audience had not the slightest notion. In particular, we accept without question that the mechanism of inheritance is particulate (through alleles, acquired from parents on a fixed odds basis) and that characters are thus not open to blending. The numerical hints offered by Galton, and the intriguing developments in cell theory happening elsewhere around the same time (Lecture 9), all taking place in ignorance of the yet-to-be-appreciated work of Mendel, only serve to emphasise the uniqueness of the time at which AMM gave his lectures.

The molecular insights of the 20th century allow us to accommodate both of the deficiencies which AMM finds in the two theories he presents. Given his musings in the lecture's final paragraph, he would have been delighted by the progress made. It is humbling, nonetheless, to realise that biologists are still asking his question (albeit with greater even-handedness of gender) and are likely to continue to do so for many years to come.

LECTURE 7

INHERITANCE

THE members of a Microscopical Society may find very legitimate cause of congratulation in the progress that is being daily made in the use and application of their favourite instrument. As regards natural history—the history of nature—it may rightly be said that the microscope has effected a veritable revolution; and this not in one branch only, but in all. In Zoology, it has rendered possible the detailed examination of forms barely perceptible, or even invisible, to the naked eye: witness the five huge volumes lately published in the reports of the *Challenger* Expedition,[108] in which are recorded Professor Haeckel's researches on the Radiolaria,[109] and those of Professor Brady on the Foraminifera. It has also revealed to us numberless facts of the utmost importance concerning the minute structure of animals whose general anatomy was previously well known: facts which in many cases have shown that our earlier ideas as to the affinities of these forms were defective or erroneous. But far more than all this, it has opened up to us the science of Embryology—the most fascinating study a man can pursue—which not merely teaches us the several stages through which the complexities of adult structure are reached, but also enables us to reconstruct the pedigrees, the genealogical histories of the various groups of animals, and proves to us that morphological resemblances are no mere accidents, but are indications of true blood relationship between one form and another.

---

108   *HMS Challenger* set out from Portsmouth in December 1872 on the world's first major oceanographic voyage. It carried 243 crew and 6 scientists and travelled 127,000 km, principally through the southern oceans, before returning in May 1876. The expedition identified 4,700 new species of life, besides discovering new features of the sea bed including the Marianas trench and the mid-Atlantic ridge. According to the *Challenger Society for Marine Science*: "[T]he scientific findings of the cruise were examined by over 100 scientists, primarily under the guidance of John Murray, who should receive the highest praise for the work's eventual publication *The Report of the Scientific Results of the Exploring Voyage of HMS Challenger* during the years 1873–76 occupied 50 volumes, each measuring about 13 by 10 inches and as 'thick as a family Bible'. They appeared between 1885 and 1895. Scientists involved with collecting and publishing the data were awarded with a specially-minted medal, the original Challenger Medal" (www.challenger-society.org.uk/History_of_the_Challenger_Expedition).

109   Haeckel's studies on radiolaria, which famously include 140 plates of painstakingly prepared drawings and paintings, were published in 1887. He produced three other reports based on *Challenger* material, dealing with *Deep-Sea Medusae* (1881), *Siphonophora* (1888) and *Deep-Sea Keratosa* (1889). Professor Brady was Henry Bowman Brady, FRS, FLS, FGS (1835–1891) who published *Report on the Foraminifera Dredged by HMS Challenger during the Years 1873–1876* in 1884 (part of *HMS Challenger Reports*, Volume 9).

In Botany, results of similar nature have been attained; and the amount of space accorded in our more recent text-books to the lower orders of plants is cogent evidence of the importance attached to the results of microscopical examination by those best qualified to pass judgment upon them. Nor has the benefit been confined to the student of biological science. To the geologist, the microscope is rapidly becoming as important, and indispensable as it has long been to the zoologist or botanist. Like his biological brethren, the modern geologist is no longer content with a knowledge, however accurate and minute, of the structure and present condition of the rock he is examining, but recognises that in the microscope he has a means which, used aright, will enable him to unravel the pedigree of the rocks, to reconstruct their past history, and to determine through what series of changes, in what order, and by what agencies, their present condition has been brought about.

The limits of time and space forbid that I should refer, save in the briefest manner, to the ever-widening applications of the microscope in other branches of knowledge, and the benefits which it is conferring on mankind. In trade and commerce the microscope is employed more frequently and relied on more fully than is generally appreciated; and in such matters as the adulteration of food, and in criminal enquiries, it often yields evidence of the most material and convincing character. There is perhaps no direction in which microscopical enquiry has advanced more rapidly of late years than its employment as a means of detecting disease, or even of determining its true nature and causation. Pathology is one of the most actively growing of modern sciences; and though its fulness of time has not yet come we may feel well assured, from the results already attained, that the scientific, and especially the microscopical, investigation of disease, will in the immediate future afford us most powerful and welcome assistance in the alleviation of human suffering. It is in considerations of this kind—presented here, I am but too well aware, in the crudest possible form— that we find the justification for the high position to which a Microscopical Society may rightly aspire. In the microscope we have perhaps the most potent instrument of research that mankind has ever possessed; and in the ever-widening circle of its influences, in its far-reaching applications, we may see opportunity for enrolling amongst our numbers men of the most varied interests and pursuits, and so gaining that free interchange of independent opinion which is one of the highest privileges and delights of civilised humanity.

One sometimes hears it said that a microscopist, being occupied with small things, is usually, perhaps of necessity, a man of small ideas. This may possibly be true in individual cases, but as a general statement I believe it to be utterly false. No better justification of this belief could be found than is afforded by the present state of our knowledge with regard to the subject I have chosen for my address.

The real nature and *modus operandi* of Inheritance are problems of the widest possible interest and importance, problems which have baffled many in the past, and which are at this moment being attacked by different observers, working from different sides, and along different lines of attack, but all relying for their evidence on microscopical observation. In dealing with the subject of Inheritance it is well to bear in mind that the problem is as yet unsolved, the question still an open one. Neither can I myself make any material contribution towards its solution: all I propose to do here is to indicate the main conditions of the problem, and to point out what appear, in the light of recent investigation, to be the most promising lines of attack. The problem itself is familiar enough, and may be expressed in its simplest form by the question—Why is a child like its father? Why is it; how does it come about that a young animal resembles and grows into the likeness of its parent? Stated thus, the problem seems definite enough. However, it is really much more complicated than appears at first sight, and it will be well to consider briefly certain preliminary matters before dealing with the more serious attempts that have been made to grapple with its difficulties.

In the first place it should be remembered that reproduction, whether of animals or plants, is effected in two principal ways; asexual and sexual; and that the phenomena of inheritance are seen in both cases. Thus, to take a familiar case, the common fresh-water Hydra reproduces either by budding, or by the formation of eggs. The former is an asexual process, the bud appearing as a hollow outgrowth from the body wall of the parent, which acquires mouth and tentacles at its distal end, and after a longer or shorter time detaches itself and becomes an independent Hydra. An egg, on the other hand, is a specialised cell of the ectoderm, or outer layer of cells of the parent, which is incapable of development of any sort until it has been fertilised by a spermatozoon, from the same or another animal. After fertilisation the egg segments, *i.e.*, divides repeatedly so as to give rise to a number of cells from which, by further growth and differentiation, an embryo and ultimately an adult Hydra is produced. The two modes of reproduction, sexual and asexual, are absolutely unlike, and yet the final results are the same; for so far as we are aware, there are no points of difference that will distinguish with certainty a Hydra produced by budding from one produced from a fertilised egg. Inasmuch as these two forms are not only similar to each other, but similar also to their parents, it follows that a Theory of Inheritance, to be of any real value, must apply equally to sexual and asexual processes of reproduction. I lay stress on this point as it is one which appears to me to have been very frequently overlooked, especially of late years.

The power of repairing mutilation, that is possessed in so marked a degree by many animals, is another phenomenon of which account must be taken in any Theory as to the real nature and mode of action of Inheritance. A Hydra may be cut into many pieces, and each piece will regenerate the missing portions and give rise to a perfect animal. One or more of the arms of a starfish may be removed, the whole of the viscera of an Antedon[110] may be turned out, the leg of a crab or the limb or tail of a newt may be cut off, and the loss will in each case be made good. Spallanzani cut off the tail of a salamander six times in succession, and Bonnet eight times;[111] while the eye-bearing tentacle of a snail has been removed twenty times; and yet after each mutilation the missing organ has been reproduced. In the more highly organised animals, such as birds and mammals, this power of repairing mutilation is much restricted: removal of parts of the epidermis is, however, readily made good; while such operations as skin-grafting, and transfusion of blood from one animal to another, show that isolated parts of even the highest animals may retain their vitality and special properties when placed under favourable conditions. More striking examples are afforded

---

110   Free-swimming, crinoid echinoderms.

111   For a comprehensive account of the experiments of Spallazani and Bonnet, including the correspondence between them, see Dinsmore CE, *International Journal of Developmental Biology* 1996; **40**: 621–627.

by such cases as those quoted by Mr. Darwin:[112] in one instance, the spur of a cock inserted into the ear of an ox lived for eight years, and grew to a length of nine inches; while in another, the tail of a pig removed from its natural position and grafted into the middle of the animal's back lived for a time and recovered sensibility.

The phenomena of Reversion are again of great importance in reference to the problem of Inheritance. It is well known that animals may transmit to their offspring characters which are not manifested in themselves: the tendency of gout and some other diseases to appear in alternate generations is perhaps the most familiar instance. In such cases we must regard the disease, or other peculiarity, as present in a latent form in the generations which it apparently skips, for how otherwise could we understand its reappearance in a later generation? The tendency of domestic animals, and more especially of cultivated flowers and fruits, to revert—either in form, colour, or other characteristics—to the ancestral wild condition, is another good illustration of what may well be termed Latent Inheritance. Readers of Mr. Darwin's works will call to mind his famous experiments on pigeons; more especially that crucial one in which he first paired a black Barb with a red Spot; then another black Barb with a white Fantail; and then paired the mongrel Barb-Spot with the mongrel Barb-Fantail, the result being that he obtained a family of birds which in colour and markings were almost identical with the blue Rock pigeon, the common ancestor of all domestic pigeons.

The tendency of cultivated and domesticated plants and animals to revert to a former ancestral condition may perhaps be illustrated mechanically in this way. Take a pack of cards, and lay it on the table; the cards will all lie on their sides, and be in a condition of stable equilibrium, so that the table may be struck or shaken without materially affecting their position: this represents the normal, i.e., the wild or ancestral condition of the race.

Now arrange the cards on their edges, resting them against one another, and so building them up into a pagoda: the resulting structure is a far more imposing one than the pack of cards when laid flat on the table, but it is also an eminently unstable one, its instability being directly proportional to the extent to which it departs from the initial condition: a very slight shake or push of the table will cause the whole structure to collapse, and revert to its condition of initial stability, the cards all falling flat on their sides as at first. So the Pouter or the Fantail are much more impressive and remarkable birds than the blue Rock, but the former are artificial productions, in a condition of great instability, and very readily revert to the ancestral condition. These are but a few of the considerations which must be kept in mind when dealing with Inheritance. Let us now consider in what way the problem may best be attacked.

Of Theories of Inheritance there are two which have attracted special attention, and which demand careful consideration. These are, first, Mr. Darwin's "Provisional Hypothesis of Pangenesis;" and secondly, the view more recently advanced by Professor Weismann, of Freiburg.

---

112   In *Origin of Species*, Chapter 1: "Variation Under Domestication".

Mr. Darwin's Theory is stated by himself as follows[113] "*It is universally admitted that the cells or units of the body increase by self-division or proliferation, retaining the same nature, and that they ultimately become converted into the various tissues and substances of the body. But besides this means of increase, I assume that the units throw off minute granules, which are dispersed throughout the whole system; that these when supplied with proper nutriment, multiply by self-division, and are ultimately developed into units like those from which they were originally derived. These granules may be called* gemmules. *They are collected from all parts of the system to constitute the sexual elements, and their development in the next generation forms a new being; but they are likewise capable of transmission in a dormant state to future generations, and may then be developed. Their development depends on their union with other partially developed or nascent cells, which precede them in the regular course of growth. Why I use the term union, will be seen when we discuss the direct action of pollen on the tissues of the mother-plant. Gemmules are supposed to be thrown off by every unit, not only during the adult state, but during each stage of development of every organism; but not necessarily during the continued existence of the same unit. Lastly, I assume that the gemmules in their dormant state have a mutual affinity for each other, leading to their aggregation into buds or into the sexual elements. Hence, it is not the reproductive organs or buds which generate new organisms, but the units of which each individual is composed. These assumptions constitute the provisional hypothesis which I have called Pangenesis. Views in many respects similar have been propounded by various authors.*"

It will be seen that Pangenesis is a mechanical theory of Inheritance; and that it recognises and faces fully the difficulties of Reversion and of Repair of Mutilations, and explains how organs may become abnormally multiplied and transposed through the gemmules developing accidentally in wrong places, as in the case of supernumerary fingers or toes, or the development of hairs or teeth in unusual situations.

Pangenesis, in spite of the ready explanation it gives of many difficulties, has never met with anything like general acceptance. Its illustrious author, however, while careful to speak of it always as a Provisional Hypothesis, regarded it with much affection; and alludes to it almost pathetically as his neglected child, for which he predicts confidently a future career of greatness. Such an opinion claims the highest respect, and compels the utmost caution in criticising the theory: yet it cannot be denied that it involves certain difficulties which seem of great weight, and which have not yet been satisfactorily met. In the first place, there is the mechanical difficulty of the extraordinary numbers of these gemmules which must be present. Gemmules must be derived from every component cell of the body: for Mr. Darwin lays much stress on the independence of these cells, quoting Virchow to the effect that "*Every single epithelial and muscular fibre cell leads a sort of parasitical existence in relation to the rest of the body ... every single bone corpuscle really possesses conditions of nutrition peculiar to itself.*" Again, Sir James Paget speaks of each cell as living its appointed time, and then dying and being cast off or absorbed; while further on Mr. Darwin

---

113   This and the subsequent quotes from Darwin are from: *Variation of Animals and Plants Under Domestication*, Volume II, Chapter XXVII: "Provisional Hypothesis of Pangenesis".

continues, *"I presume that no physiologist doubts that, for instance, each bone-corpuscle*[114] *of the finger differs from the corresponding corpuscle in the corresponding joint of the toe; and there can hardly be a doubt that even those on the two sides of the body differ, though almost identical in nature."*

Again, these gemmules must not only be formed from every cell, but must be present in enormous numbers from every cell, and at every period of life: for it is well known that a portion of a leaf of Begonia or Asplenium can reproduce the whole plant; and to do this there must, on the theory of Pangenesis, be present in this portion of the leaf gemmules from every part of the plant. So again, the repair of mutilations can only be possible through the presence of a great reserve stock of gemmules of every kind, from all parts, Moreover. it must be borne in mind that the component cells of the body are not simple, homogeneous structures, but that, as is daily becoming more and more evident, some at any rate of them have an exceedingly complex structure, consisting of parts of very different composition, and discharging very different functions. Hence it must follow that many kinds of gemmules, each in enormous numbers, must be required from a single cell in many cases. Inasmuch as the body structure of the young and of the adult animal are different, there must also be different sets of gemmules for the several stages of existence: and it becomes a matter of the utmost difficulty to form the slightest conception of how these different sets take up the running in successive stages of development. If we remember that gemmules from all parts, from each component cell of the body, have not only to be formed in great numbers, but have also to find their way about the body to be collected together from the most remote parts, and planted in due proportion in each of the ova or spermatozoa of the parent, we become fairly staggered at the magnitude of the operations we are asked to believe that each animal performs with such apparent ease.

It must also be remembered that Pangenesis requires that besides the active gemmules there must be enormous numbers of latent gemmules, corresponding to ancestral characters, present in each egg and spermatozoon: for Pangenesis explains such instances of Reversion as the production of a pigeon practically identical with the blue Rock from a Barb-Fantail and a Barb-Spot as due to the development of ancestral blue Rock gemmules, which must be supposed to be present in all pigeons in sufficient numbers to produce fully formed offspring, though they usually remain in a latent condition. Considering the enormous number of generations that must have intervened between the original ancestral blue Rock and the present Barb or Fantail, and that each member of each of these generations must be supposed to have possessed these ancestral germs in sufficient numbers to cause Reversion if an opportunity occurred, the magnitude of the operation and the numbers of such germs originally present become simply inconceivable.

A further difficulty is found in the consideration that Pangenesis, involving as we have seen the presence of gemmules corresponding to the different periods of life of the parent, fails altogether to explain the inheritance of the characters of old age, or of any period beyond that at which the ova or spermatozoa were discharged from

---

114   The word *corpuscle* is problematic, especially as *gemmule* has been used to mean a unit of inheritance, but the easiest interpretation seems to be *cell* in the sense that we now use that word. AMM reverts to *cell* in the following paragraphs.

the parent. It is hardly sufficient to say in answer to this that "in all the changes of structure which regularly supervene in old age, we probably see the effects of deteriorated growth, and not of true development," for the objection may apply not merely to the period of old age, but to three-fourths or more of the entire life of the animal.

One further objection may be alluded to: On the theory of Pangenesis we should certainly expect that the removal of a part, such as the tail of a sheep or horse, especially when effected in early life before the breeding period has been reached, would lead at any rate to diminution of size of the part in the offspring: for surely the removal at an early age of the source from which the gemmules arise ought to have at least some effect on the transmission, through the gemmules, of this part to the offspring; yet it is well known that such mutilations do not tend to be inherited.[115]

Considerations such as these show clearly that whatever may be the ultimate fate of the Theory of Pangenesis, it is not yet in a position to command acceptance: indeed, some of the objections seem of so important and fundamental a nature as to compel us to regard them as fatal to the Theory, at any rate in its present form. Quite recently, Mr Francis Galton has published the results of a series of most laborious statistical enquiries, undertaken with the view of ascertaining whether inheritance takes place according to definite laws, and if so to determine as accurately as possible what these laws are. His researches, which are of the greatest possible interest and importance, and must exercise great influence on future speculations, lead him to a view which he refers to as Particulate Inheritance, and which may be described as an aggravated form of the Theory of Pangenesis, propounded by his illustrious kinsman. Mr. Galton states his view thus:[116] *"All living beings are individuals in one aspect, and composite in another. They are stable fabrics of an inconceivably large number of cells, each of which has in some sense a separate life of its own, and which have combined under influences that are the subjects of much speculation, but are as yet little understood. We seem to inherit, bit by bit, this element from one progenitor, that from another, under conditions that will be more clearly expressed as we proceed, while the several bits are themselves liable to some little change during the process of transmission. Inheritance may therefore be described as largely, if not wholly, particulate."* Farther on, he compares the process of inheritance to the construction of a modern building out of the corresponding parts of the ruined edifices of former days. *"This simile,"* he says, *"gives a rude though true idea of the exact meaning of Particulate Inheritance, namely, that each piece of the new structure is derived from a corresponding piece of some older one, as a lintel was derived from a lintel, a column from a column, a piece of wall from a piece of wall. We appear then to be severally built up out of a host of minute particles of whose nature we know nothing, any one of which may be derived from any one progenitor, but which are usually transmitted in aggregates, considerable groups being derived from the same progenitor. It would seem that while the embryo is developing itself, the particles more or less qualified for each new post wait as it were in*

---

115   This is clearly the anti-Lamarckian objection.

116   The following three quotes from Galton are from his most famous work *Natural Inheritance* which many consider to have founded the science of biometrics. Curiously, the publication date of that book (by Macmillan) is 1889, the year after this lecture was given to the Manchester Microscopical Society. AMM says that Galton published his results *quite recently*, so either Galton's manuscript became available in advance of publication or AMM revised the content of the transcribed version of his lecture.

competition to obtain it. Also, that the particle that succeeds must owe its success partly to accident of position, and partly to being better qualified than any equally well placed competitor to gain a lodgment. Thus the step by step development of the embryo cannot fail to be influenced by an incalculable number of small and most unknown circumstances."

These views are boldly expressed but they are also distinctly crude, and the metaphor, which Mr. Galton is dangerously fond of using, rather confuses than aids the explanation. Of more real value are his attempts to determine the numerical ratio in which characters, such as height, colour of eyes, etc., tend to be transmitted to successive generations. On this point Mr. Galton comes to the following very definite and important conclusions: *"The average contributions of each separate ancestor to the heritage of the child were determined apparently within narrow limits, for a couple of generations at least. The results proved to be very simple: they assign an average of one quarter from each parent, and one-sixteenth from each grandparent. According to this geometrical scale continued indefinitely backwards, the total heritage of the child would be accounted for."* Results of this kind are of the greatest possible value, and open up a most promising field for further enquiry.[117]

I turn now to a consideration of the important series of researches by Professor Weismann, which have of late years attracted so much attention. These are of especial interest to microscopists, because the data on which Prof. Weismann bases his arguments are obtained from a careful study, with the most refined histological methods, of the minute structure of the egg, and of the changes which it undergoes during development. This is a perfectly philosophical standpoint to adopt; for the egg, say of a hen, has the power of developing into a chick without any further assistance from the parent, provided only that certain conditions of temperature and moisture are fulfilled: and it follows that the problem of heredity is centred in the egg, and that it is therefore reasonable to hope that patient investigation of the egg-structure may throw some light on the question.

Weismann in the first instance lays special stress on the fundamental difference between the unicellular animals, or Protozoa, on the one hand, and the Metazoa, or multicellular animals, on the other. In Protozoa there is no distinction between body cell and germ cell: the entire animal is but one single cell. Reproduction is effected by simple division of this cell, and every individual peculiarity in the parent must

---

117   This final sentence, and the preceding dissatisfaction with a *crude metaphor* illustrate the importance AMM attaches to rigorous objectivity in approaching a tricky question. Neither Francis Galton (at the time when this was written) nor AMM knew anything of Mendel's work, yet the proportions stated are what we would now refer to as representing simple Mendelian ratios. According to NW Graham (*A Life of Sir Francis Galton*, Oxford: Oxford University Press, 2001), *"Galton knew about chromosomes and their segregation as he cited an excellent review by John McKendrick on contemporary cell biology, but the significance of these 'chromatin filaments' was still unknown. Nevertheless, Galton succeeded in elaborating what may be described as the second best theory of heredity's approach."* McKendrick's review (On the modern cell theory and phenomenon of fecundation, *Proceedings of the Philosophical Society of Glasgow* 1887–1888; **19**: 71–125) is dated 1887–1888 but Galton seems not to have known of it when he penned these words quoted by AMM. There is no mention of chromosomes anywhere in AMM's Lectures but hints of them can be found in Lecture 9.

therefore be transmitted directly to the offspring. Heredity therefore in Protozoa is no problem at all, but a simple and direct consequence of the mode of reproduction. In the Metazoa however the case is different: here the adult animal consists not of a single cell, but of many cells arranged variously so as to form the epithelial, muscular, nervous, and other component tissues of the animal. Of these cells certain ones are early distinguished as genital cells, and to these the power of reproduction, at any rate sexual reproduction, is confined, and in them are centred the hereditary tendencies of the whole organism.

The problem is to explain how it is that in the Metazoa, one particular cell, the ovum, should have acquired and retained this special power of transmitting the characters of the entire animal. This problem Weismann proceeds to attack. He calls special attention to the general agreement among competent observers, that the part of the cell directly concerned in the transmission of hereditary features is the nucleus; and he brings forward observations of his own in support of this view. He assumes the presence in the nucleus of a special substance to which he gives the name *germ-plasma* and to which he supposes the power of hereditary transmission to be confined.[118] He maintains that this germ-plasma is of exceedingly complex structure, and that it has the power of indefinite growth without loss of its essential characters. He further supposes that the germ-plasma of an egg is not wholly employed in building up the body of the embryo, or young animal, but that a certain portion of it remains unchanged, and produces the germ cells of the succeeding generation. In this way the germ-plasma is supposed to pass unchanged from one generation to another, and this *continuity of the germ-plasma* is regarded by Weismann as the fundamental cause of heredity. It cannot be said that this explanation is a satisfactory one. In the first place it is not really an explanation of inheritance at all; for unlike Darwin's theory of Pangenesis, it does not attempt to explain the actual *modus operandi* of inheritance, but merely localises the power of transmitting hereditary characters to the germ-plasma; and asserts that the power is due to the special and utterly unknown molecular constitution of this germ-plasma, the very existence of which is altogether hypothetical.

A more valuable part of Weismann's work concerns the share which sexual reproduction takes, in his opinion, as the direct cause of specific variation. Inasmuch as the germ-plasma is supposed to be of constant composition, and to be transmitted unchanged from generation to generation, it follows that characters acquired by the parent cannot be transmitted to the offspring. This statement, that acquired characters are not inherited, is one that has attracted much attention, and that at the present

---

118   In the HL copy of *Biological Lectures and Addresses* (see Author's apologium) a marginal note adjacent to this sentence, probably pencilled by Duckworth, says: "as in R Owen's theory of Parthenogenesis 1849". This refers to: "*Parthenogenesis, or the Successive Production of Procreating Individuals from a Single Ovum*. A discourse, introductory to the Hunterian lectures on generation and development, for the year 1849, delivered at the Royal College of Surgeons of England", published by John Van Voorst, London (see *http://darwin-online.org.uk/converted/Ancillary/1849_Owen_A649.html*). In it, Owen uses the terms *germ-cell*, *germ-mass* and *germinal vesicle* but not germ-plasm(a), so that term can be correctly attributed to Weismann.

time is being most keenly debated.[119] Should it prove to be true, it becomes necessary to look elsewhere for the cause of specific variation in animals: and this, Weismann maintains, is to be found in the mingling of germ-plasma from two separate animals in the act of fertilisation of the ovum by a spermatozoon.

Limits, both of time and space, forbid that I should follow these arguments further. We are at present very ignorant concerning the nature, properties, and reactions of living things, while life itself remains as great a mystery as ever. Perhaps it will not be until we have gained some clue to the mystery that we shall be able to understand what is the real nature of inheritance: in the meantime we may heartily welcome all earnest efforts towards the solution of the problem.

---

119   Yet another sceptical reference to Lamarckism. Weismann vehemently rejected the inheritance of acquired characteristics.

LECTURE 8

THE SHAPES AND SIZES OF ANIMALS

This 1889 President's address to the Manchester Microscopical Society raises questions which, though mildly interesting, would scarcely detain us for long today, at least not in the same way. (Some matters are also discussed in Lecture 13.) Interest in morphometrics blossomed again in the early 20th century with the work of D'Arcy Thompson and, in a more publicly approachable way, JBS Haldane, but the emphasis was more on body symmetry, proportion and energetics than on size for its own sake.

Despite the impressive observational skill of late 19th-century microscopists, perceptions of animal development and approaches to taxonomy were radically different then as compared with today. This makes it very difficult to provide a critique of the lecture with any degree of fairness to its author. I have tried in my comments to offer a modern perspective, but doing so tends to render many of AMM's arguments redundant and make his examples seem excessively selective. (Readers with a greater knowledge of embryology than I have may find more of value and interest in AMM's observations: if this is the case, I am quite content to appeal to the second of the justifications he offers in the first paragraph!)

There are hints, especially in the later part of the lecture, of a more physiological and physical approach to questions of the evolution of size and shape, which are more interesting to a reader of today. When looking at animal morphologies and the pressures under which they may have evolved we tend to head straight for arguments grounded in the accessibility of nutrients and gasses, in questions of energetic physiology, in the physics of gravitational and fluid environments, in adaptations to thermal and osmotic stresses, in the avoidance of predation, and such like. AMM is clearly comfortable dealing with these kinds of matters (frequently prompted by the work of Herbert Spencer) and it is something of a disappointment that he doesn't give them greater prominence from the start of the lecture.

Perhaps one of the most remarkable points in this lecture occurs towards the end, in the prophetic reference to an extinction threat to elephants. At the time of writing (the last day of 2016, coincidentally the 123rd anniversary of AMM's death on Scafell), the Chinese government has just announced a halt to all trade in ivory in order to assist with elephant conservation. AMM speaks of 100,000 killed each year; the current estimate is that some 350,000 African elephant remain, but that number is 30% less than it was 7 years ago. One wonders whether AMM would have anticipated their survival even this far.

A year before delivering this lecture, AMM gave a public lecture on *Elephants* to the Ancoats Brotherhood (9 December, 1888). Its exact content is not known.

LECTURE 8

THE SHAPES AND SIZES OF ANIMALS

I T may, I believe, be laid down as a general rule, that when a man deliberately selects a particular subject for conversation or for more serious consideration, his choice is made, consciously or unconsciously, for one of the following reasons. Either it is because he has himself paid attention to the subject, and has something to say about it which he thinks will be novel or at any rate interesting to his hearers; or it is because he is addressing others better instructed in the subject than himself, whose opinions he is desirous to obtain; or finally, it may be that he introduces the subject with the express though probably unavowed purpose of finding out what his own opinions are about it.

I confess at once that it is this last motive that has determined my choice of a theme for the Presidential Address, and I make no apology for this. It must have happened many times to each of us, that ideas have occurred unexpectedly on subjects to which we had previously paid but little attention; ideas, which though recognised at once as crude and disjointed, are yet felt instinctively to contain germs of interest, worthy of future development. Such ideas should not be let slip: it is well to docket them, and without attempting too soon to frame a consistent notion of their real bearing, or of the conclusions to which they may lead, it is well also to keep a look-out for any additional facts or ideas bearing on the subject, to take due note of them, and after a time to turn out one's box, go over one's notes, and take stock of one's material. At this stage conversation and discussion with others will probably afford material assistance: and in bringing before you this evening a subject concerning which my state of mind is exactly expressed by saying that I want to find out what I think about it, I believe I am utilising in a very proper and legitimate manner the advantages which such a Society as ours confers upon its members.

The problem that I desire to deal with, that of the forms and dimensions, or in more popular language, the shapes and sizes of animals, may be stated thus: Is it a mere matter of chance that animals, say butterflies and birds, have certain characteristic shapes? Clearly it is not altogether so: it is plain to every one that the shapes of animals are correlated with their habits, and that if it were a purely haphazard business there would be no reason why butterflies should so much resemble one another, or why birds should be constructed on one common plan. If then it is not mere matter of chance, can we determine in any way what are the causes which govern the shapes of animals, and what are the laws in accordance with which their effects are produced?

So with size: for each animal we have a certain standard of size, which is rarely very greatly departed from. Such names as cat, pigeon, cockroach, convey to our minds not merely impressions of animals of certain shape, structure, and habits, but also of tolerably fixed dimensions. A cockroach as big as a cat would at once arrest our attention as unusual. The problem we have to deal with is to find out, if possible, the causes which regulate the dimensions of animals, and determine that there shall be for each kind of animal a certain average size. These are most elementary considerations, but they will serve to show us what our subject is, and that it is one well worthy of attention.

With regard to the shapes of animals, we find that among the simplest animals, or Protozoa, it is characteristic of the more primitive genera that there should be no definite or consistent shape. This is well seen in Amoeba, which consists of a minute speck of protoplasm[120] equally contractile in all directions; protrusions of the protoplasm, known as pseudopodia, being put out from any part of the surface, and in any direction. In Amoeba and in all forms which exhibit similar "amoeboid" movements, there is no distinction of ends, sides, and surfaces, such as we are familiar with in the higher animals. Anterior and posterior end, right and left sides, dorsal and ventral surfaces, are terms which have no meaning in reference to an Amoeba, for any part of the animal may go first in locomotion, and when crawling the animal moves along on whatever part of its surface happens to be in contact with foreign bodies. The only distinction in such an animal is between inside and outside, and even this is not permanent in all cases. In the higher Protozoa the body presents a clear distinction between its outer or superficial layer, which is clearer, firmer, and more contractile; and the inner or central part, which is more fluid, less contractile, and usually less transparent; the former being named ectosarc, the latter endosarc. In the simpler forms however as may be seen in many Amoeba, this distinction between ectosarc and endosarc is not a constant or permanent one; the protoplasm of the whole body exhibits constant flowing movements, by which parts that at one time are at the surface, at another are carried into the interior: in such cases the ectosarc is merely the layer of protoplasm that at any one moment is at the surface; and if this differs in appearance from the more deeply placed protoplasm, such difference is perhaps due to the effects of contact with the water in which the animal dwells, rather than to any fundamental distinction in structure or composition of the protoplasm itself.

Even amongst the Amoeboid Protozoa it is however possible to speak of a distinct shape, for when perfectly at rest they all tend to withdraw their processes, and to assume more or less definitely a spherical form. This spherical shape is very characteristic either of the normal condition or of the resting state of a large number of Protozoa, and deserves further notice. When widely departed from, such departure is commonly associated with induration of part of the surface either as a mere thickening of the ectosarc, or in the form of a definite shell; a condition which is clearly a more modified one, and usually involves and is associated with further differentiation, such as the presence of a definite mouth, and usually of distinct oral and aboral ends to the body.

---

120   This word fell out of technical use from roughly the start of the second half of the 20th century. It had two, related meanings: it was a general term for the living matter of which organisms are made—the substance of life, whatever that may be—and it was also used to describe the fluid material found inside cells. AMM is clearly using it in the latter sense, for which we would now use the word cytoplasm. Protoplasm carried the implication of an amorphous colloid which was not properly understood. The histologists in AMM's audience would have known that cells contained organelles but would have been ignorant of the cytoskeleton and of the highly organized intracellular environment with which we are now familiar. See also Lecture 9.

Concerning the spherical forms, it may be noted that they are all aquatic and free swimming aquatic, because in the case of animals with soft body-substance, it is only in water that the body of the animal can be sufficiently supported on all sides to enable the spherical shape to be retained free swimming, because a habit of crawling leads, as we shall see directly, to definite modification of external shape, involving a distinction between dorsal and ventral surfaces, and almost invariably a further distinction between anterior and posterior ends. In the typically spherical animal, all parts of the surface are equidistant from the centre, and are alike in all respects; and such an animal will float suspended in the water with any part of the surface uppermost, or any part undermost.

There is much reason for thinking that the spherical shape is not merely the simplest which an animal can offer, but is also the most primitive. In evidence of this we may quote the spherical shape of the ovum, which is characteristic of all the higher animals as regards the earliest stage of its formation, and of most as regards its fully matured condition. If we are right in regarding the embryological development of an animal as a recapitulation of its ancestral history,[121] then the earliest developmental condition—i.e., the ovum, or egg—must represent the most primitive ancestral phase, and the significance of the spherical shape so characteristic of the earliest condition becomes at once apparent.[122]

A further argument in favour of the primitive nature of the spherical form may be drawn from the development of the more modified forms of cells, even in adult animals. Thus, the shapes of the epithelial cells vary greatly, according to the part from which they are taken. Cells from the surface-layer of the epidermis, such as those lining the mouth or covering the hand, are thin scales fitted together edge to edge, and with their flat surfaces parallel to the surface which they cover. Cells from the epithelial lining of the stomach or intestine on the other hand, are columnar or rod-like in form, being placed side by side with their long axes vertical to the surface they clothe. Yet a section through the epidermis shows that its deepest layer consists of spherical cells, which gradually approach the surface as those lying over them get rubbed away, and which, as they move towards the surface, get flattened more and more until ultimately they become converted into the scale-like cells. Thus, each scale-like or pavement epithelial cell is in the first instance, in its earliest stage of existence, a spherical cell. So also with the columnar cells of the stomach or intestine. Each such cell is formed in the deeper layers of the epithelium as a spherical cell, and gradually becomes elongated into a columnar cell as it approaches the surface. The spherical cell is therefore the link connecting the scale-like and the columnar cells; an indifferent or primitive form, from which either of the more modified forms may be derived, and which is really the earliest stage in the developmental history of both. It would be easy to multiply instances of this kind, but I have said enough to show that there is really strong ground for holding that the spherical form is to be regarded as a primitive one; perhaps as the most primitive form met with amongst animals.

---

121   This topic is dealt with at length in Lectures 10 and 13.

122   AMM is using the concept of primitivity to imply both simplicity and fundamentality of origin. For a related discussion concerning the use of *high* and *low* to describe the status of organisms, see footnote 62 (Lecture 3).

The truly spherical shape, in which all parts of the surface are alike and of equal value, is only seen in the unicellular animals, or Protozoa, and in the individual cells of higher animals. The early embryonic phases of many of the higher animals, known as Morula and Blastula, are also spherical, the former being a solid heap of spherical and polygonal cells resulting from the repeated division of the fertilised ovum; while the blastula is a later stage, having the form of a hollow ball with a wall composed of a single layer of cells surrounding a cavity filled with fluid. Neither morula nor blastula however is absolutely spherical in the sense in which I have used the word above, for the boundary lines between the individual cells must be of different value to the cells themselves, so that all parts of the surface cannot be identical.[123] Moreover, the component cells very usually, perhaps always, present differences of size or structure by which upper and lower hemispheres may be marked off from each other, and by which the true spherical symmetry becomes still further disturbed. Volvox and Pandorina[124] may be quoted as examples of permanent blastula in which the component cells present no such differences, but they are forms the animal nature of which is still extremely doubtful.

Leaving the spherical forms, the next characteristic shape we meet with among animals is that known as radially symmetrical, of which the most typical instances are met with in the group of *Coelenterates*; an ordinary jelly-fish affording as excellent an example as one could wish to find. A sphere is said to have an infinite number of axes, all equal to one another; but a jelly-fish may be described, if the mathematicians will pardon the phrase, as having of axes a number that can only be expressed as being one more than infinity, for its bell-shaped body, besides having an infinite number of transverse equal axes, has one definite longitudinal axis round which all the parts are symmetrically arranged. Watch a jelly-fish swimming in still water, and you will note that while locomotion is always effected in the direction of the main or longitudinal axis of the animal, the rounded end of the bell going first, the open mouth of the bell last, yet that it is a matter of indifference which part of the rim of the bell is uppermost. The animal, in fact, may be said to have anterior and posterior ends, but no distinction between dorsal and ventral surfaces, or between right and left sides; and this is the characteristic arrangement in a radially symmetrical animal.

If the spherical form is primitive, then the radially symmetrical form must be derived from it, and of this we have direct evidence in the fact that every radially symmetrical animal is developed directly or indirectly from a spherical egg. Concerning the actual historical mode of derivation of the radial from the spherical form however there has been much discussion, and the question cannot yet be regarded as settled. The chief difficulty arises from the fact that in actual development there are at least two quite different ways in which the spherical ovum may give rise to a radial larva,

---

123 The meaning of this statement is unclear. It seems to confuse the sphericity of the morula or blastula with that of the cells from which they are composed. Alternatively, it is pointing out that the surface of the structure, being made of adjacent cells, is not completely smooth.

124 Chlorophytic green algae which form spheroidal or sack-shaped colonies. Regarding their *animal nature*, chlorophytes are most certainly alive but would now be classified within the plant kingdom.

and it has not yet been determined which of these modes is the more primitive, and in what way one of them could have been derived from the other.

The more usual mode of development is as follows: The ovum, after fertilisation, divides into two cells; each of these again divides, giving four in all; the process is repeated until a solid heap of cells, the morula, is produced; then this becomes converted into a blastula by the cells moving away from the centre and becoming arranged so as to form a spherical ball, consisting of a single layer of cells enclosing a central space filled with fluid. The blastula now becomes flattened on one side and the flattened side becomes doubled up within the rounded part, so that the larva now assumes the form of a hemispherical cup, the walls of which consist of two layers of cells, outer and inner, between which is a narrow chink-like space containing fluid, which is really the last disappearing remnant of the blastula cavity of the earlier stage. The cup-shaped larva is spoken of as a gastrula. A gastrula developing in this fashion is said to be formed by invagination. Such an invaginate gastrula is of very wide occurrence, occurring as an early larval stage in members of all the large groups of the animal kingdom above the Protozoa i.e., in Sponges, Coelenterates, Echinoderms, Worms, Molluscs, Arthropods, and Vertebrates.[125]

The second mode, referred to above, in which a gastrula is formed is by what is called delamination. The starting-point, the egg, is the same as before, and so also is the gastrula itself; for the delaminate and invaginate gastrulae, though formed in entirely different ways, cannot always be distinguished from each other. In the development of the delaminate gastrula the egg segments, giving rise to a solid heap of cells, the morula; and this becomes a blastula as before. Each cell of the blastula now divides into inner and outer parts, so that the blastula wall becomes double, consisting of outer and inner layers of cells, surrounding a central cavity filled with fluid. By perforation of one pole the cavity is placed in communication with the exterior and the embryo becomes a gastrula.

---

125   This paragraph and subsequent ones, though they indicate an awareness of more than one developmental route to the formation of a gastrula, show that there was no appreciation of the fundamental distinction between the two great animal superphyla: protostomes and deuterostomes. Leaving aside sponges and coelenterates, we would place "worms", molluscs and arthropods in the former group (where the start of gastrulation, the blastopore, forms the mouth) and echinoderms and vertebrates in the latter (where it forms the anus). The process of gastrulation described in this paragraph applies best to that of the protostomes whilst that in the next paragraph seems more applicable to deuterostomes, although the *perforation of the pole* is hard to interpret. The subsequent difference, which does seem to have been appreciated, is largely in the mechanism of formation of the fluid cavity or coelom. Rather frustratingly, AMM fails to identify any animals in the second group so we are unable to interpret fully his understanding of the processes and their distribution. The *Oxford English Dictionary* identifies the first use of *deuterostome* in 1959 and *protostome* (with modern meaning) in 1958, but it is not clear when the importance of the distinction was first appreciated taxonomically.

It is by no means easy to determine which of these two forms is the more primitive.[126] The invaginate gastrula is much more widely distributed in the animal kingdom, occurring, as we have seen, in all the large groups, while the delaminate gastrula is much more restricted. On the other hand it is not easy to see how the invaginate gastrula first came into existence, for it is by no means clear what advantage a spherical blastula-like animal gets by becoming flattened on one side, or what further advantage is conferred by a very slight depression or cupping of this flattened surface. Such a depression is useful enough after it has reached such proportions as to give rise to a sac-like cavity suitable for the reception and digestion of food, but the early stages of its formation are useless, and could not have been preserved for such purpose. In the case of the delaminate gastrula however there is no such difficulty, and it is possible to construct a hypothetical series of forms which may well represent the ancestral series in the pedigree of the gastrula, each step marking a distinct advance in organisation, and being a sufficiently definite improvement to justify its perpetuation; and the whole series corresponding to the successive stages of development of the delaminate gastrula of the present day.

Starting with the blastula stage, a hollow ball whose wall is but one cell thick, we note that the inner and outer ends of each of the cells are exposed to very different environment. The outer ends being on the surface of the blastula can come in contact with and gain cognisance of the outer world, while the inner ends facing towards the blastula cavity can have no direct contact or concern, except with bodies that have passed inwards through the outer parts of the cells. Hence merely as a consequence of this arrangement, physiological differentiation will be set up between the outer and inner ends of each cell; the outer parts of the cells will become the seats of sensation and of locomotive activity, while the inner ends, freed from these functions, apply themselves to other purposes and become specially nutritive or digestive in function. The next stage is a simple one. The differences between the outer and inner ends of each cell once established will tend to increase as each part of the cell learns to discharge its special functions more efficiently: a mechanical separation of the two parts of the cell is but a slight further differentiation; each cell dividing into two—an outer cell, sensitive and locomotive and probably respiratory in function, and an inner cell specially digestive in purpose.

So far we have supposed the food to consist of small particles captured by the outer parts of the cell, or the surface cells where two layers are established, and passed inwards to the inner parts, or the inner cells, to be digested. The act of digestion is still intra-cellular—i.e., is effected entirely within the substance of the cells just as in an Amoeba. If now we suppose particles of larger size to be taken in as food, we can well imagine how these might be passed on by the inner cells into the central

---

126   This word now becomes problematical for we are asked to decide which of the two developmental mechanisms came first. The protostome/deuterostome split, which is thought to have occurred at least 558 million years ago, tells us nothing about the developmental process taking place in their common bilaterian ancestor: it could be either or neither, so one does not need to be "more primitive" than the other. The earliest bilaterian in the fossil record, *Kimberella*, from 555 million years ago, appears to be a protostome. This problem rather bedevils much of the rest of this lecture.

cavity, which will then become a digestive cavity; the inner cells pouring out into this cavity the secretions which dissolve the food, and which it is their special purpose to manufacture. The formation of a mouth by thinning away of the wall at one point, will be a manifest advantage, as it will avoid the necessity of the food particles having to pass through both layers of cells in order to reach the digestive cavity; and on its appearance the gastrula is completed.[127]

The Theory of Natural Selection requires that each stage in the gradual evolution of a complex organ or system should be a distinct, if slight, advance on the stage immediately preceding it; an advance so distinct as to confer on its possessor an appreciable advantage in the struggle for existence. This condition is often overlooked; we are apt to assume, though most erroneously, that if it can be shown that the ultimate stage is more advantageous than the initial or earlier condition, then the whole problem of the evolution of the organ in question is solved.[128] It is indeed seldom that we are able to refer to so complete a series of intermediate stages as that given above in the case of the delaminate gastrula, each step being but a very slight advance beyond the previous condition, and yet each step conferring on its possessor a distinct and tangible advantage. The fact that such a series of forms can be pointed out, every one of which is repeated in the life-history of certain jelly-fish, is a strong argument in favour of the primitive nature of the delaminate gastrula; while, as already noticed, it tells strongly against the claims of its rival, the invaginate gastrula, that it is at present not possible to point out the progressive advantage gained by the successive stages of gradual flattening and gradual invagination through which the gastrula stage is acquired. Indeed, we seem here driven to suppose a much more rapid change to have occurred than is commonly recognised as possible. Perhaps the real explanation may be that the delaminate gastrula is the older form historically, and that the formation of a gastrula by invagination is merely an embryological device to save time and facilitate the course of individual development.[129]

The formation of the mouth, which in the delaminate gastrula is the final stage of development, is an event of first-rate importance, both from the morphological and physiological standpoints. Now in the invaginate gastrula a mouth is, by the very mode of development, present from the first commencement of the process of invagination, and it may be that the advantage gained by the early formation of this important organ, which at once obviates the necessity of the food having to traverse

---

127   This appears to us as a strange perspective on the matter. Physiologists now view digestion in the gut of multicellular animals as taking place *external* to the body, the gut chamber providing a restricted environment in which the actions of exocrinally secreted enzymes can be concentrated. Rather than evolving to avoid the necessity of passing food particles "through" the body, it is simpler to imagine the development of a digestive epithelial surface, capable of absorption as well as secretion.

128   AMM's interpretation here, as elsewhere in the lectures, is clearly grounded in his admirably evangelical desire to avoid teleology.

129   Notwithstanding the limited understanding of developmental processes already pointed out, the concept of there being *an embryological device to save time* etc. appears superficial. It is an uncharacteristically lazy intellectual approach, especially if he still wishes to avoid connotations of purpose in animal design. Why not extend the thought and wonder why there is any intermediate development at all?

the ectoderm cells in order to reach the digestive layer or endoderm, it may be that this advantage has led to the substitution in actual or individual development of the invaginate for the historically delaminate older type of gastrula formation.

Turning from this somewhat lengthy digression to our more immediate subject, we find that radial symmetry, seen in its most typical form in the gastrula, is confined to aquatic forms.[130] The reason for this is the same as in the case of the still more primitive spherical form—*i.e.*, that it is only in the case of animals, whether young or adult, which live immersed in fluid, that the relations between the animal and the surrounding medium are such as to allow of the animal having identical relations to the environment, whichever part of its circumference happens to be uppermost or undermost. It is also a matter of common observation that radially symmetrical animals are not merely all aquatic, but are almost entirely marine.[131] The reason for the paucity of radially symmetrical forms in the fresh-water fauna appears to be that they are weak swimmers, depending for locomotion on the action of cilia when the animals are of small size, and having no special locomotive organs when of larger dimensions. Fresh-water animals, at any rate such as dwell in rivers and streams, have to be able to hold their own against the currents in which they live; nay more, they must not merely be able to hold their own but also to make their way up stream as well as down, or else in the long run they will be carried slowly but steadily lower and lower down the river, until ultimately they become swept out to sea. For this reason weakly swimming animals cannot, unless under exceptional circumstances, establish themselves in fresh water.[132]

While the typically radial animal is a free swimming form, there are a large number that in the adult condition are attached either temporarily or permanently. Of these the common fresh-water Hydra, and the whole of the great group of Hydrozoa,[133] known popularly as zoophytes, are familiar examples. The majority of these attached radial animals reproduce by budding, the buds usually remaining attached to their parent, and so giving rise to plant-like colonies. In many of these, and especially in the higher group of Coelenterates, the Actinozoa,[134] of which sea anemones and corals are instances, a curious modification of the typical radial symmetry is manifested, to which the term biradiate symmetry is commonly applied. In a biradiate animal, while the radial symmetry is well preserved, there is superadded to it a further change in the shape and arrangement of certain of the internal organs, whereby a definite plane of symmetry is established, on either side of which the organs are perfectly similarly

---

130   Despite appearances, earthworms (for example) do not exhibit radial symmetry.

131   Fresh water protozoa such as Heliozoa would contradict this (not to mention archaea and bacteria).

132   This argument, which was also used in Lecture 6 and is made further use of later in this lecture, appears very flimsy. Strong currents occur in sea water and there are large bodies of fresh water (ponds and lakes) with very little current.

133   Part of the large phylum of (mostly marine) invertebrates which includes the cnidaria or jellyfish. The term *zoophyte* implies a passing resemblance to a plant, as for example with the sea anemones.

134   A now obsolete term for a group which included jellyfish, echinoderms and rotifers, as well as sea anemones and corals.

and symmetrically arranged, but which is the only plane by which the animal can be so divided.

The simplest case of biradiate symmetry would be of this kind: Imagine a Hydra with the body, as usual, cylindrical, i.e., circular in transverse section, and with the mouth also circular in outline; and, for the sake of simplicity, imagine the tentacles to be absent. Such an animal has any number of planes of symmetry, for any plane of division passing along the whole length of the animal and along its axis will divide the Hydra into two perfectly symmetrical halves. Now, imagine the mouth of our Hydra to become oval or elliptical in outline instead of circular; there will now be only two planes of symmetry, which will divide the animal into exactly similar and corresponding halves; one of these planes passing along the longer diameter of the elliptical mouth, the other along its shorter diameter. Next imagine the mouth, instead of being elliptical, to be ovoid or egg-shaped in outline, with a larger and a smaller end. There is now only one single plane of symmetry possible, namely, that passing along the longer diameter of the mouth opening, for any other plane will divide the mouth, and therefore the animal, into two unlike halves.

The origin of this biradiate symmetry is a little obscure. There are reasons for thinking that it first arose in colonial forms, such as Alcyonium[135] or Pennatula,[136] inasmuch as in these colonial forms it is very well marked, and furthermore the plane of symmetry of the individual polypes always has a definite relation to the axis of the entire colony, and the differences between the two sides of the animal on which the biradiate symmetry depends seems to be associated with a special provision for securing a rapid and efficient circulation of water, not merely through the individual polypes themselves, but throughout the whole colony. This explanation seems fairly satisfactory in most cases. It must be noted however that it involves the descent of solitary forms, such as Cerianthus[137] and many other Anemones in which biradiate symmetry is well marked, from colonial ancestors, a line of ancestry for which there is but little independent evidence. Much greater difficulty is offered by the Ctenophora, in which biradiate symmetry is usually well established, and by some Medusae, in which the number of tentacles arising from the margin of the bell may be reduced to two, or even to a single one; in these latter cases however the biradiate symmetry is probably independently acquired, as theoretically it might readily be by any radiate animal.

We have next to consider the type of animal shape spoken of as bilaterally symmetrical, which must be carefully distinguished from the biradiated symmetry we have just been describing. In biradiate symmetry, as in a sea anemone, there is one divisional plane or plane of symmetry, by which the animal can be divided into identical halves; this plane however concerns the internal organs only, and has no constant relation to the movements of the animal. In cases of bilateral symmetry on the other hand, the animal, as in a worm,[138] a lobster, or a frog, is divided by a median vertical plane into symmetrical right and left halves, while furthermore a distinction

---

135   Soft corals (e.g. Dead Men's Fingers).

136   Sea pens (types of cnidaria).

137   Tube anemones.

138   As elsewhere, including in other lectures, this group is ill defined, although some clarity is provided later on (see footnote 161). See also footnote 130.

may be readily made between dorsal and ventral surfaces and between anterior and posterior ends.

Just as the radiate and biradiate shapes are associated with free-swimming habits, or else with an attached condition, so is the presence of bilateral symmetry similarly connected in its earlier phases with the habit of crawling along the sea-bottom. A most instructive series of gradations is shown by the simpler Turbellarian[139] worms, commencing with such forms as Anonymus,[140] in which the body is greatly flattened and almost circular in outline, and in which the mouth is almost in the centre of the ventral or oral surface, while a row of eye spots occurs all round the edge of the animal, though more thickly set at the two extremities. Starting from such a form as Anonymus, which may be compared to a very flat jelly-fish, like Aurelia, which instead of swimming freely in the water, has taken to crawling about on the sea-bottom mouth downwards, we find two diverging series in both of which the body gradually becomes more and more elongated and vermiform in shape, while the sense organs tend to become concentrated at the anterior end. In one series, that of the Turbellaria Acotylea, the mouth gradually moves backwards as the shape of the body becomes more markedly oval and elongated; in the other series, the Turbellaria Cotylea, which receive their name from the presence of a muscular sucker on the ventral surface, the mouth, starting as in the Acotylea from a central position, gradually shifts further and further forwards until it ultimately, in the genus Prosthiostomum, becomes placed quite at the anterior end of the body.

There is little room for doubt that, just as among epithelial cells the spherical form is the primitive one from which both the columnar and the squamous forms have been derived, so also in the series of Turbellarian worms, those with the body approximately circular in outline and with a centrally placed mouth are really the primitive ones from which the more modified Cotylea and Acotylea have alike sprung. It is very significant that these more primitive Turbellarians should present many points of affinity with the radiate animals, such as Coelenterates, not merely in general shape but in the position of the mouth and central gastral chamber, the radial arrangement of the diverticula of the gastral chamber by which nutriment is distributed to all parts of the animal, the disposition of the sense organs all round the margin of the animal, and the position and relations of the nervous system and reproductive organs. These resemblances are too close and too fundamental to be accidental, and they lend much support to the view hinted at above, that bilateral animals are descended from radiate ancestors, that bilateral symmetry is something additional to and imposed on the radiate symmetry, and that this further modification is a direct consequence of the animals having exchanged their pelagic free-swimming habits for crawling ones; a change that would at once lead to the establishment of a difference, both structural and physiological, between the ventral or oral surface along which locomotion is effected, and the opposite or dorsal surface; while the further differentiation between anterior and posterior ends would very soon follow as a necessary consequence of this same crawling habit. It is very possible also that the Turbellarians are themselves the simplest group in which the crawling habit has been

---

139   Non-parasitic planaria (flat worms).

140   The identity of this turbellarian is unknown. The others mentioned in this paragraph are well recognized.

acquired, and are directly descended from Coelenterate ancestors, a view that finds favour with many zoologists, but which can at present hardly be regarded as more than an hypothesis.

With regard to the mechanical origin of bilateral symmetry suggested above, Herbert Spencer, whose writings on the fundamental laws governing animal form and structure are of the greatest possible interest, speaks as follows: "*Where the movements subject the body to different forces at its two ends, different forces on its under and upper surfaces, and like forces along its two sides, there arises a corresponding form, unlike at its extremities, unlike above and below, but having its two sides alike.*"[141]

We have seen above that there are reasons of very great weight for regarding the radiate type as more primitive than the bilaterally symmetrical one, and further than this, for regarding the latter as directly descended from the former. The group of Echinodermata, including the starfish, brittle stars, sea urchins, and their allies, warn us however that we must not generalise too hastily. About the radiate symmetry of an Echinoderm there can be no possible doubt; a starfish is as markedly, as conspicuously, radiate as any animal in existence, indeed it has been by older writers taken as the type of radiate animals. Take one of the ordinary five-fingered starfish for instance: note its shape; the central disc-like body produced into five equal and symmetrically arranged arms, the mouth placed centrally on the lower, or oral surface, the anus subcentrally on the dorsal surface. As regards internal structure, the radiate symmetry is equally well marked; the muscular, skeletal, digestive, circulatory, nervous, and reproductive systems all extend radially and symmetrically along the arms. There are no anterior or posterior ends, right or left sides to the animal, for in its dilatory ramblings a starfish moves indifferently in any direction, any one of its five arms leading.

Bearing in mind what has been said above as to the primitive nature of radiate symmetry, and of the relation between it and bilateral symmetry, bearing in mind also that Echinodermata are not merely all aquatic, but are exclusively marine, and that there are the most cogent reasons for regarding the marine fauna as the primitive one, from which both the fresh-water and terrestrial have sprung, we should I think naturally conclude that the radiate symmetry of Echinodermata is primitive, and that a starfish is a radiate animal, which has adopted crawling rather than pelagic habits, presumably for convenience in obtaining food, and in which consequently a distinction between ventral and dorsal surfaces has been established; but that the further structural modification by which anterior and posterior ends, right and left sides, become differentiated, has not yet appeared in it. If we want further evidence in support of this primitive character of Echinodermata, we may obtain it from the past history of the group, for Echinoderms are, geologically considered, a group of extreme antiquity, and a group in which the characteristic radiate symmetry is as marked in the older as in the more recent members. If however we consider the actual embryological development of a starfish or other Echinoderm, we find difficulties in the way of the view sketched out above, difficulties which have not yet been overcome and which may not improbably prove fatal.[142]

---

141   *The Principles of Biology*, Vol II (revised and enlarged edn; London: Williams and Nothgate, 1899), Part IV, Chapter XIV: 'The General Shapes of Animals'.

142   This comment and the final sentence of this paragraph are intriguing, given that we would now group starfish (echinoderms) amongst the deuterostomes. Perhaps AMM

A starfish lays small eggs, from which a radially symmetrical larva, a gastrula,[143] is developed. This larva however soon acquires a very marked and unmistakable bilateral symmetry; ventral and dorsal surfaces, anterior and posterior ends, right and left sides may readily be distinguished in it, and the internal organisation shows the bilateral symmetry as clearly as does the external shape. This bilateral symmetry is not confined to starfish, but is present in the larval stages of all other Echinoderms as well. The importance of the point is at once apparent: it shows us that the radiate symmetry of the adult Echinoderm is not directly continuous with, and may indeed not be the same thing as, the radiate symmetry of the early larva, for between the two radiately symmetrical stages a bilaterally symmetrical stage is intercalated. If the developmental history of an Echinoderm is a true recapitulation of the pedigree of the race, then the history can only be interpreted as meaning that Echinoderms are descended from bilaterally symmetrical ancestors, and that the radiate symmetry of the adult Echinoderm is secondary, and of later origin. The matter is one of great interest, especially when we bear in mind that the relations of Echinoderms with other groups of animals are at present entirely unknown to us, and that consequently any light that can be obtained from a study of embryology would be peculiarly welcome.

It is impossible to discuss the question at all adequately here, but it is perhaps worthwhile pointing out that though the adult starfish is an animal showing radiate symmetry in a marked manner, yet that the radiate symmetry of a starfish, or indeed of any other Echinoderm, differs in some important respects from the radiate symmetry of the Coelenterate. Thus, in the first place, the symmetry of an Echinoderm is pentamerous, the typical number of arms in a starfish being five, and five being the typical number of corresponding parts met with in the other groups of Echinoderms also. This may seem an unimportant point, but it becomes significant when we note that though the actual number of corresponding parts in a Coelenterate varies very greatly in different forms, yet that it is almost invariably either four or six, or some multiple of these numbers, while five or any multiple of five is unknown. Then again, in the actual development of an Echinoderm the radial symmetry of the adult is first shown by a set of organs, the ambulacral system, which is absolutely and entirely unrepresented in Coelenterates; and it is apparently on this radially arranged ambulacral system that the radial symmetry of other parts and systems is based. When further we bear in mind that the whole structure of an Echinoderm is altogether different to that of a Coelenterate, and in many respects very much more complex, any real comparison between the two groups becomes very nearly impossible, and we find it less anomalous than it at first appeared to regard the adult symmetry of an Echinoderm as something quite distinct from, and acquired perfectly independently of, that of a Coelenterate.

Limits both of space and time forbid that I should pursue further the discussion of the shapes of animals. Where once bilateral symmetry is established however the further modifications seen in the higher forms become comparatively easy to follow. The development of a head, with accompanying concentration of the nervous

---

was conscious that his developmental story, described above, did not completely fit the facts.

143   This is now a general term for an early, multicellular embryo in the which the gut has started to form, not restricted to starfish.

system and sense organs, are merely further developments of processes and tendencies which we have seen already established in the Turbellarian worms, while the formation of limbs, perhaps shadowed forth in the parapodia of Chaetopods,[144] is the most important step in the upward progress to the highest groups of animals.

Bilateral symmetry is characteristic of all the higher groups, though it may be masked or modified by further development, as the twisting of the body of a snail or other gastropod, the asymmetrical form of the tail of a hermit crab, or the shifting of the eye in a sole. Speaking generally we may say that the forms of the higher animals are derived from those of the lower bilaterally symmetrical worms, by exaggerating the differences between one part of the body and another already present in these latter. Thus the differences between the dorsal and ventral surfaces or rather halves of the body, and between the anterior and posterior ends of the body, gradually become intensified, attaining their maximum in birds and mammals, the two highest groups of animals.

I can hardly conclude this part of my address more fittingly than by the following quotation from Herbert Spencer who, in his "Principles of Biology," has discussed in most philosophical fashion, and with far greater thoroughness than I can pretend to here, the laws regulating the shapes of animals. *"The one ultimate principle,"* says Spencer,[145] *"that in any organism equal amounts of growth take place in those directions in which the incident forces are equal, serves us a key to the phenomena of morphological differentiation. By it we are furnished with interpretations of those likenesses and unlikenesses of parts which are exhibited in the several kinds of symmetry; and when we take into account inherited effects wrought under ancestral conditions, contrasted in various ways with present conditions, we are enabled to comprehend, in a general way, the actions by which animals have been moulded into the shapes they possess."*

Passing from the consideration of the shapes to that of the sizes of animals, is very like turning from a well-made road into a ploughed field, across which progression becomes not only slow but difficult and irregular. Hitherto the problems concerned with the sizes or magnitudes of animals have received but very scant attention, and we are not only ignorant of the principles and laws that govern them but of the directions in which to seek for these principles. Indeed we have at present but a very limited number or range of facts on which to base our arguments. Still the questions are of much interest, and it is certainly worth while enquiring into them, even though we may not be able to make much progress to-night.

In the natural or wild state the size of each kind of animal in the adult condition is fairly well defined, and often very sharply so. The words cat, rabbit, sparrow, convey to our minds the impression of animals, not merely of certain appearances, habits, and structure, but also of a certain well understood and fairly constant size. The limits of variability[146] are much wider in some cases than in others. Speaking generally they are much wider in the case of aquatic, and especially of marine, than of terrestrial animals. If we say of an animal that it is as large as a fox, we know fairly exactly what is meant; but to speak of anything being as large as a salmon, would convey a very vague

---

144   Annelid worms, including the oligochaetes and polychaetes.
145   See footnote 141.
146   Presumably he is referring to the extent of variation amongst individuals within species.

notion of magnitude. As a standard of size, the salmon is indeed but little better than the traditional "lump of chalk."[147] The same is true of most other fish, and indeed is characteristic of aquatic animals in general. The explanation seems to be that in terrestrial animals the period of growth is practically limited to the earlier stages of existence; while aquatic animals continue to grow for a much longer period, or indeed throughout their entire lives.[148] Why an animal should stop growing on reaching a given size is a very difficult question to answer, but one that, if time permits, I will return to later on.

As regards the actual dimensions attained, here again the aquatic animals lead the way. Of all animals now existing, whales are incomparably the largest; next to the aquatic come the terrestrial forms, with the elephant in the forefront; while last and smallest of all, come the aerial, or flying animals.[149] The actual size seems here to be associated with the density of the medium in which the animal lives. in water an animal has to support but a very small part of its weight by its own muscular effort, for more than half the weight of a fish or whale is water, and of the solid components the fats are lighter than water, so that the specific gravity of such an animal is not much in excess of the water in which it dwells. Flying animals live under very different circumstances, for here every part of the body is considerably heavier than the air and great muscular efforts are necessary to sustain the animal during flight. Large size or great weight of the body becomes therefore impossible.[150]

Again, we may lay it down as a general rule that the largest animals belong to the higher, or even the highest groups, i.e., that great size is associated with great complexity of organisation. Exceptions are readily met with, but on the whole the statement is correct. Vertebrates are clearly on the whole of much larger size than any group of Invertebrates. Amongst Vertebrates, mammals rank as the highest group and to them belong the largest animals now living, both aquatic and terrestrial. So also amongst Invertebrates; in the important group of Mollusca, the highest forms are undoubtedly the Cephalopoda, and it is amongst these that the largest members of the group occur.

It has been suggested recently that the rule is a stricter one than has been hitherto recognised; and evidence has been quoted in support of the further statement that the ancestral forms or progenitors of the higher groups, such as mammals or birds, were of distinctly small size, i.e., much below the average stature attained by their descendants—the present members of the groups in question. The direct influence of size on structure has as yet been but very imperfectly investigated, but there are many cases known in which among animals of the same zoological group the larger forms are distinctly of more complicated structure than their smaller allies, and in which

---

147　For which, let us read *length of a piece of string*.

148　This seems to be a gross generalization for which convincing evidence would be required. More importantly, the sentence elides body size and the duration of growth. These are far from being the same phenomenon, especially as size is not defined (length, weight?).

149　This generalization seems to ignore non-flying insects and other terrestrial invertebrates and to ignore the large size of some birds.

150　The point made is sound but we would now express it in terms of the energetics of flight. We should bear in mind that AMM had never seen an aircraft.

there are very valid grounds for holding that the greater complication of structure is connected casually with the increased dimensions.

We shall perhaps best deal with the several points just mentioned by taking the large groups into which the animal kingdom is divided one by one, and noticing with regard to each, the principal facts concerning the sizes of the several members of the group.

Protozoa, the simplest of all animals, are defined as those forms which, not merely in the earliest phases of their existence, but throughout their entire lives, remain in the condition of single cells. This unicellular nature is associated with great simplicity of organisation and with extremely small size. The differentiation of parts or organs within a single cell can only proceed up to certain limits, and if of great bulk parts of the cell would be too far removed from the surface to obtain nourishment or to get rid of excretory matter.[151] What the actual limits of size are among unicellular animals we perhaps do not know accurately. Stentor,[152] which may attain a length of 1/20 inch,[153] is usually regarded as one of the largest. Individual cells in higher animals may however attain much larger dimensions: thus, if histologists are right in regarding a single muscle fibre as formed by differentiation within a single cell, then we must grant that a cell may be some inches in length; while, if the statement be true that a nerve fibre is merely a process of a single cell, it will follow that a single cell may in a man extend from the spinal cord, perhaps from the brain itself, to the extremity of the fingers or toes—i.e., may attain a length of some feet, or in an animal the size of a whale as much as 60 feet[154] or more. Such nerve fibres are however exceedingly slender.

Among the largest cells known to us are the eggs of reptiles, of Elasmobranch fishes, and, largest of all, those of birds. Embryology shows us that the yolk of a bird's egg, which is the only part from which the embryo is directly developed, is morphologically a single cell, and is indeed in the earlier stages of its formation indistinguishable from the ordinary epithelial cells covering the surface of the ovary. The largest eggs are those laid by the Struthious birds, and the yolk of the egg of an ostrich is perhaps the largest individual cell known to us, though it was very greatly exceeded in recent geological times by that of the gigantic Aepyornis.[155] It must however be remembered that the large size of the yolk of an egg is due mainly to its distension by the granules of food-yolk imbedded in it, which cannot properly be regarded as part of the living substance of the cell.

In the next group—that of the Sponges—it is difficult to speak of the size of the individual animals, owing to the very general habit of forming colonies by continuous gemmation,[156] in which colonies the outlines of the component individuals are impossible to determine. The solitary sponges are usually of small size, not exceeding an inch or two in height, but some forms, as the beautiful Euplectella,[157] attain a

---

151   Or achieve adequate gas exchange.
152   A class of horn-shaped protozoa.
153   About 1.25 mm
154   18 metres
155   The extinct Elephant Bird of Madagascar.
156   Budding.
157   The glass sponges, including Venus' Flower Basket.

height of a foot or more. Among Coelenterates the same habit of colony formation prevails, but the boundaries and hence the size of the component members of the colony can almost always be determined. The majority of Coelenterates are of small size: the individual zooids of a hydroid colony are commonly but a fraction of an inch, rarely much over an inch in length, though occasionally attaining a larger size. The entire colonies often reach great dimensions; the branching zooids may form brush-like masses some feet across, while the reef-building corals reach a far larger size.

Speaking generally the Anthozoa, which are the higher group of the two, are of larger size than the Hydrozoa; and it is also among the more highly organised groups that the largest individuals, the giant Coelenterates, are met with. Thus, of the two groups of jelly-fish, the Hydromedusae, which belong to the Hydrozoa, are small, while the Scyphomedusae,[158] which in many points of structure and also in some peculiarities of development are more nearly allied to the Anthozoa, are of much greater average size and occasionally reach extraordinary dimensions, individuals measuring as much as seven feet in diameter with tentacles nearly fifty feet in length having been met with. Sea anemones again, the typical Anthozoa, are much larger than the hydroids; a height of three or four inches with a diameter of one or two inches is not at all uncommon, while from tropical seas anemones have been described over three feet in diameter.

The Echinodermata are as a rule of small size and are measured by inches. The vermiform Holothurians[159] are sometimes greatly elongated, some species of Synapta[160] measuring five feet or more in length, while the extinct Crinoids had stems up to seventy feet long.

Among the heterogeneous and in many respects unnatural assemblage of animals grouped together by zoologists under the name Vermes,[161] the average dimensions are distinctly small. Turbellarian worms,[162] perhaps the most primitive members of the group, average less than an inch in length, though some forms may measure four to six inches. Trematodes[163] are similarly small; and the great length attained by some Tapeworms; thirty feet or more, is due rather to their peculiar mode of asexual reproduction than to increase in individual dimensions.

Nemertines[164] may attain an enormous length, fifteen or twenty feet being not uncommonly exceeded. Amongst Annelids the average length is certainly below six inches; but individual genera, as Halla,[165] often exceed three feet, and the giant earthworm of Australia measures six feet or more.

Rotifers are a very interesting group;, invariably of small size, and often actually microscopic, they yet exhibit very great complexity of organisation. A Rotifer may be actually smaller than a Stentor, and yet while the latter is but a single cell exhibiting but very slight differentiation of parts, the former is an animal with well developed

---

158 True jellyfish.
159 Sea cucumbers, members of the Echinoderm phylum rather than worms.
160 Characterized by calcareous anchor plates.
161 These paragraphs help to elucidate the compass of the *worm* group.
162 Free-living flatworms.
163 Parasitic flatworms including flukes.
164 Proboscis, ribbon and bootlace worms.
165 A marine polychaete.

cutaneous and muscular systems, a large body cavity, an alimentary canal in which the various regions are modified in special manner for special purposes, a definite system of excretory organs, a very perfect nervous system, with well developed and sometimes highly specialised sense organs; and all these various parts formed of specially differentiated cells. Comparison of a Rotifer with a Protozoon shows us very forcibly that although it may be true that larger animals are, on the whole, more highly organised than smaller ones, yet it is not magnitude alone that determines structure.

Arthropoda, though an extraordinarily numerous group, far exceeding in number of species all the other groups put together, yet do not present any unusually great diversity of size. The majority of arthropods are small, and a length of four inches is a long way above the average, while many large and very numerous groups such as Entomostraca[166] are minute, indeed almost microscopic, in their dimensions. Crabs may reach a large size, but this is almost entirely due to great elongation of their legs.

Amongst Mollusca a much larger size may be attained. Of Lamellibranchs the genus Tridacna[167] affords the giants, shells of these huge clams having been found measuring two feet across and weighing upwards of five hundred pounds, the animal itself exceeding twenty pounds in weight. Gastropods also reach a large size, but are far exceeded by the Cephalopods,[168] or cuttle-fish, which group yields the largest invertebrates known to exist.

From time to time the bodies of enormous cuttle-fish are cast up on the shores in various parts of the world. Two or three very large specimens have occurred on the West Coast of Ireland, others are described from New Zealand, but by far the largest and also the most numerous cases are from the East Coast of North America, more especially the Coast of Newfoundland. Some of the specimens are of really gigantic proportions, a length of body of as much as twenty feet having been measured, while the long arms, or tentacles as they are often named, were thirty or even forty feet in length.

The Gastropods yield some very instructive series of forms, illustrating in a remarkable way the influence of size on structure. Of these one of the most interesting is that afforded by the genera Aplysia, Doris, Eolis, and Pontolimax, all four of which belong to the group of Opisthobranchiate Gastropods, or sea slugs.[169] Aplysia, the sea hare, is an animal of some size, four or five inches in length,[170] and has a very well-developed gill covered over and protected by a thin uncalcified shield-like shell. Doris is of smaller size, has no shell, and has its gill processes less developed than in Aplysia but still grouped together in a tuft on the hinder part or its back. In Eolis, there are a number of elongated papillae covering the back of the animal, some of which, at any rate, act as respiratory organs. Pontolimax,[171] finally, is a very small

---

166   An arthropod group, no longer recognized, which included Branchiopoda, Cephalocarida, Ostracoda and Maxillopoda.

167   Giant Clam.

168   Includes squid and octopus, as well as cuttlefish.

169   Broadly, the nudibranchs.

170   Much larger ones are known, up to 75 cm.

171   Now Limapontia.

slug-like animal only a twelfth of an inch in length, and having a perfectly smooth dorsal surface devoid of respiratory processes of any kind.

Herbert Spencer has pointed out very clearly how in two animals of the same zoological type, but of markedly different size, the smaller one may be able to do without respiratory organs, while to the larger one, merely in consequence of its larger size, such organs are absolutely essential. Suppose you have two animals of identical shape, but of different size: for the sake of simplicity let us suppose the animals to be spherical, and let the diameter of the smaller animal be one inch, that of the larger two inches. The extent of surface in the two will be proportionate to the squares of their linear dimensions—i.e., in this case, the larger animal will have a surface four times as extensive as the smaller one. But the mass of bulk of the animals will be proportionate to the cubes of their linear dimensions—i.e., the larger animal will be eight times the bulk of the smaller one. Hence the larger animal, with a bulk or mass eight times that of the smaller one; will have a surface only four times as extensive. This fact, that as animals increase in size their bulk or mass increases at a much faster ratio than their surface, explains how it is that a small animal, such as Pontolimax, may find its body surface sufficient for the interchange of gases that constitutes respiration, while in a larger animal Doris, of similar shape and constitution, the body surface may be altogether insufficient and special respiratory organs may become necessary. Again, in a small animal no part of the body is very far removed from either the surface or from the digestive organs, hence each part is able to effect its respiratory interchange of gases and to obtain its due supply of nutriment directly, without the intermediation of a system of circulatory organs or blood-vessels. In a larger animal however such blood-vessels are absolutely necessary, for without them the more deeply lying parts would be unable to obtain the oxygen which is necessary for their vital activity, or to get rid of the carbonic acid and other poisonous products of that activity; while the nutrition of the surface and more remote parts of the body could not be properly kept up unless there were direct communication between the digestive organs and these parts. Such considerations must suffice to illustrate in what way mere increase of size may involve and necessitate greater complexity of structure.

Amongst Vertebrates we have already noticed the extreme individual variability of size seen amongst fish, in which apparently growth continues throughout the whole period of life. The actual limits of size of fish are probably only imperfectly known to us. The blue shark, Carcharias,[172] attains a length of twenty-five feet; specimens of Carcharadon[173] have been measured over forty feet in length, while of the genus Rhinodon[174] examples of fifty, sixty, or even seventy feet in length have been described. This is very probably the limit of size reached by fish at the present day, but judging from the fossil teeth of Carcharodon, Cestracion, and other forms, sharks of these genera must have existed in tertiary times more than twice the dimensions of any now living.

It is difficult to speak with any certainty about the size of Amphibians, but the existing genera are all small. Some of the Salamanders attain a length of four feet or more, but the majority of recent Amphibians are of much smaller size. Fossil

---

172   Sand Tiger Shark.
173   Great White Shark.
174   Whale Shark.

forms occur of much larger dimensions, some, such as the Labyrinthodonts, being veritable giants. It is however impossible in most cases to speak with certainty as to the Amphibian character of these fossils, but we are probably right in regarding recent Amphibians as diminutive and pigmy representatives of a group formerly of much larger size.

Reptiles are in much the same position, many of the largest groups being now extinct. Of these the Plesiosauri and Ichthyosauri attained lengths up to twenty feet, while the flying Pterodactyls were also of large size, some measuring as much as twenty feet in expanse of wing; but these dimensions were greatly exceeded by other forms. It has been estimated that Mosasaurus was as much as seventy feet in length, while in the genus Brontosaurus, Professor Marsh believes we have a truly terrestrial reptile eighty feet in length, thirty feet high, and estimated to have weighed not less than twenty tons. Of recent reptiles the Crocodiles and some of the Chelonians attain large dimensions, while snakes are described up to thirty feet in length and as thick as a man's body.

It is however amongst Mammals that we find as already noticed the real giants, and it is interesting to note, that in different geological periods different groups of Mammals have worked up to a maximum of size and then disappeared. Thus the Edentates[175] at one time gave rise to giant forms culminating in the huge Glyptodon and still larger Megatherium, but the existing Edentates like existing Amphibians are a very puny race compared to their forefathers. Ungulates seem in past times to have had a greater number of large forms than at present, though even now they are among terrestrial animals second only to the Elephants. It is interesting also to note that there seems to be an actual limit to the size of a terrestrial mammal, which has been approached more than once by separate groups. It can hardly be an accident that the Megatherium attained dimensions very closely comparable to those of an elephant, while the biggest fossil Ungulates were not far short of this size.

It has been suggested that the limits of size are due to the nature of the materials of which animals are constructed, and that the difficulty in increasing that size is a mechanical one; that just as it would be impossible to construct a Forth Bridge of stone, for in order to get sufficient strength it would be necessary to employ so great a quantity of material that the structure would be crushed by its own weight, so the bones, ligaments, and muscles of which the animal frame is constructed will permit of size up to that of an elephant but not beyond. It is not hard to find points in the anatomy of the elephant that support this view. The several joints of each limb, which in smaller quadrupeds are in the natural standing position bent on one another, often at considerable angles, are in the elephant placed vertically one above another, an arrangement that enables them to support with less muscular effort the enormous weight of the body and head. The massive pillar-like form of the legs, which take up the greater part of the space below the body, are also indications that the limits of size are being approached, and when we find that though approached closely time after time, the size of the elephant has not been exceeded in past times by terrestrial animals, save by a few members of its own group, and with the possible but as yet doubtful exception of Brontosaurus and some other reptiles, we are probably not far

---

175   Anteaters, sloths and armadillos.

wrong in assuming that this size is somewhere near the mechanical limit imposed by the strength of the materials of which the animal frame is composed.

With aquatic animals in which the body is supported on all sides, and the specific gravity of the entire animal is not greatly in excess of that of the water in which it lives, the case is very different; and among the whales we meet with genera which attain the length of ninety feet, and whose weight is estimated at upwards of one hundred tons, figures which according to some authorities may in exceptional cases be considerably exceeded. Whales are not merely the biggest animals that live, they are probably also the biggest that ever have lived, for we have no satisfactory evidence of larger forms at any period in the world's history.

It is worth while also to note that the giants, whether aquatic or terrestrial, are doomed to extinction. Elephants have been tolerated because of the readiness with which they can be tamed and made to place their great strength at the service of man; but it is many centuries since the African elephant was so employed, and no one of the African tribes at the present day makes the slightest attempt to so use him; while with regard to the Indian elephant, though large numbers are still employed, experts are becoming more and more doubtful as to the profitable nature of this mode of labour. Elephants require much attention, they consume enormous quantities of food, and they are subject to a variety of ailments, which more or less incapacitate them for their work, so that the elephant seems in real danger of being shouldered out of existence by the steam engine and hydraulic jack, which are more reliable, easier to apply, and on the whole less costly. Enormous numbers of elephants, both African and Indian, are killed each year for the sake of the ivory of which their tusks consist; it is difficult to form anything like a correct estimate of the actual number, but competent authorities state that it cannot be less than 100,000 annually. This terrible destruction is admittedly thinning their numbers, and elephants are on all hands recognised as doomed to destruction; and with the whales, which are yearly becoming scarcer, to form the last and greatest victims of a ruthlessly advancing civilization.

A final problem, that time only permits me to hint at but which has been keenly discussed by many writers, is this. Why is it, how does it come about, that some animals are so much bigger than others? To take a concrete case: why is a cow bigger than a sheep? The two animals belong to closely allied groups; we see them side by side in our fields, eating the same food and digesting it in the same characteristic fashion; why then should there be so great and constant a difference in size? Herbert Spencer, who discusses the problem at some length,[176] is inclined to explain it by the size of the animal at birth, and attempts to establish the position that those animals which are larger at the time of birth or of hatching are those which are larger also when adult. It is true that an ostrich lays a bigger egg than a hen, and a hen than a sparrow; but it is very easy to show that the relation is not a general one, and that size at birth has no necessary relation to size in the adult condition. For example, a crayfish and a lobster are two closely allied animals, and yet the crayfish lays the larger egg of the two, though the adult is not more than a quarter the length of a lobster. Or again, were Spencer's contention right, then an eight months' child, being born of smaller size than the average, should not attain to average stature. It is more probable

---

176   *The Principles of Biology*, Vol. I (revised and enlarged edn; London: Williams and Nothgate, 1899), Part II, Chapter I: 'Growth'.

that the real explanation is a much more complicated one, and that there is for each animal a certain average size which is most advantageous to the animal when living wild in its natural condition. Natural Selection will then tend to preserve this average size by placing at a disadvantage those individuals which depart from it conspicuously, whether in the way of excess or diminution.

In connection with this point, reference must be made to Mr. Galton's law of Regression, enunciated in his recently published and most important book on Natural Inheritance.[177] It is impossible to discuss the subject further on the present occasion, and I will conclude by giving the Law in Mr. Galton's own words, referring my readers for further details to his most fascinating and suggestive work, *"If the word 'peculiarity' be used to signify the difference between the amount of any faculty possessed by a man, and the average of that possessed by the population at large, then the Law of Regression may be described as follows: Each peculiarity in a man is shared by his kinsmen, but on the average in a less degree. It is reduced to a definite fraction of its amount, quite independently of what its amount might be. The fraction differs in different orders of kinship, becoming smaller as they are more remote."* In other words, according to Mr. Galton's Law of Regression, any tendency to individual deviation—say, from the normal stature—whether in the way of excess or of diminution, is counteracted by an inexorable law, which acting to a definite degree in each succeeding generation, constrains return to the normal height.[178]

---

177   Francis Galton, *Natural Inheritance* (London: Macmillan, 1889), Chapter XII: 'Summary'. See also footnote 116, Lecture 7.

178   For a discussion of Galton's Law of Regression, including the role of Karl Pearson in developing its application in statistical analysis, see Gillham NW, *A Life of Sir Francis Galton* (Oxford: Oxford University Press, 2001).

LECTURE 9

SOME RECENT DEVELOPMENTS OF THE CELL THEORY

Where Lecture 8 discussed matters which, to our eyes, seem to be of little moment, Lecture 9 deals with one of the most important issues in biology, both then and now—the processes of cell division and their relationship with the mechanism of inheritance.

The development of the *Cell Theory* was one of the most important developments in early- to mid-19th-century biology, and AMM gives a helpfully concise account of how it developed and the roles of some of the key players. We are so used to accepting the cell as the principal unit of function and continuity in life that it is hard to imagine a time when this was not so. The theory was first explicitly formulated by Matthias Schleiden and Theodor Schwann, in separate papers published in 1839, following pioneering work by Anton van Leeuwenhoek, Lazzaro Spallanzani, Robert Hooke, Robert Brown, Jan Purkyné and Johannes Müller. It was later developed by Remak and Rodolf Virchow and exploited by Henle, Kölliker, Weismann and others as a ready means of understanding observations in development, reproduction, embryology and inheritance. Its emergence rested on improvements in microscopy and the increasing availability of high-quality instruments, and was subsequently enhanced by reliable techniques for making thin sections of tissues.

Lectures 8 and 9 were given by AMM to the same audience just one year apart. The difference, in terms of style, intensity of inquiry and evaluation of evidence, is remarkable. As in some previous lectures, he uses his extraordinary facility with language to conjure up simple, direct and persuasive images of complex processes (one could almost turn away from the text at any point and sketch a diagram), whilst at the same time assembling a wealth of evidence from the latest research to underpin his explanations. The result is an impressively detailed potted history of cell science; one might almost call it a status report.

The descriptions of processes are essentially anatomical and mechanical, and only towards the end of the lecture do they become interpreted against the requirements of inheritance. The investigators cited by AMM clearly realized the crucial importance of what they were seeing, yet they lacked the biochemical tools to interpret it in a completely meaningful way. Optical microscopy allowed them to make impressively accurate observations of fine structures but not always to understand composition or function. Nevertheless, the quality of their work and attention to detail is demonstrated by the extent to which the terminology they coined is used in today's descriptions of mitosis and meiosis.

Of all the transitional points in biological understanding represented by AMM's lectures, this must surely represent one of the most significant. He relishes the detail and gives no quarter in demanding that interpretations fit the facts. Hints such as the ...*the mathematical precision* [of] *the division of the chromatin into equal halves* ... and the implied limitations of ... *the reagents at present in use* ... give a strong sense of progress being made, and the fertility of current theories, even if the target was yet to be fully identified. One can sense a premonition of 19th-century descriptive anatomy preparing itself for 20th-century biochemistry and genetics. The plethora of cited work accentuates the feeling, while the final paragraph expresses it directly.

My annotations identify the named authors of cited work where that has proved possible. Comprehensive accounts of cell theory and the development of embryology, including events subsequent to AMM's time, can be found in: Richmond ML, *Journal of the History of Biology* 1989; **22**(2): 243–276; Laubichler MD and Davidson EH, *Developmental Biology* 2008; **314**(1): 1–11; Churchill FB, *August Weismann: Development, Heredity, and Evolution* (Cambridge, MA: Harvard University Press, 2015).

SOME RECENT DEVELOPMENTS OF THE CELL THEORY

M Y address last year was concerned with big rather than with little animals, and might perhaps with greater propriety have been delivered to a macroscopical rather than to a microscopical society. For this however I propose to make amends to-night; and to the subject-matter of my address the most ardent microscopist could hardly take exception, however legitimate may be his dissatisfaction at the mode of its presentation. For these more recent developments of the cell theory not merely depend absolutely on the use of the microscope, but require for their elucidation the very highest powers obtainable, and the most refined methods of modern histological research. In what I have to say tonight I can make no claim to originality: my aim is rather to give a summary of recent progress along certain important lines of investigation, and a statement of the present position of problems recognised by all as of the highest interest and importance.

A few words concerning the origin and the earlier phases of the cell theory may fitly preface my remarks. Like all great theories it is impossible to date precisely its first enunciation. Commonly stated to have been founded in 1839, first for plants by Schleiden, and almost directly afterwards extended to animals by Schwann, the cell theory in its main outlines is to be found lurking in an incomplete unformed condition, rather hinted at than clearly expressed, in the earlier writings of Robert Brown, Dutrochet, Von Mohl, and others. These earlier and less precise attempts were however confined to vegetable tissues. As regards animals, to which I propose to limit myself this evening—from no desire to minimise the importance of the sister science of Botany, but merely because I am unfortunately less familiar with it and unable to speak from my own knowledge and observation—as regards animal tissues, Schwann is rightly regarded as the founder of the cell theory.

By the cell theory, Schwann meant that the bodies of all animals, as also of all pants, are formed of cells: that it is by evolution of, and changes in these cells, that the tissues of which the various parts of the body consist are formed, and further, *"that the differences in the properties of the different tissues and organs of animals and plants depend on differences in the chemical and physical activities of the constituent cells."*[179] Concerning the mode of formation of cells, Schwann held that they might arise independently in a surrounding matrix or blastema. In 1845 however Goodsir first promulgated the doctrine that cells never originate without pre-existing cells, from which they arise by fission;[180] a view strongly supported by Remak, Kölliker, and Virchow, and soon universally accepted. The next great step was von Baer's discovery of the mammalian

---

179 Source not known.
180 Goodsir J and Goodsir HDS, *Anatomical and Pathological Observations* (Edinburgh, 1845), Chapter 1: 'Centres of Nutrition'.

egg or ovum,[181] and his recognition of the fact that it was a single nucleated cell; a discovery that threw an entirely new light on Harvey's dictum, *omne vivum ex ovo*.[182]

The cell theory was now firmly established. Histology, or the microscopical study of the animal body, showed that all its parts and tissues were really built up of cells, as a wall is of bricks; while embryology furnished even more cogent proof, showing that the ovum or egg of all animals is one single cell, and that from this single cell by repeated division all the component cells of the adult animal are derived. Thus expressed the theory is the grandest generalisation and the most firmly established fact in all morphology and the division of the animal kingdom into Protozoa and Metazoa – *i.e.*, into animals which on the one hand remain single cells all their lives, and on the other hand commence as single cells or ova but speedily become multicellular, takes rank as the most fundamental and most natural of all zoological distinctions.

The cell theory early gained acceptance as regards its main conclusions, but concerning the detailed structure of the unit or cell discussions quickly arose, which the progress of knowledge has served rather to widen than to restrict. Schwann's idea of a cell was *"a vesicle closed by a solid membrane, containing a liquid in which floats a nucleus enclosing a nucleolus, and in which also one may find small granular bodies."*[183] This definition was soon challenged. The cell wall was first attacked, and shown by Nägeli in 1845 to be not essential to a cell. Others speedily confirmed Nägeli on this point, and in 1857 we find Leydig defining a cell as *"a soft substance containing a nucleus."*[184] Next the nucleus received attention. Max Schultze, one of the most careful of observers, denied its existence in Amoeba porrecta; other investigators quickly followed on the same line, and by Haeckel a distinct sub-kingdom, the Protista, was proposed for the reception of forms such as Protamoeba, Protomyxa, and Protomonas,[185] which were supposed to be devoid of nucleus and were regarded as possibly representing the parent forms from which animals and plants have alike descended. Further research has however tended to check rather than to confirm these statements; and the use of special reagents and more refined histological methods has shown that nuclei are present in many forms, such as the Foraminifera, in which their existence was at first denied.

The more recent investigations have also shown that the nucleus is not as was at first supposed a structure always presenting the same characters, but that it may vary greatly in form, structure, and relations, even amongst the simplest animals, or Protozoa. In the young of many forms no nucleus can be detected, while the adult possesses a conspicuous one. Some forms, as Amoeba, have a single nucleus; others, as Opalina, or Arcella,[186] have many nuclei. In Paramoecium there are two nuclear

---

181   In 1826.

182   "All life comes from the egg."

183   Source not located.

184   In *Lehrbuch der Histologie des Menschen und der Tiere*.

185   Protamoeba and Protomyxa are now grouped in the phylum Amoebozoa but Protomonas is actually a Methylobacterium. As a prokaryote, the latter has no nucleus.

186   Opalina is an alga; Arcella is a freshwater amoeba.

structures, the nucleus proper and the paranucleus,[187] which have different properties and very different functions. Then as to shape, the nucleus is in most cases spherical or approximately; in Vorticella it is greatly elongated; in Stentor it is elongated, and furthermore constricted in a moniliform manner; and in such forms as Dendrosoma it presents an extraordinarily branched and extremely complex condition. A still more curious state is seen in Opalinopsis, an Infusorian[188] living parasitically in the liver of the squid. Here the nucleus appears at times as a much branched reticular structure extending the whole length of the animal; at other times the reticular nucleus breaks up or becomes "pulverised" into an immense number of extremely minute particles, which become diffused through the protoplasm of the animal, and would most certainly escape detection but for a knowledge of their previous condition. These few illustrations will serve to show how extraordinarily variable the nucleus may be even amongst the Protozoa. I shall return later on to the condition of the nucleus in higher animals, but propose first to deal with the cell-body itself.

The cell-body or cell contents—*i.e.*, the whole cell except the cell membrane and the nucleus—were spoken of by Dujardin as consisting of "sarcode," which he defined[189] as *"a kind of mucus endowed with spontaneous movement and contractility."* His idea of the physiological activity of the cell-body was very early recognised as important, and Schwann himself expressed the conviction that the study of the cell-body or cell-substance was essential to a true knowledge of the processes of life. The term "protoplasm" that has since come into general use for the cell-substance[190] was first applied to the cell-contents of plants, and was adopted by Max Schultze for animals as well,[191] when in 1861 he maintained the fundamental identity of the living matter of animals and plants, however high or however low in grade[192] they might be.

In 1868 Huxley, in his celebrated lecture at Edinburgh on "The Physical Basis of Life,"[193] laid great stress on the fundamental unity of the living matter of animals and plants, maintaining that it is right to speak of life as a property of protoplasm; and that just as the properties of water—*e.g.*, convertibility under given conditions of temperature and pressure into steam—result from the nature and disposition of its component molecules, so do the properties of protoplasm result from the nature and disposition of its molecules. This idea, really originating with Schwann himself, that the riddle, the mystery of life, was to be solved by study of the molecular constitution of living matter or protoplasm was a great conception and in all probability a correct one. However for the moment it

---

187   Now called macronucleus and micronucleus; both contain the complete genome of the organism but they have different roles during reproduction.

188   Now a loose collective term for a variety of aquatic protists, mostly found in fresh water.

189   Probably in 1835, but the source is not known.

190   See footnote 120 (Lecture 8).

191   Exact source unknown but possibly: *Das Protoplasma der Rhizopoden und der Pflanzenzellen; ein Beiträg zur Theorie der Zelle* (1863).

192   See footnote 62 (Lecture 3).

193   Published in America in 1869; see https://archive.org/details/onphysicalbasis-ooohuxlrich.

checked rather than aided the progress of biological science; it was too big a jump. Our knowledge of molecular physics is far too incomplete to enable us to tackle the problem of life from this side, with even the remotest prospect of success; while the habit which this conception almost necessarily engendered, of speaking of protoplasm as a substance of a certain chemical constitution, containing so much carbon, hydrogen, nitrogen, oxygen, phosphorus, and so on, as a substance allied to the albumens, perhaps most akin to white of egg, by diverting attention from other and more promising lines of enquiry, threatened to hamper rather than to further real progress.

The reaction soon came. Physiologists reminded themselves that it is the essence of protoplasm that it should be alive, and that their wisest course was to study directly the phenomena of life as manifested by it. The more conspicuous of these phenomena, contractility and irritability, the powers of movement and of sensation, first attracted attention. More recently attention has been specially directed to a further problem, resulting from the consideration that protoplasm is never idle, never at rest, but is always wearing itself away, incessantly wasting. Every living animal or plant wastes—i.e., loses weight; and this in all its parts and at all times, though at unequal rates. The loss of weight must mean loss of actual matter, and this loss is due to the breaking down of the body substance into simpler chemical bodies or excretory matters, of which carbonic acid and urea are the most important and the most characteristic.[194]

But this is not all. A dead body also wastes away, and the products of its wasting are much the same as the excretory matters formed by a living animal. The difference is that the living animal or plant has the power of repair, renewal, or regeneration of its living substance, while the dead body has no such power.[195] This renewal is effected at the expense of the food. Food is essential, or we starve and die; and it has been well said that hunger is the essential and diagnostic character of living things. This power of building up living from non-living matter is peculiar to protoplasm, of which we have seen that the cell-bodies, whether of animals or plants, consist. The process is partly a chemical one, a building up of simpler into more complex bodies; but in part it is something more, something peculiar, the true nature of which we do not understand. The final touch, the conversion of the dead food into living brain, muscle, etc., is effected by a process which we name assimilation, but of which the modus operandi is at present absolutely unknown.

This conception of protoplasm or the living matter of animals and plants, as undergoing incessant change or metabolism as it is termed, is one of much importance. Living protoplasm has been compared to a fountain in which the form remains constant though each component particle of water is in constant movement. In protoplasm as in the fountain, we distinguish two main processes, an uphill or anabolic process, as the water rises to the crest of the wave, or the food is being built up into the living tissues; and a downhill or katabolic process, as the water falls from the crest

194   This may be generally the case but whether it is as important as AMM implies is less clear. Lower weight may be due to water loss and/or gradual loss of cells and tissue mass; these are rather different from the processes of constant metabolism (*protoplasm is never idle*) taking place in cells.

195   *Death* is the subject of Lecture 12.

back into the basin, or as the living brain, muscle, etc., become broken down into the various excretory products. The up-hill or anabolic processes are synthetic and require or absorb energy. The water of the fountain will not rise of itself, but must be forced above the level of the water of the basin; and similarly the digestion and assimilation of food, the building-up of protoplasm, demand an expenditure of energy on the part of the animal.[196]

On the other hand, the downhill or katabolic processes are analytic[197] and are sources of energy; and just as the falling water of the fountain may be used to drive a wheel or other machine, so in the living body the katabolic changes, like a series of explosions, give out energy which as muscular movement or mental activity or in other ways may be employed to do the work of the body. And just as in a fountain, the energy required to raise the water a given height is equal to the energy liberated by the water in falling from the same height; and if work is to be done by the fountain, the water must be allowed to fall to a lower level than that from which it was raised; so also in the living protoplasm of an animal, if work, muscular, mental, or of other kind is to be performed, the katabolic fall must be greater than the synthetic rise; or in other words the excretory matters must be of simpler constitution than the food taken in.

Considerations of this kind which form the basis of modern physiology, lead us to regard the protoplasm of which the cell-bodies consist, in rather a different manner, and from a new standpoint. Protoplasm may, to adopt the simile given above, be regarded as the topmost point, the crest of the physiological wave, consisting of matter in a condition of extremely unstable equilibrium, and varying greatly in structure and in composition in different cells, or even in the same cell at different times. The essential character however of protoplasm is that it is alive; and although we are unable to formulate precisely what we mean by life, there is at any rate one very definite idea which we associate with living matter and with it alone, and this is the power which living matter possesses of building itself up, renewing itself from the dead matter which is taken in as food. To define protoplasm is difficult, perhaps impossible, but the one essential thing to remember about it is that it is alive.[198] To speak of dead protoplasm is a misnomer.

Starting from this new position attempts were soon made to determine whether the protoplasm of which cell-bodies consist possesses any structure, and more especially whether there are any structural changes which occur normally in cell-bodies, in association with vital acts. It was soon found that the cell-body or cell-substance is not necessarily homogeneous. In many, perhaps in most cases, a more or less pronounced reticular structure is present; the protoplasm consisting of a network of

---

196   This is a slightly clumsy analogy but its use anticipates biochemical energetics, besides relating to the question of whether matter is dead or alive (see above and below). This and the next paragraph express principles embodied in the Laws of Thermodynamics.

197   In the sense of breaking things into their component parts.

198   This statement illustrates the close association between the two usages of the word protoplasm, as indicated in Lecture 8. However, from here on, the word is clearly used to mean the fluid content of a cell.

firmer strands, the meshes of which are filled with a fluid or gelatinous substance in which are often present minute particles or granules.[199]

Furthermore the action of reagents of various kinds shows that the network or reticulum is of different character, presumably of different composition, to the substance filling the meshes, and that in some cases the network itself may vary in different parts. Examples of such reticular protoplasm are frequent among Protozoa—as for example in the genera Noctiluca,[200] Trachelius,[201] and many other Infusoria; while extreme cases of reticulation or vacuolation are seen among the Heliozoa,[202] Radiolaria,[203] and many of the Foraminifera[204]—e.g., Globigerina, in which the protoplasm has a characteristic bubbly or frothy appearance from the great number and size of the vacuoles. The contents of the meshes are usually fluid, or semi-fluid, and a study of their behaviour under various conditions, with a comparison of the mode of formation of fat cells, suggests strongly that the reticulum or network is the active part of the protoplasm, and that the substance filling the meshes is of secondary importance to it. Of this view, a most suggestive application has been made with regard to the structure of striated muscular fibre, which appears to consist of a reticulum, the strands of which are arranged in regular geometric pattern, and the meshes filled with a more fluid sarcous substance. It has been suggested that the reticulum is the active agent by contraction of which shortening of the muscle is produced; and that the reticulum of a striated muscle fibre cell is equivalent to the reticulum of a Protozoan or of any ordinary cell, differing mainly in its more regular arrangement.

The changes that actually go on within the protoplasm of the cell-body in different phases of its activity have been studied most closely, and with the most definite results in secreting cells, either from special glands, such as the pancreas[205] or salivary gland, or with still greater success in the epithelial cells lining the digestive stomach of certain insects. These secreting cells have their outer ends bathed in lymph,[206] which diffuses from the neighbouring blood-vessels, while their inner ends form the bounding surface either of the gland or of the alimentary canal itself. The cells separate from the lymph certain products which they modify in various ways, and finally discharge at their free surface as the special secretion of the gland. Careful histological examination has shown that during rest the gland-cell enlarges, the special secretions accumulating in the meshes of the reticulum as a fluid or

---

199   This is a clear reference to what we would now call the cytoskeleton, seen clearly only with the advent of advanced staining techniques and electron and confocal microscopy.

200   A marine dinoflagellate.

201   An alga.

202   A freshwater protozoan.

203   Protozoa with calcareous skeletons.

204   Amoeboid protists, often with external test (shell).

205   This undoubtedly means the exocrine pancreas, secreting digestive enzymes. The first hormone (secretin, from the gut) was not discovered until 1902; the sugar-regulating role of the pancreas was suggested by the work of Minkowski in 1889 but the hormones involved were not known until the 1920s.

206   This is presumably what we would now call extracellular fluid.

semi-fluid substance, often granular in appearance. During digestion, this secreted matter is discharged from the cells at their free surface, the meshes of the network becoming more or less completely emptied.[207] As to the part played by the network itself the evidence is incomplete, but there is some reason to think that the secreted substance is formed in part at the expense of the reticulum. The discharge of the secretion does not necessarily involve the death of the cell, which may fill up again, time after time, with the secreted matter.[208]

This very brief and imperfect account must suffice to indicate the lines along which investigation is at present advancing, in the attempt to determine the nature of the processes which go on within active living protoplasm.

Turning now to the nucleus, we find that an extraordinary number of minute researches have been made of late years, more especially with a view to determine as far as possible the part played by the nucleus in the acts of fertilisation and reproduction, in the hope of obtaining some clue to the attractive but bewildering problem of heredity.[209] The important part played by the nucleus in initiating, indeed determining, the act of cell-division, was first definitely established by F. E. Schulze; in Amoeba polypodia. He describes the successive changes as follows:[210] The nucleus first elongates, then becomes dumb-bell shaped; then the bridge between the two knobs becomes thinner and thinner, and finally breaks, so that there are now two separate nuclei, formed by division of the single original one. Next, the body of the Amoeba begins to elongate in the same direction as the nucleus did previously; then follow constriction, and finally the division of the Amoeba into two parts, each containing one of the two nuclei already formed. The entire process occupied ten minutes; division of the nucleus taking about a minute and a half, and the remaining eight and a half minutes being occupied by division of the body of the Amoeba. A similar process of direct nuclear division,[211] as it is called, has been described by Ranvier in the leucocytes of the Axolotl, in which case the process occupied an hour and a half; by Waldeyer in Infusoria; and since then by many other writers.

Of recent years attention has been more specially directed to a far more complicated series of changes, which have been seen by many observers to occur in the nucleus during division, and which are spoken of as karyokinesis or mitosis,[212] or indirect nuclear division, in contrast with the former or direct mode. To understand these more complicated processes, we must first describe the structure of the nucleus

---

207   Presumably secretory vacuoles.

208   A modern interpretation of these two sentences is not evident. Exocrine secretion does not occur at the expense of the secreting cells, nor is cell death involved.

209   N. B. Waldeyer named the chromosome in 1888 but Rabl and others had given good descriptions of them before that—see later comments and notes on the highlighted individuals.

210   Source not known, although Schultze's drawings have been reproduced in several descriptive works on amoeba.

211   This is later contrasted with *indirect* division in which the nucleus divides first followed by the cytoplasm—our familiar view of mitosis. (AMM subsequently queries whether the two processes are distinct.)

212   The *Oxford English Dictionary* gives the earliest use of these words as 1882 (by von Sachs) and 1887 (*Journal of the Royal Microscopical Society*), respectively.

as determined by more recent and more detailed investigations. It now appears that in the cell-nucleus there may normally be distinguished four elements: the network or reticulum, the nucleoli, the nuclear membrane, and the nucleoplasm.

The network or reticulum consists of finer or coarser threads, which differ in their arrangement in different cells. In epithelial cells from a Chironomus[213] larva, Balbiani found the nuclear reticulum to be one single complexly coiled thread. In a large number of other cases it has been found by Rabl[214] and others that the typical arrangement is as follows: Two kinds of threads may be distinguished; thicker primary threads and thinner secondary ones.[215] The primary threads may be twenty or more in number: each is folded on itself so as to form a loop, and the several looped threads are placed in a symmetrical manner in the nucleus; the loops being arranged around one pole of the nucleus, and the free ends of the threads interlacing at the other pole. The secondary threads form a fine network, connecting together the primary threads, and may be so numerous as to conceal more or less completely the definite arrangement of these latter. The nucleoli are spherical bodies of various size, which stain deeply with the ordinary colouring reagents. It is not certain whether they are in all cases corresponding structures. Sometimes they are merely nodes or local thickenings of the network of threads, while at other times they appear to be quite independent of the network. The general tendency at present is to regard them as non-essential, or at any rate secondary structures; and it is held by many that their purpose is to serve as a store of reserve matter for the nutrition of the other constituents of the nucleus.

The nuclear membrane has recently attracted renewed attention. By many it is regarded merely as a denser and superficial part of the nuclear reticulum; while by others it is held that it is, at any rate in certain cases, a continuous membrane, though authorities are not agreed as to whether it belongs really to the nucleus itself, or is rather to be regarded as a limiting layer of the cell protoplasm, immediately surrounding the nucleus.

The nucleoplasm or nuclear sap, which fills up the whole of the rest of the nucleus, is an albuminous coagulable liquid. A fine reticulum has been described in some cases as traversing the nucleoplasm, and apparently independent of the main nuclear reticulum.

As regards the chemical constitution of the various parts of the nucleus, very little is as yet known. The network and the nucleoli take up stains more readily than the rest of the nucleus: the substance of which they consist is hence contrasted as chromatin with the non-staining nucleoplasm. In some cases the threads of the reticulum appear to consist of rows of minute chromatin granules arranged in an achromatin basis.

We are now in a position to consider the changes which occur in the nucleus during the process of indirect nuclear division or mitosis. The first change is that the secondary threads of the reticulum disappear, apparently through being absorbed into the primary threads. The primary threads thus become much more conspicuous, and their arrangement in loops is clearly seen. The threads forming the loops next become

---

213   A non-biting midge.
214   *Über Zelltheilung, Morphologisches Jahrbuch* 1885; **10** (Part 2): 214–330.
215   This and the subsequent sentences seem to be referring to cytoskeletal (microtubular) structures as well as to chromosomes.

somewhat shorter and thicker;[216] they may also by transverse division become more numerous. There is some reason for thinking that in cells of the same kind from the same animal, the number of threads is constant, but this has not yet been proved to be a general rule. In the epidermal cells of the Salamander twenty-four loops are said to be constantly present.[217] The next stage is an extremely important one, for the discovery of which we are indebted to Flemming. Each of the looped threads splits longitudinally along its whole length into two parallel threads. The pair of threads formed by the splitting of a single primary thread are spoken of as " sister-threads." As they are of exactly equal size, and as each of the primary threads becomes divided into a pair in this way, the upshot of the process is that the whole of the chromatin becomes divided into two precisely equal portions, a process the full significance of which will become apparent immediately. About or shortly after the time of the splitting of the primary threads into sister-threads, a structure known as the nuclear spindle appears. This is a fusiform figure bounded by a number of exceedingly fine and feebly staining threads, It is said to be formed not from the nucleus, but from the cell protoplasm immediately surrounding the nucleus. At the time of its first appearance it has no clear relation to the looped threads of the nucleus, but very soon the two structures take up definite positions with regard to each other, the looped threads being grouped in a ring round the equator of the spindle, with the loops directed inwards towards the centre and the free ends outwards. At the same time the spindle itself becomes more clearly marked; at its poles are two rounded bodies, the "pole-bodies" of van Beneden,[218] from which fine threads radiate outwards in all directions through the protoplasm of the cell-body. The nuclear membrane disappears about this time, though what exactly happens to it is unknown. It was formerly supposed to give rise, at any rate in part, to the nuclear spindle, but it is now generally agreed that the spindle is extra-nuclear and a product of the cell protoplasm alone. The chromatin loops now begin to move from the equator along the threads of the spindle, towards its poles. This takes place in a perfectly regular manner, the sister-threads of each pair moving towards opposite poles. As the two sister-threads of each pair are precisely equal, the two groups of threads around the two poles of the spindle will contain precisely equal quantities of chromatin. The two groups separate completely from each other, and each group becomes arranged in a manner corresponding to that of the primary – threads in the original nucleus. In this way two daughter-nuclei are formed by division of the mother-nucleus: each daughter-nucleus containing half the chromatin of the mother-nucleus. After completion of the daughter-nuclei, the protoplasm of the cell-body divides, precisely as in the process of direct division, and the division of the mother-cell into two daughter-cells is complete.

The above description of the phenomena of mitosis or indirect nuclear division, represents the generally accepted view as to the sequence of events, but several points

---

216   Presumably chromosomal condensation.

217   The various species of salamander vary considerably in their chromosome number but some do indeed have 24 as the diploid number.

218   The centrosomes, to which the nuclear microtubules are attached. Van Beneden's major work was *Recherches sur la composition et la signification de l'œuf* (1868). *Pole bodies* are clearly distinct from the *polar bodies* whose presence and ejection from the dividing cell are discussed later.

in it are still very imperfectly understood. One of the most important of these is the origin of the nuclear spindle and its pole-bodies. Van Beneden, whose long-continued and extraordinarily careful observations entitle him to speak with great authority, maintains that the pole-bodies are of primary importance. He says that the pole-bodies appear before the spindle; that they arise distinctly in the cell protoplasm, quite independently of the nucleus; and that they appear to be centres from which the spindle threads diverge: the spindle threads themselves he describes as "apparently muscular," and as governing the movements of the chromatin threads. Van Beneden would therefore regard the pole-bodies as the most important structures, and as really determining cell division rather than the nuclei. Other observers however are not yet prepared to accept this view, but regard the nuclei as the determining agents in all cases of cell division.

A still more important problem is as to whether the direct and indirect processes of nuclear division are entirely independent, or whether they are not really related to each other, and if so in what manner. On this point very keen discussion has been and is still being carried on. As regards the relative frequency of the two processes, direct nuclear division has only been seen in comparatively few cases: in leucocytes by Rabl and Flemming; in the intestinal epithelium of Crustacea by Frenzel;[219] in regenerating tissues by Fraisse;[220] in the Infusorian Euplotes[221] by Möbius; in the early stage of spermatozoon formation in Amphibia, and in a few other cases. On the other hand indirect nuclear division, or mitosis, has been seen in both animal and vegetable cells of almost every kind, and in processes both normal and pathological, and would certainly appear to be by far the more usual method.

With reference to a possible relation between the two methods, Waldeyer has strenuously maintained that there is one fundamental form of nuclear division, presenting a series of gradations from the simple direct method described by Remak up to the extreme complication seen in typical cases of mitosis; and in favour of this view there is a considerable amount of evidence already accumulated. Thus Waldeyer himself has shown that if a frog's epidermis, in which mitosis usually occurs, be treated with silver nitrate, direct nuclear division can alone be demonstrated, from which he argues with great justice that the distinction between the two methods may in other cases also be only apparent, and due to the particular mode of histological treatment adopted. Pfitzner has succeeded in staining both the nucleoplasm and chromatin simultaneously, and maintains that mitosis concerns not merely the chromatin but the entire nucleus; and further that the cell protoplasm takes no part in the process. Bütschli, Hertwig, Schewiakoff, and others have shown that in Protozoa, in which all the phenomena of mitosis occur, the nuclear membrane may remain complete the whole time until the final division of the nucleus into the two daughter-nuclei; while Boveri and others have found that in segmenting eggs of various animals the various stages of mitosis may sometimes be very clear, and at others very obscure. Finally,

219   The original source is not known but the following work, which slightly post-dates AMM's lecture, is referenced in *Proceedings of the Academy of Natural Sciences* May–Sep 1902; **LIV**: Frenzel J, *Biologisches Zentralblatt* 1891; **11**: 558.

220   Fraisse P, *Die Regeneration von Geweben und Organen bei den Wirbelthieren, besonders bei Amphibien und Reptilien* (Kassel and Berlin, 1885).

221   A ciliated Hypotrich protozoan.

Carnoy asserts that mitosis, so far from being a uniform process in all cases, presents so many varieties that no one series can be viewed as essential; he even denies that the longitudinal splitting of the chromatin threads into sister-threads is universal: a point on which Waldeyer differs from him emphatically.

On the whole it would appear that the balance of evidence is against the existence of a sharp distinction between the direct and indirect methods of nuclear division; and in favour of there being one common form, subject to considerable variation in detail. Much however yet remains to be done before the question can be regarded as settled. We are at present absolutely in the dark as to the real meaning and significance of the chromatin threads; and it is especially important that the true relation of van Beneden's pole-bodies and of the nuclear spindle should be clearly determined, for until this is done, no theory of nuclear division can be held to be definitely established. Perhaps the most striking points brought to light by the recent researches on nuclear division are the extreme complication of a process formerly regarded as a perfectly simple one, and the mathematical precision with which the division of the chromatin into equal halves is effected in typical cases.

One of the most interesting problems arising from the advance of our knowledge concerning the minute changes that occur during nuclear division, is the relation of the phenomena of mitosis to the act of fertilisation of the egg. By the older writers it was supposed that either prior to, or as a consequence of the act of fertilisation, the nucleus of the egg, or germinal vesicle as it is commonly termed, disappears completely. It is now however known that part of the egg-nucleus persists and fuses with part of the spermatozoon, this fusion constituting the act of fertilisation. Auerbach was the first to show, in 1874, that in the egg of Ascaris two nuclei may be seen shortly after the spermatozoa reach the egg, and that these two nuclei fuse together. In 1875 Oscar Hertwig stated, as the result of a careful series of observations on the eggs of Echini, that one of the two nuclei seen by Auerbach is the head of the spermatozoon, and that the other is part of the nucleus or germinal vesicle of the egg, probably the nucleolus or germinal spot. In 1875 and 1876 van Beneden saw the fusion of two nuclei, or pronuclei as he named them, in the ovum of the rabbit, as a result of fertilisation; one of these pronuclei he found to arise from the egg nucleus, though not as Hertwig had supposed from the nucleolus; the other, or male pronucleus, van Beneden recognised as in some way connected with the spermatozoon, though he failed to trace it directly to the head of the spermatozoon as Hertwig had done in the sea urchins. These discoveries naturally attracted great attention, as they for the first time rendered possible a theory of fertilisation based on the changes actually known to occur during the process. A number of other investigators, prominent among them being Fol, Greeff,[222] Selenka,[223] Flemming, Hensen, and Boveri, studied the phenomena in various groups of animals and definitely established the fact that the essence of the act of fertilisation consists in fusion of part of the egg nucleus with the head or nucleus of the spermatozoon—i.e., fusion of the

---

222 Source uncertain but possibly Greeff R, *Ueber den Bau und die Entwicklung der Echinodermen. (Sitzungsb. Gesellsch. z. Beford, d. gesammten Naturwiss. zu Marburg,* 1876, Nr. 5.)

223 Selenka E, *Studien über Entwicklungsgeschichte der Thiere, Das Opossum (Didelphis virginiana),* Vol 4 (CW Kreidel: Wiesbaden, 1887), 101–172.

male and female pronuclei. Concerning the details of this fusion, and especially its relations to the processes of mitosis or karyokinesis described above, many elaborate and careful investigations have been made of late years which have brought to light points of extraordinary interest.

Before describing these in detail it is necessary to say a few words concerning a phenomenon closely connected with fertilisation, though no part of the actual process itself; I mean the extrusion from the egg, prior to fertilisation and independent of it, of the so-called polar bodies. Carus[224] first noticed, in 1828, minute globules on the surface of an egg which took no part in the formation of the embryo. These were described carefully by Friedrich Müller in 1848 as seen by him in the snail's egg, and were called by him directive corpuscles because of the constant relation they appeared to have to the first plane of segmentation; he showed clearly that they were derived from the egg itself. Hensen was the first to show that the extrusion of the directive corpuscles or polar bodies occurred independently of fertilisation, for he found that in the rabbit and guinea-pig the polar bodies were formed while the eggs were still in the Graafian follicles, before their discharge from the ovary. This has been confirmed by many other investigators, who have shown that as a rule two polar bodies are extruded in succession; that both are usually extruded previous to fertilisation; but that in some cases as in the lamprey and the frog, the first polar body may be extruded prior to fertilisation; and the second after or during the process; while in some cases, as in Ascaris, according to van Beneden, both polar bodies are extruded after the entrance of the spermatozoon. In 1875 Butschli showed that in Nematodes the extrusion of polar bodies is accompanied by the formation of a nuclear spindle, and by other changes similar to those seen in indirect nuclear division. He in consequence suggested the view that the formation of a polar body is really a division of the egg cell into two very unequal portions, and is accompanied by the ordinary nuclear changes seen in typical cases of mitosis. This view has since been widely adopted, though authorities are not yet agreed as to whether any part of the protoplasm of the egg is extruded with the daughter nucleus in the polar body. The point is of considerable interest, for on one view the whole process would be one of cell division, on the other merely nuclear division with extrusion of one of the daughter nuclei. The balance of opinion inclines strongly towards the former of these views, but the point cannot be considered definitely settled as yet.

The changes that occur in the egg nucleus during the formation of polar bodies have been recently studied with great care by many observers. In the frog's egg Oscar Schultze has shown that at the time of ripening of the egg the nucleus undergoes important changes. Previous to this time it has been of very large size, as much as a third the diameter of the egg itself, and consists of a well-marked nuclear membrane enclosing fluid nucleoplasm in which float a number of chromatin globules of varying size and apparently unconnected with one another. At the time of ripening of the egg, shortly before it leaves the ovary, the nuclear membrane becomes crumpled and indistinct; the whole nucleus shrivels very greatly, its fluid contents being diffused through the egg; the majority of the chromatin globules disappear, but a group of very small ones near the centre persist; these unite together to form a convoluted varicose

---

224   Identity uncertain, possibly Carl Gustav Carus (1789–1869), https://en.wikipedia.org/wiki/Carl_Gustav_Carus.

chromatin thread which soon breaks up into loops; then a nuclear spindle appears with which the loops soon become connected as in ordinary mitosis. The spindle with the associated loops, which together are of exceedingly small size, move to the surface of the ovum; the loops and the spindle now divide in the ordinary way into two equal halves which become the daughter nuclei. One of the daughter nuclei is extruded as the first polar body, while the other retreats into the interior of the egg. The formation of the second polar body is a repetition of the process by which the first was formed. As the amount of chromatin is halved at each nuclear division, the female pronucleus, which is the portion of the egg nucleus left after the formation of the second polar body, will contain exactly one-fourth of the amount of chromatin present in the egg nucleus at the formation of the nuclear spindle. Van Beneden's account of the process in Ascaris agrees exactly with this. At each division of the nucleus to form a polar body the chromatin, which is present rather in the form of spherules than of threads, is exactly halved in amount.

Concerning the fusion of the male and female pro-nuclei in the act of fertilisation some points of great interest have recently been brought to light. In sonic cases, as shown by Nussbaum[225] in Ascaris, a direct fusion of the two pronuclei is described as occurring, while in other forms and by other observers the process is said to be of a more complicated character. In the Nematodes, genus Leptodera, Nussbaum says that the two pronuclei, male and female, take up a position parallel to the long axis of the egg, which is ovoid in shape, and then fuse together lengthways. The first segmentation plane is a longitudinal one and passes along the axis of the fused pronuclei, so that each of the two cells formed by the first cleft contains one-half of the male pronucleus and one-half of the female pro-nucleus. Inasmuch as all the cells of the body of the adult animal are derived by division from the two primary ones, it follows, as Nussbaum points out, that if this equal division of male and female nuclear elements obtains in the later stages of cell division, each cell of the adult animal will possess a nucleus, one half of which is derived from the father and the other half from the mother.

This suggestion, the bearing of which on theories of heredity is of the greatest importance, has received most striking confirmation from the extraordinarily minute observations of van Beneden on the eggs of Ascaris. These investigations, made in the years 1883 and 1884, rank deservedly among the most remarkable of microscopical triumphs. Van Beneden finds that after extrusion of the polar bodies, and entrance of the spermatozoon into the egg, the two pronuclei, male and female, which are precisely equal in all respects, come very close together but do not fuse directly. Each pronucleus bears at first a single much convoluted and varicose thread of chromatin, which soon divides transversely into two, each of which becomes bent into a U-shaped loop. There are then four loops in all, two male and two female. Each loop now splits longitudinally into two sister threads. A spindle figure with pole-bodies and polar rays at its apices now appears, and the outlines of the pronuclei, previously distinct, disappear. The chromatin loops, of which owing to the longitudinal splitting there are now eight, four male and four female, take up a position at the equator of the spindle. The two sister threads of each pair now separate, one moving towards one pole of the

---

225   Uncertain but possibly: Nussbaum M, Ueber die Veränderungen der Geschlechts-produkte bis zu Eifurchung. *Archiv für mikroskopische Anatomie* 1884; **XXIII**.

spindle the other towards the opposite pole; so that at each pole of the spindle there is a group consisting of two male threads and two female threads. Each group now forms a daughter nucleus, and then the entire egg divides into the first two segmentation cells. This account agrees with Nussbaum's as regards the equal division of male and female threads between the first two segmentation cells; but differs inasmuch as according to Nussbaum, the two pronuclei fuse directly, while, according to van Beneden, there is no fusion of the pronuclei, but merely a re-arrangement of the chromatin threads. Zacharias and Hertwig have suggested that fusion really does occur but may have been overlooked, but for the present the point cannot be considered settled. It is possible indeed that the details are not the same in all cases.

The equal division of male and female elements in the first segmentation, a point in which Nussbaum and van Beneden agree and which has been confirmed by others, is of the highest possible interest and importance. If it should prove to occur in the later as well as in the earliest stages of development, then, as pointed out above, it will follow that the nucleus of each cell in the body of the adult animal will contain male and female elements derived from the male and female pro nuclei—i.e., from the father and mother—in precisely equal amounts. In other words, each cell of the adult body may be spoken of as hermaphrodite. If this is true, then the egg, which is merely an epithelial cell indistinguishable in its early stages from the surrounding cells, must itself be hermaphrodite. The further suggestion at once presents itself: Is not the extrusion of the polar bodies a casting out of the male elements of the egg?

This is a view which in one form or another has commended itself to many embryologists. Balfour first suggested that the extrusion of the polar bodies was a device for ridding the egg of its male elements, and so ensuring that cross fertilisation must occur, the advantage of which as regards vigour of offspring is well known. Minot has independently suggested and strongly supported the same view. A serious difficulty however has been raised to this view by Weismann, who points out that if the male element in the egg is got rid of completely by extrusion of the polar bodies, then it becomes very difficult to understand the transmission of male ancestral characters through the mother. It is well known that children may inherit peculiarities of their grandfather on the mother's side, and this, according to Weismann, should not be possible if the extrusion of the polar bodies removes all male elements from the egg. We must suppose that a small remnant is or may be left behind; but then, asks Weismann, why should any be eliminated at all if it is not necessary that all should be got rid of?

The matter is very far from a simple one. Weismann himself has shown that in various parthenogenetic crustacea, Polyphemus, Moina, Daphnia, and others, only one polar body is extruded; and Blochmann in 1887 announced that in Aphis the parthenogenetic eggs extrude one polar body only, while those that require fertilisation extrude two. This seems to indicate that the two polar bodies may have very different value in spite of the close similarity in their modes of formation; for the extrusion of one polar body seems to leave the egg still capable of developing without fertilisation; while after two polar bodies are extruded, fertilisation is necessary for development to occur.

Weismann's explanation of the process is a complicated one. Regarding the nucleus as specially concerned with heredity, he distinguishes in the nuclear substance two kinds of matter, which he names histogenic plasma and ancestral plasma respectively. The former or histogenic plasma he regards as specially concerned with the growth, nutrition, and shaping of cells; while the second or ancestral plasma is supposed to be specially concerned with reproduction and heredity. He then assumes that the histogenic plasma having completed its work when the egg has reached maturity, it is convenient to get rid of it, which is done by extruding it bodily from the egg as the first polar body. The remaining half of the nuclear substance is the ancestral plasma; and with regard to the second polar body, Weismann argues that if the whole of the ancestral plasma of the egg were retained, and were to fuse with the whole of the ancestral plasma of the spermatozoon, then the total amount of ancestral plasma would be doubled at each generation. This he argues would cause inconvenience, and so the device is adopted of getting rid of half of this ancestral plasma in the form of the second polar body.

Apart from its extreme complexity, this theory of Weismann's is open to grave objection. In the first place the supposed difference between histogenic plasma and ancestral plasma is a pure assumption, in support of which no direct evidence has been advanced. Secondly, it is very difficult to understand why, on Weismann's view, the histogenic plasma should be exactly half the nucleus. Thirdly, it cannot be overlooked that the modes of formation of the first and of the second polar bodies are precisely similar to each other, and also that they agree precisely with the changes that occur in the nucleus of an ordinary epithelial cell during indirect division, similarities which become very difficult to understand if, as Weismann supposes, the two polar bodies are of totally different nature, and do not correspond to the halves of the nucleus of a dividing epithelial cell.[226] Lastly, it may be noted that cases have been described in which more than two polar bodies have been extruded.

On the whole therefore, while acknowledging the extreme ingenuity of Weismann's theory and the service which its publication has rendered by stimulating investigation and discussion, it cannot be said that the theory is in accordance with all the facts it seeks to explain, or that it helps us very materially towards an understanding of these facts. So far direct observation has failed to show any difference in structure or in mode of formation of the two polar bodies, or any important respects in which their development differs from the phenomena of ordinary mitosis. These processes are so complicated and require such extraordinary care and patience for their successful investigation, that it is but natural to assume that they have some deep and far-reaching significance; and it is well therefore to remember that they are in no way specially concerned with the processes of sexual reproduction, but occur as characteristically in the act of division of an ordinary epithelial cell as in the formation of a polar body by a ripe ovum.

It is well also to bear in mind that although the arrangements and divisions of the chromatin threads naturally attract special attention, it is by no means certain that these threads are the main factors in the act of nuclear division. Indeed the comparison of the direct method of division with the indirect method which we have discussed above, rather suggests that the nucleoplasm or nuclear substance itself is the really essential element, and that the chromatin threads need not even be present,

---

226 AMM's objection here is based on his sense (reiterated in the next paragraph) that the polar bodies are the same and do not represent different types of *plasma*. The logic of this is that he also finds the distinction between the two types of plasma difficult to accept (as well as unsupported by evidence).

at any rate in a form capable of being brought into view by the reagents at present in use.

And so it would appear from these more recent researches, of which time has only permitted me to give a brief and most imperfect summary, that the cell theory, great and important as it is most undoubtedly, is rather the commencement of a great movement, a fresh starting-point from which to begin investigations anew, than a complete scheme or final explanation; and the one great lesson for us to learn is that processes of apparently the simplest kind are really of an extremely complicated nature, and will well repay the most minute and attentive study: for a right understanding of the changes that occur during the act of division of an ordinary epithelial cell, and of the causes determining those changes, would throw most welcome light on the more complicated processes accompanying the ripening and fertilisation of the egg, which microscopists of all nationalities are at present studying with such intense earnestness.

LECTURE 10

ANIMAL PEDIGREES

This lecture to the Birmingham Natural History Society was based on that of Lecture 13: CFM explains in his Preface to the collection his reasons for including both. The Birmingham society was formed in 1858 *"by a group of amateur Victorian naturalists"* (http://bnhsoc.org.uk/) and still exists as an organization open to all comers. AMM also delivered a lecture with this title at 3.30 p.m. on 6 December 1891 in the New Islington Public Hall, Ancoats, Manchester, during the tenth series of Sunday Lectures to Men and Women, part of the fifteenth Recreation in Ancoats event organized by the Ancoats Brotherhood.

These non-academic audiences explain the accessible style of the opening remarks and the gently ironic humour with which the discussion of pedigrees is suffused. But this leads swiftly into deeper matters, starting with the Recapitulation Theory, through the role of yolk in embryo maturation, on to degeneration and finally to the incompleteness of current ideas. If this really is the material of one public lecture, it can be imagined that the audience found the combination of embryology, palaeontology and developmental interpretation across a range of animal examples, both living and extinct, not to mention the frequent insertion of authoritative names, challenging.

To modern eyes, AMM seems to trip himself up when talking about degeneration, even if we scrupulously avoid attaching any pejorative meaning to the word, and one wonders whether the audience of the time felt any discomfort in his phraseology. Throughout the Lectures, AMM strives for objectivity and rationality and for an interpretation of biology firmly grounded in the Darwinian theory of natural selection. Yet here we see development (and embryology) described as a progression towards perfection and ultimate forms. It may well be that he felt himself restricted, consciously or otherwise, by the terminological conventions of the day and the need to make his theoretical explanations accessible. It is noticeable that Lecture 13, which was to an academic audience, uses teleological language much less frequently.

This Lecture perhaps illustrates as clearly as any other the transition that was taking place in the *language* of biology as well as in its ideas and interpretation. A post-modern analysis might suggest that these are one and the same thing.

ANIMAL PEDIGREES

THERE are few things in which men take greater pride or from which they derive more solid enjoyment, than in tracing out and publishing to the world their pedigrees; and we must acknowledge that the proceeding itself, and the satisfaction obtained from it, are entirely legitimate. For we all have pedigrees; for two or three generations each one of us could give his descent, trace his pedigree, offhand; and if we fail in attempting to go further back we know that this is from lack of knowledge, not of facts.

Would we learn these facts, we know that there are those whose profession it is to supplement our deficiencies of memory or of information on these points, and who are prepared for a sum of half-a-crown,[227] to provide the enquirer with a duly attested pedigree dating from the time of the Conquest. For a guinea a Roman emperor can be obtained; while the avaricious in such matters, who are prepared to spend a five-pound note, may satisfy themselves, and for all we know, truthfully, of their descent from a Pharaoh of the 19th dynasty.

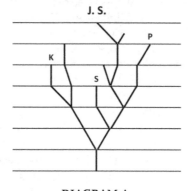

DIAGRAM A.

The mode of construction of such pedigrees, or genealogical trees, is familiar to us all from our school days. We begin by ruling a series of horizontal lines, which we agree shall represent successive generations. Then, assuming that we are of those whose aspirations are satisfied by a two-and-sixpenny ancestry, we commence with the dawn of respectability in the year 1066, and gradually trace upwards from this date the line representing our descent from the selected progenitor. We indicate in capitals, or in italics, any kings, or statesmen, or poets, or other eminent people whose memories we like to think derive renewed lustre from association with ourselves; and to include a sufficient number of these we trace the side branches of our tree for some distance. Finally, on the topmost twig, and in largest letters, we write—John Smith.

---

227 Two shillings and sixpence (12.5 new pence).

We all have such pedigrees, whether we can declare them or not; and for the prince and the pauper they are of equal length, dating back not merely to the Norman Invasion of England, but to the first appearance of man on the earth.

A special point with regard to these genealogical trees, and one to which attention may well be directed, is their absolute truthfulness. Suppose for example, you have three young friends, two of whom, Tom and Dick, are brothers, while Harry, the third, is their first cousin; then the diagram, or genealogical tree, representing their mutual relationships will be as follows, the horizontal lines marking successive generations:—

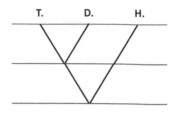

DIAGRAM B.

Go back one generation, and Tom and Dick's lines meet, for they are sons of one father. Before Harry's line joins in it is necessary to go back one generation further—*i.e.*, to the grandfather of our young friends, who was one and the same person for all three, and who forms the true link or bond of union between them. If once these relationships are determined correctly, and the diagram constructed aright, then it expresses facts, and facts only, and can never he disturbed. It matters not what the occupations or residences of the three may be; they may never see one another, never even suspect one another's existence. It makes no difference how many people may be living contemporaneously, how many may have preceded, how many may he born in after ages; this little bit of history remains untouched and absolutely true for all time. It is for this reason that genealogy or blood relationship affords the only satisfactory basis for a classification of men; and we shall find that the same considerations apply to the lower animals as well.

One further matter of a preliminary nature requires mention. It is customary in preparing genealogical tables to construct them as is done above, starting from some one more or less remote ancestor, and following upwards the branches representing his descendants, so that the whole diagram takes the form of an upright tree. It should be noted however that in a certain sense the diagram would be more correct if it were inverted. In tracing back human pedigrees there is a marked tendency to follow out one special line of descent, and to concentrate attention on one particular ancestor from whom we desire to make our line arise, ignoring the fact that there were many other contemporary ancestors from any of whom our path might with equal truth have commenced.

A man has two parents, four grandparents, eight great-grandparents—*i.e.*, in tracing back his pedigree from the present time the number of his ancestors in each generation is double those of the generation that succeeded it in time. This is graphically expressed in the following diagram:

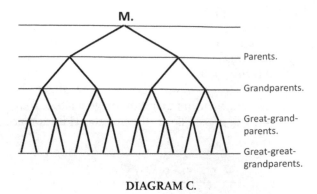

M.

Parents.

Grandparents.

Great-grand-
parents.

Great-great-
grandparents.

DIAGRAM C.

which takes, as noted above, the form of an inverted, not an erect, tree.[228] If we allow three generations to a century there will have been twenty-five generations between the Norman Invasion and the present time: so that a man now living may be descended not merely from one ancestor who came over to England in 1066, but directly and equally from over sixteen million ancestors who lived at or about that date. I say advisedly "may be descended;" for unless we assume that many of these ancestors were identical individuals, we shall find that the existence of a single man to-day involves the existence, a thousand years ago, of over a thousand millions of ancestors: and at the commencement of the Christian era of nearly seventy thousand millions of millions of ancestors a state of things which would involve serious reconsideration of the dimensions of the earth.

Genealogical trees, such as I have described, we are all familiar with. Furthermore we know that the principles employed in constructing them are not confined to Smith and Jones and the kings and queens of England, but apply to the lower animals as well. The pedigrees of racehorses, and of other artificially bred animals, such as cattle, sheep, pigs, dogs, pigeons, poultry, etc., have for many years past been kept with the most scrupulous care; and there are men who could tell you in detail the pedigree of the winner of last year's Derby or Leger, who would be sorely perplexed if asked for their own, and would perhaps prefer that the results of researches on this point should not be made too public.

We all recognise that the cats and rabbits and dogs of to-day did not come into existence spontaneously, but are descended from the cats, rabbits, and dogs of preceding generations, decades, or centuries; and that the same applies to birds, to butterflies, to sea anemones, or to any other animals we like to think of. We have now merely to enlarge our sphere of action, to widen our boundaries with regard to such genealogies, and we find ourselves face to face with the great problem with which naturalists are confronted, and which they are attacking on every side and by all means in their power.

We recognise that Diagram A represents correctly the relation between man and man: and we admit that it is equally true when applied to horses, to cows, to dogs, or to canaries. In other words, we acknowledge that the principle on which

---

228   The point about the plethora of ancestors is well made. Why the tree should be depicted one way up rather than the other is less clear.

the diagram is constructed is true in all cases in which historical or documentary evidence is forthcoming.

Can we not go further than this? Is this written testimony essential? Would the facts be in any way altered if no documentary evidence were forthcoming? Do we not agree that all animals have had pedigrees of this kind; and is it not worth enquiring whether we cannot reconstruct, unravel these pedigrees, even in cases where of necessity documentary evidence cannot be obtained?

Again, to return for a moment to the human argument. So far our horizontal lines have been used to indicate successive generations, and the relation, admitted by all, has been that each generation has sprung from the preceding generation, and has in its turn given birth to the next succeeding one.

Supposing now that we widen our boundaries, and agree that the intervals between successive horizontal lines shall indicate, not generations but longer intervals, say centuries, the relations will remain unaltered.

	XIX Century.
	XVIII Century.
	XVII Century.
* * * *	XVI Century.
	XV Century.
	XIV Century.

If for example we fix our attention on the sixteenth century, we find that the men of that century did not arise spontaneously, but were the direct descendants of those of the preceding or fifteenth century; furthermore we all admit that from the men of the sixteenth century those of the seventeenth, eighteenth, and nineteenth centuries have all directly sprung. And what holds good with regard to the men of the sixteenth century applies equally well to the horses, the cats, the dogs, the birds, the butterflies, the starfishes of that time. Now widen the intervals still further: let them represent not merely centuries but thousands, tens of thousands, of years; let them finally indicate the great geologic periods, and the argument will still hold:

	Quaternary and Recent.
	Pliocene.
* * * *	Miocene.
	Eocene.
	Cretaceous.

The animals, and for that matter the plants too, of the Miocene age, did not come into existence irrespective of pre-existing animals and plants, but were the direct lineal descendants of the Eocene animals and plants; and the forefathers of those of Pliocene and Recent times.

	Kainozoic.
* * * *	Mesozoic.
	Palaeozoic.

So also were the animals of Mesozoic times the children of those of the Palaeozoic age, and the parents of the Kainozoic[229] fauna. And so this idea of continuity of life from its earliest dawn on the earth, through age after age, down to the present time, forces itself upon us: an idea involving the further conception of the evolution of existing animals from unlike ancestors of former times: an idea constraining us to admit that animals, like men, have pedigrees, and that between the animals of all ages a kinship, a blood relationship, exists.

The recognition of this kinship is the determining feature of the Natural History of to-day. The reconstruction of these pedigrees is the great work of the future: the rewriting of the past histories, not merely of one or two groups for a limited number of generations, but of all animals and for all time:—a formidable, but an entrancing problem; and whatever misgivings I may have as to my power of presenting it aright, no apology is needed for asking your attention to a consideration of the means at our disposal for attacking it, of the evidence on which we rely in our attempts to reconstruct the past histories, to determine the pedigrees of animals.

On the present occasion, it is not with the whole evidence, but with one special side of it, that we shall be concerned, that namely which is derived from a study of the development of existing animals. Every one knows that animals in the earlier stages of their existence differ greatly in form, in structure, and in habits from the adult condition; a lung-breathing frog for example commences its life as a gill-breathing tadpole; and a butterfly passes its infancy and youth as a caterpillar. It is clear that these developmental stages, and the order of their occurrence, can be no mere accidents; for all the individuals of any particular species of frog, or of butterfly, pass through the same series of changes. It is not however until recent years that naturalists have realised that each animal is constrained to develop along definitely determined lines; and that the successive stages in its life-history are forced on an animal in accordance with a law, the determination of which ranks as one of the greatest achievements of biological science.

The doctrine of Descent, or of Evolution, teaches us that as individual animals arise, not spontaneously, but by direct descent from pre-existing animals, so also is it with species, with families, and with larger groups of animals, and so also has it been for all time; that as the animals of succeeding generations are related together, so also are those of successive geologic periods; that all animals living or that have lived are united together by blood relationship of varying nearness or remoteness; and that every animal now in existence has a pedigree stretching back, not merely for ten or a hundred generations, but through all geologic time since the dawn of life on this globe.

---

229   Cainozoic or Caenozoic.

The study of Development, in its turn, has revealed to us that each animal bears the mark of its ancestry, and is compelled to discover its parentage in its own development; that the phases through which an animal passes in its progress from the egg to the adult are no accidental freaks, no mere matters of developmental convenience, but represent more or less closely, in more or less modified manner, the successive ancestral stages through which the present condition has been acquired. Evolution tells us that each animal has had a pedigree in the past. Embryology reveals to us this ancestry, because every animal in its own development repeats its history, climbs up its own genealogical tree. Such is the Recapitulation Theory,[230] hinted at by Agassiz,[231] and suggested more directly in the writings of Von Baer, but first clearly enunciated by Fritz Müller, and since elaborated by many, notably by Balfour and Ernst Haeckel.

A few illustrations from different groups of animals will best explain the practical bearings of the theory, and the aid which it affords to the zoologist in his attempts to reconstruct the pedigrees of animals; while these will also serve to illustrate certain of the difficulties which have arisen in the attempt to interpret individual development by the light of past history; difficulties which I propose to consider at greater length.

A very simple example of recapitulation is afforded by the eyes of the sole, flounder, plaice, turbot, and their allies. These "flat fish" have their bodies greatly compressed laterally, and the two surfaces, really the right and left sides of the animal, unlike, one being white or nearly so, and the other coloured. The flat fish has two eyes, but these, in place of being situated as in other fish one on each side of the head, are both on the coloured side. The advantage to the fish is clear, for a flat fish when at rest lies on the sea bottom, with its white surface downwards and the coloured one upwards. In such a position an eye situated on the white surface could be of no use to the fish, and might even become a source of danger, owing to its liability to injury from stones or other hard bodies on the sea bottom.

No one would maintain that flat fish were specially created as such. The totality of their organisation shows clearly enough that they are true fish, akin to others in which the eyes are symmetrically placed one on each side of the head, in the position they normally hold among vertebrates. We must therefore suppose that flat fish are descended from other fish in which the eyes are normally situated.

The Recapitulation Theory supplies a ready test; and on employing it—*i.e.*, on studying the development of the flat fish—we obtain a conclusive answer. A young flounder or other fish, on leaving the egg, is shaped just as any ordinary fish, and has the two eyes placed symmetrically on the right and left sides of the head. As the young fish increases in size, the shape gradually approaches that of the adult; the body increases in height and becomes flattened laterally, the median, dorsal, and ventral fins becoming greatly developed at the same time; and the fish now begins to adopt the habit of the adult of lying on one side on the sea bottom. Another change occurs: the eye of the side on which the fish lies, usually the left side in a flounder,

---

230   This is the main topic of Lecture 13 and is also central to Lecture 17. See also the book review interlude and Lecture 20.

231   Louis Agassiz.

becomes shifted slightly forwards, then rotated on to the top of the head, and finally twisted completely over to the opposite or right side.[232]

Crabs differ markedly from their allies, the lobsters, in the small size and rudimentary condition of their abdomen or "tail." Development however affords abundant evidence of the descent of crabs from macrurous ancestors. A crab leaves the egg in what is termed the zoea condition, possessing a long and clearly jointed abdomen; and throughout all the earlier stages of existence the abdomen remains at least as long as the body. At the megalopa stage the shape and proportions are very similar to those of a lobster or other macrurous decapod. It is only in the last stages of development when the shape, though not the size, of the adult crab is attained, that the abdomen becomes relatively smaller, and is turned forwards out of sight beneath the hinder part of the thorax.

Molluscs afford excellent illustrations of recapitulation. The typical gasteropod[233] has a large spirally coiled shell; the limpet however has a large conical shell, which in the adult gives no sign of spiral twisting, although the structure of the animal shows clearly its affinity to forms with spiral shells. Development solves the riddle at once, telling us that in its early stages the limpet embryo has a spiral shell, which is lost on the formation subsequently of the conical shell of the adult.

Recapitulation is not confined to the higher groups of animals, and the Protozoa themselves yield most instructive examples. A very striking case is that of Orbitolites,[234] one of the most complex of the porcellanous Foraminifera, in which each individual during its own growth and development passes through the series of stages by which the cyclical or discoidal type of shell was derived from the simpler spiral form.

The fully formed Orbitolite shell is a thin calcareous disc. It is hollow, and the central cavity is divided into chambers by concentric partitions or septa. These chambers are further subdivided by incomplete radial partitions. The concentric partitions are perforated by numerous holes, which place the chambers in communication with one another; and the outermost or marginal chamber communicates with the outer world through a series of holes round the edge or rim of the disc. During life all the cavities are filled with a slimy protoplasm, very similar to that of which an Amoeba consists; through the perforations in the septa the protoplasm of one chamber communicates freely with that of the neighbouring chambers; and through the marginal apertures at the rim of the shell pseudopodia can be protruded and food captured and digested.

An Orbitolite grows by addition of new chambers round the margin of the shell. The protoplasm, becoming too abundant to be contained within the cavities of the shell, protrudes as a rim all round the margin of the shell, and by deposition of calcareous matter gives rise to successive new chambers to the shell.

The discoidal shape of shell, so characteristic of Orbitolites, is very unusual amongst Foraminifera, and it is a matter of great interest to determine in what way it has been acquired. Bearing in mind what has just been said as to the mode of growth

---

232   For 21st-century genetic explanations of flatfish morphology and of their taxonomic position among the teleosts, see Chen S et al., *Nature Genetics* 2014; **46**: 253–260, and Shao C et al., *Nature Genetics* 2017; **49**: 119–124. See also Lecture 17.

233   Now *gastropod*.

234   An extinct order of Foraminifera, known from the Eocene.

of the shell, it is clear that the oldest part—*i.e.*, that which alone was present in the young animal—is the central portion, and that the successive concentric rings are younger and younger as we pass outwards towards the circumference, the marginal chamber being the youngest and latest formed of the whole series. Now if we look at the central or oldest part of an Orbitolite we find that it has not the concentric arrangement of the peripheral part, but is coiled spirally like the majority of Foraminifera. The central or oldest turns of this spiral are not chambered; the outer turns are divided by partitions into chambers, and these chambers, as we follow the spiral round, become wider and wider, so as to overlap and wrap round the older part of the shell, at first partially but ultimately completely; the first chamber that completely surrounds the shell marking the transition from the spiral to the discoidal type.

We thus find that the discoidal Orbitolite shell commences its development as a spiral shell, and acquires the discoidal character merely through an exaggerated mode of growth on the part of the spiral. The Recapitulation Theory tells us that this is to be interpreted as meaning that the discoidal shells are descended from spiral ancestors, and the close agreement between a young Orbitolite and an adult Peneroplis suggests that either Peneroplis itself, or forms closely allied to it, were the actual ancestors.

The Orbitolite is peculiarly instructive, owing to the fact that the addition of new chambers during growth takes place in such a way as to leave the older parts of the shell unaltered and fully exposed to view, so that simple inspection of an adult shell reveals the whole course of development, and shows us not merely the anatomy but the embryology as well. It is as though a kitten were to develop into a cat, not by interstitial growth in all its parts, but by the addition of successive lengths to its nose, its ears, its legs, and its tail; the additions being cleverly effected so as to leave the original kitten unaltered in the middle, and fully exposed to view the whole time.

The above examples, selected almost haphazard, will suffice to illustrate the Theory of Recapitulation. The proof of the theory depends chiefly on its universal applicability to all animals, whether high or low in the zoological scale, and to all their parts and organs. It derives also strong support from the ready explanation which it gives of many otherwise unintelligible points.

Of these latter, familiar and most instructive instances are afforded by rudimentary organs—*i.e.*, structures which, like the outer digits of the horse's leg, or the intrinsic muscles of the ear of a man,[235] are present in the adult in an incompletely developed form, and in a condition in which they can be of no use to their possessors; or else structures which are present in the embryo, but disappear completely before the adult condition is attained; for example, the teeth of whalebone whales or the gill-slits present in the neck during the embryonic phases of all higher vertebrates.

Natural selection explains the preservation of useful variations, but will not account for the formation and perpetuation of useless organs, and rudiments such as those mentioned above would be unintelligible but for Recapitulation, which solves the problem at once, showing that these organs, though now useless, must have been

---

235   Thanks to Brenda Luck for identifying these as the helicis major, helicis minor, tragicus, antitragicus, transversus auriculae and obliquus auriculae muscles. In some other primates they change the orientation of the outer ear (pinna) to improve the directional focus of hearing. Some humans can move the pinna slightly (waggle the ears) but this seems to give no auditory benefit.

of functional value to the ancestors of their present possessors, and that their appearance in the ontogeny[236] of existing forms is due to repetition of ancestral characters.

Rudimentary organs are extremely common, especially among the higher groups of animals, and their presence and significance are now well understood. Man himself affords numerous and excellent examples, not merely in his bodily structure, but by his speech, dress, and customs. For the silent letter b in the word "doubt," the g in "reign," or the w of "answer," or the buttons on his elastic-side boots are as true examples of rudiments, unintelligible but for their past history, as are the ear muscles he possesses but cannot use; or the gill-clefts, which are functional in fishes and tadpoles, and are present, though useless, in the embryos of all higher vertebrates; which in their early stages the hare and the tortoise alike possess, and which are shared with them by cats and by kings.

The fossil remains of animals and of plants yield results of the greatest importance when studied in the light of the Recapitulation Theory, I have thought it well to ask special attention to these, even at the risk of repeating what has been said elsewhere and by others, for it seems to me that zoologists are too apt nowadays to neglect palaeontology, while palaeontologists have a tendency to regard embryology as something beyond their own ken, and concerning them but little. Moreover there are certain points arising from a study of fossils which, I venture to think, may possibly commend themselves to some of our members as suitable subjects for practical investigation.

The elder Agassiz was the first to point out, in 1858, the remarkable agreement between the embryonic growth of animals and their palaeontological history. He called attention to the resemblance between certain stages in the growth of young fish and their fossil representatives, and attempted to establish, with regard to fish, a correspondence between their palaeontological sequence and the successive stages of embryonic development. He then extended his observations to other groups of animals, and stated his conclusions in these words[237] "*It may therefore be considered as a general fact, very likely to be more fully illustrated as investigations cover a wider ground, that the phases of development of all living animals correspond to the order of succession of their extinct representatives in past geological times.*"

This point of view is of great importance. If the development of an animal is really a repetition or recapitulation of its ancestral history, then it is clear that the agreement or parallelism which Agassiz insists on between the embryological and palaeontological records must hold good, and a most important field of work is thus opened up to us. It is sometimes urged however that such work is necessarily unfruitful and inconclusive, because of the scantiness of our knowledge concerning life in the earlier geologic periods, or as it is commonly termed, the imperfection of the geological record. I have elsewhere[238] protested against this objection, and would repeat my protest here. The actual number of fossils already obtained, especially from the more recent formations, is prodigious; and what we have to do is to make the

---

236   The *Oxford English Dictionary* gives 1872 as the date when this word was first used (*Microscopical Journal*, July, 85). It was used by Weismann, Haeckel and Fritz Müller in the mid-1870s.

237   *Essay on Classification*, 1857.

238   Lectures 1 and 4. Lecture 16 examines the objection in detail.

most of the material already accumulated, rather than to fold our hands and idly lament the absence of forms that perhaps never existed.

It is true that with all groups the chances are not equal. But by judicious selection of groups in which long series of specimens can be obtained, and in which the hard skeletal parts, which alone can be suitably preserved as fossils, afford reliable indications of zoological affinity, it is possible to test directly this alleged correspondence between the palaeontological and embryological histories; while in some instances a single lucky specimen may afford us, on a particular point, all the evidence we require. Many serious attempts have already been made to work out in detail this comparison between fossils and the developmental stages of living forms, and the results obtained are most promising.

Following the lines laid down by his father, Alexander Agassiz has made a detailed comparison between the fossil series and the embryonic phases of recent forms in the case of the Echinoids or Sea Urchins, a group peculiarly well adapted for such an investigation, as the fossil representatives are extremely numerous and well preserved, and the existing members well known and comparatively few in number. Agassiz shows that the two records in this case agree remarkably closely; more especially in the independent evidence they give of the origin of the asymmetrical forms from more regular ancestors. The young Clypeastroid[239] for example has an ovoid test, a small number of coronal plates, few and large primary tubercles and spines, simple straight ambulacral areas, and no petaloid ambulacra; in fact has none of the characteristic features of the adult Clypeastroid, while the characters it does possess are those of geologically older and preceding forms. So again, in the group of Echinidae,[240] the members of the comparatively recent polyporous group, in which each ambulacral plate bears more than three pairs of ambulacral pores, commence their existence in the older and more primitive oligoporous condition, and become polyporous through fusion of originally distinct ambulacral plates.

Agassiz gives many other examples, and from a careful consideration of the entire group, arrives at the conclusion that *"comparing the embryonic development with the palaeontological one, we find a remarkable similarity;"* and again, *"the comparison of the Echini which have appeared since the Lias with the young stages of growth of the principal families of recent Echini, shows a most striking coincidence, amounting almost to identity, between the successive fossil genera and the various stages of growth."* [241]

In this connection Agassiz makes a suggestion of much interest. We are apt he says to assume, and perhaps rightly, that enormous periods of time have elapsed during the conversion of genus into genus, but the fact that these very changes can be repeated before our eyes in a few days' or even hours' time, during the development of the individual animal, may perhaps afford us a hint that such enormous periods are not really necessary in historical development, and that transformation of one form to a widely different one may, under favourable circumstances, be effected with considerable rapidity.

---

239   The Sand Dollar.

240   Sea Urchins.

241   *Paleontological and Embryological Development.* Address by Alexander Agassiz, *Science* 1880; 1(12): 142–149.

The Echinoids, and other groups of Echinoderms as well, have been worked at from the same standpoint and with the same results by Neumayr, in whom we have recently lost one of the most gifted and painstaking of palaeontologists.

As an example of the extreme value in certain cases of a single fossil specimen, the singular fossil bird Archaeopteryx may be referred to. In recent birds the meta-carpal bones of the wing are firmly fused with one another and with the distal row of carpal or wrist bones, but in development the metacarpals are at first and for some time distinct. The first specimen of Archaeopteryx discovered, which is now in the British Museum, showed that in it this distinctness was preserved in the adult—i.e., that what is now an embryonic character in recent birds was formerly an adult one.[242]

Another very excellent illustration of the parallelism between the palaeontolog-ical and the developmental series is afforded by the antlers of deer, which as is well known are shed annually, and grow again of increased size and complexity in each succeeding year. In the case of the red deer, Cervus elaphus, the antlers are shed in the spring, usually between the months of February and April; during the summer the new antlers sprout out, and growing rapidly attain their full size at the pairing season in August or September: they persist throughout the winter, and are shed in the following spring. The antlers of the first year are small and unbranched; those of the second year are larger and branched; in the antlers of the third year three tynes or points are present; in the fourth year four points, and so on until the full size of the antler and the full number of points are attained.

The geological history of antlers has been worked out by Professor Gaudry and by Professor Boyd Dawkins, and is of great interest. In the Lower Miocene and earlier deposits no antlers have been found. In the genus Procervulus from the Middle Miocene, a pair of small, erect, branched, but non-deciduous antlers were present, intermediate in many respects between the antlers of deer and the horns of antelopes. From slightly later deposits a stag has been found with forked deciduous antlers, which however do not appear to have had more than two points. In Upper Miocene times antlered ruminants were more abundant, and the antlers themselves larger and more complex while from Pliocene deposits very numerous fossils have been obtained showing a gradual increase in the size of the antlers and the number of their branches down to the present time.

Antlers are therefore, geologically considered, very recent acquisitions: at their first appearance they were small, and either simple or branched once only; while in succeeding ages they gradually increased in size and in complexity. The palaeonto-logical series thus agrees with the developmental series of stages through which the antlers of a stag pass at the present day before attaining their full dimensions.

There is another point of view from which fossils acquire special interest in connection with the Recapitulation Theory. If the theory is correct, it must apply not merely to the animals now living on the earth, but to all animals that ever have lived; and it becomes a matter of considerable interest to enquire whether we have any evidence whereby we can test this point, and determine whether or not the

---

242   The first Archaeopteryx feather was discovered in 1860 and the first more or less complete skeleton (the British Museum fossil) in 1861, in Langenaltheim, Bavaria, Germany. See the frontispiece preceding the second set of lectures and further discussion in Lectures 13 and 19.

fossil animals in their own individual development repeated the characters of their ancestors.

At first sight the enquiry does not seem a promising one, for it may well be asked what possibility there is of determining the embryology or mode of development of animals which are only known to us through the chance preservation of their bones or shells as fossils. In most cases it is true that such determination is impossible, but in some groups as for example the Trilobites, great numbers of well preserved specimens have been obtained, not merely of adults, but of young forms in various stages of growth; and the study of these young forms has already yielded results of considerable interest. According to Barrande, to whom our knowledge of these early stages is mainly due, four chief types of development may be recognised, differing from one another much as existing Crustaceans do in the relative size and perfection of the three regions, head, thorax, and tail, into which the body is divided.

Evidence of a very different kind, and often of far greater value, is afforded by the study of shells, whether of Mollusca or of Foraminifera. Such shells, like those of Orbitolites already noticed, have no power of interstitial growth, and increase in size can only be effected by the addition of new shelly matter to the part already in existence. In most instances these additions take place in such manner that the older parts of the shell are retained unaltered in the adult; and examination of the adult or fully formed shell will then reveal the several stages through which the shell passed in its development. In such a shell for instance as Nautilus[243] or Ammonites,[244] the central chamber is the oldest or first formed one, to which the remaining chambers are added in succession. If therefore the development of the shell is a repetition of ancestral history, the central chamber should represent the palaeontologically oldest form, and the remaining chambers, in succession, forms of more and more recent origin.

Ammonite shells present, more especially in their sutures and in the markings and sculpturing of their surface, characters that are easily recognised. Upwards of four thousand species are known to us, of many of which large numbers of specimens can be obtained, in excellent preservation. The group consequently is a very suitable one to study from our present standpoint; and the enquiry gains additional interest from the fact that Ammonites are an entirely extinct group of animals, no single species having survived the cretaceous period, so that our only chance of learning anything about their embryology is to study the fossil shells themselves.

Würtenberger, who has made a special study of the Jurassic Ammonites, has shown that there is the same correspondence between historic and embryonic development that obtains among living animals. In the middle Jurassic deposits, for instance, the older Ammonites are flattened and disc-like, with numerous ribs; in later forms the shell bears rows of tubercles near the outer side of the spiral, and later still a second inner row of tubercles as well, while the ribs gradually become less conspicuous and ultimately disappear. In forms from more recent deposits the outer row of tubercles disappears, and then the inner row, the shell becoming smooth, swollen, and almost spherical. On taking one of these smooth spherical shells, such

---

243   A pelagic, cephalopod mollusc with a characteristic flat spiral shell.
244   An extinct order of cephalopod molluscs with a flat spiral shell, possibly ancestral to Nautilus.

as Aspidoceras cyclotum, and breaking away the outer turns of the spiral so as to expose the more central and older turns, Würtenberger found first an inner and then an outer row of tubercles appearing, which nearer the centre disappeared, and in the oldest part of the shell were replaced by the ribs characteristic of the earlier, and presumably ancestral forms. Results such as these open up to us a new field of inquiry, which if energetically worked must yield results of great interest and importance.

In order to understand fossils aright, and to derive from them the full amount of information they are capable of yielding us, it is necessary that we should have a thorough knowledge of the development of their living descendants; and more especially that we should be fully acquainted with the several stages of formation of the shells or other hard parts of the recent forms, which in their fossil representatives are, with rare exceptions, the only parts sufficiently well preserved to give trustworthy evidence.

Embryologists have too often confined themselves to the earlier stages of development, and have unduly neglected the later stages, and more especially the later stages of the skeletal structures. By so doing they have failed to afford to palaeontologists the aid which they are peculiarly qualified to give, and which to the palaeontologist would be of the utmost value. Fortunately the mistake is now recognised, and serious efforts are being made to remove the reproach.

We must now turn to another side of the question. Although it is undoubtedly true that development is to be regarded as a recapitulation of ancestral phases, and that the embryonic history of an animal presents to us a record of the race history; yet it is also an undoubted fact, recognised by all writers on embryology, that the record so obtained is neither a complete nor a straightforward one. It is indeed a history, but a history of which entire chapters are lost, while in those that remain many pages are misplaced and others are so blurred as to be illegible; words, sentences, or entire paragraphs are omitted, and worse still alterations or spurious additions have been freely introduced by later hands, and at times so cunningly as to defy detection.

Very slight consideration will show that development cannot in all cases be strictly a recapitulation of ancestral stages. It is well known that closely allied animals may differ markedly in their mode of development. The common frog is at first a tadpole, breathing by gills, a stage which is entirely omitted by the West Indian Hylodes.[245] A crayfish, a lobster, and a prawn are allied animals, yet they leave the egg in totally different forms. Some developmental stages, as the pupa condition of insects, or the stage in the development of a dogfish in which the oesophagus is imperforate, cannot possibly be ancestral stages. Or again, a chick embryo of, say the fourth day, is clearly not an animal capable of independent existence and therefore cannot correctly represent any ancestral condition, an objection which applies to the developmental history of many, perhaps of most, animals.

Haeckel long ago urged the necessity of distinguishing in actual development between those characters which are really historical and inherited and those which are acquired or spurious additions to the record. The former he termed palingenetic or ancestral characters, the latter cenogenetic or acquired. The distinction is undoubtedly a true one, but an exceedingly difficult one to draw in practice. The causes which prevent development from being a strict recapitulation of ancestral characters, the

---

245   Tree Toad or Torrent Frog.

modes in which these came about, and the influence which they respectively exert, are matters which are greatly exercising embryologists; and the attempt to determine them has as yet met with only partial success.

The most potent and the most widely spread of these disturbing causes arise from the necessity of supplying the embryo with nutriment. This acts in two ways. If the amount of nutritive matter with the egg is small, then the young animal must hatch early, and in a condition in which it is able to obtain food for itself.[246] In such cases there is of necessity a long period of larval life, during which natural selection may act so as to introduce modifications of the ancestral history, spurious additions to the text.

If on the other hand the egg contain within itself a considerable quantity of nutrient matter, then the period of hatching can be postponed until this nutrient matter has been used up. The consequence is that the embryo hatches at a much later stage of its development, and if the amount of food material is sufficient may even leave the egg in the form of the parent. In such cases the earlier developmental phases are often greatly condensed and abbreviated; and as the embryo does not lead a free existence, and has no need to exert itself to obtain food, it commonly happens that these stages are passed through in a very modified form, the embryo being, as in a four-day chick, in a condition in which it is clearly incapable of independent existence.

The nutrition of the embryo prior to hatching is most usually effected by granules of nutrient matter, known as food yolk, and embedded in the protoplasm of the egg itself; and it is on the relative abundance of these granules that the size of the egg chiefly depends.

Large size of eggs implies diminution of their number, and hence in that of the offspring; and it can be well understood, that while some species derive advantage in the struggle for existence by producing the maximum number of young, to others it is of greater importance that the young on hatching should be of considerable size and strength, and so better able to begin the world on their own account. In other words, some animals may gain by producing a large number of small eggs, others by producing a smaller number of eggs of larger size—i.e., provided with more food yolk.[247]

The immediate effect of a large amount of food yolk is to mechanically retard the processes of development; the ultimate result is to greatly shorten the time occupied by development. This apparent paradox is readily explained. A small egg, such as that of Amphioxus,[248] starts its development rapidly, and in about eighteen hours gives rise to a free-swimming larva, capable of independent existence, with a digestive cavity and nervous system already formed; while a large egg, like that of the hen, hampered by the great mass of food yolk with which it is distended, has in the same time made but very slight progress. From this time however other considerations begin to tell. Amphioxus has been able to make this rapid start owing to its relative freedom from food yolk. This freedom now becomes a retarding influence, for the

---

246 Discussed at length in Lecture 6.
247 This is the r versus K argument (see footnote 66, Lecture 4) but applied to the provision of yolk rather than parental investment.
248 The lancelet or branchiostome; a marine, non-vertebrate chordate.

larva, containing within itself but a very scanty supply of nutriment, must devote much of its energies to hunting for and to digesting its food, and hence its further development will proceed more slowly.[249]

The chick embryo on the other hand has an abundant supply of food in the egg itself; it has no occasion to spend time in searching for food, but can devote its whole energies to the further stages of its development. Hence, except in the earliest stages, the chick develops more rapidly than Amphioxus, and attains its adult form in a much shorter time.

The tendency of abundant food yolk to lead to shortening or abbreviation of the ancestral history, and even to the entire omission of important stages, is well known. The embryo of forms well provided with yolk takes short cuts in its development, jumps from branch to branch of its genealogical tree, instead of climbing steadily upwards.

An excellent illustration of the influence of food yolk on development is afforded by the life histories of frogs.[250] The common frog, *Rana temporaria*, lays as is well known eggs of small size, about 1/16 inch[251] in diameter. A small egg can only contain a limited amount of food yolk, and hence the young frog can only accomplish a small part of its developmental history within the egg, and must then hatch in

---

249   This is a strange paragraph, as are the next two. It is not immediately clear (a) why yolk should hamper development, mechanically or otherwise, or (b) why the *rate* of development has to be elided with the *degree* of development? A far simpler explanation would be that the presence of yolk allows hatching at a *later* stage of development and that a more complex animal (bird) takes longer to reach a point at which it has the capacity for independent existence. (Some later passages in this Lecture and others suggest that this may be what was meant.) The rate at which development takes place (perhaps the speed with which mitosis and differentiation can occur) seems to be a different matter altogether, unless embryology is somehow viewed as a race to hatching. AMM invokes this *yolk effect* in later lectures (11, 13, 17, 18) and it was clearly important in his interpretation of embryology. Its origin is not in doubt. His friend and mentor Francis Maitland Balfour had devised, in 1875, a "law" relating the cleavage rate of an egg to its yolk content: "*In eggs in which the distribution of food material is not uniform, segmentation does not take place with equal rapidity through all parts of the egg, but its rapidity is, roughly speaking, inversely proportional to the quantity of food material*" (Balfour FM, A comparison of the early stages in the development of vertebrates. *Quarterly Journal of Microscopical Science* 1875; **15**; 207–226). According to BK Hall's scientific biography of Balfour (*Journal of Experimental Zoology, Molecular and Developmental Evolution*, 2003; **299B**: 3–8), he was encapsulating a principle well recognised at the time. His insight was to understand the difference between animals whose eggs divided completely (the *primitive* condition, with rapid cleavage) and those with large amounts of yolk (a secondary condition with incomplete division, slow cleavage and inward migration of mesodermal cells from a primitive streak). A key point here is that the yolk itself divides in the former (microlecithal) but not in the latter (macrolecithal).

250   One of AMM's popular publications was *The Frog. An Introduction to Anatomy, Histology and Embryology* which had reached its fifth edition by 1894.

251   1.6 mm.

order to obtain food from without. Consequently a frog hatches not as a frog but as a tadpole—*i.e.*, at the fish stage in the ancestral history of frogs. At the time of hatching there are no limbs and no lungs; the heart, the alimentary canal, and the nervous system are in an extremely imperfect condition; while other organs of the adult frog, such as the kidneys, have not yet commenced to appear. The frog has therefore to effect the greater part of its development after the time of hatching.

In the little West Indian frog, Hylodes, the course of events is very different. This frog which is of small size—less than a couple of inches in length—lays its eggs not in water but on the leaves of plants. The eggs are large, having a diameter of about 1/5 inch[252]—i.e., are about three times the diameter and twenty-seven times the bulk of the eggs of the common English frog. The large size of the egg is caused as we have seen by great abundance of food yolk; and the consequence of this large supply of food yolk in the egg of Hylodes is that the frog is enabled to complete the whole of its development before hatching, and emerges from the egg capsule in a form differing from the adult merely in the possession of a rudimentary stump of a tail; and even this disappears before the close of the first day of its existence.

A further and direct consequence of this development within the egg is that the successive stages are hurried over, and are at best but imperfectly recapitulated. Thus although Hylodes passes through what may be called a tadpole stage, it never develops gills; so that were Hylodes the only frog known to us, it is very likely that we should have arrived at other conclusions concerning the pedigree of frogs.

The influence of food yolk on the development of animals is closely analogous to that of capital in human undertakings. A new industry, for example that of pen-making, has often been started by a man working by hand and alone, making and selling his own wares; if he succeed in the struggle for existence, it soon becomes necessary for him to call in others to assist him, and to subdivide the work; hand labour is soon superseded by machines, involving further differentiation of labour; the earlier machines are replaced by more perfect and more costly ones; factories are built, agents engaged, and in the end a whole army of workpeople employed. In later times a man commencing the same business with very limited means will start at the same level as the original founder, and will have to work his way upwards through much the same stages—i.e., will repeat the pedigree of the industry. The capitalist on the other hand is *enabled*, like Hylodes, to omit these earlier stages, and after a brief period of incubation, to start business with large factories equipped with the most recent appliances, and with a complete staff of work-people—i.e., to spring into existence fully fledged.[253]

There is no doubt that abundance of food yolk is a direct and very frequent cause of the omission of ancestral stages from individual development; but it must not be viewed as the sole cause. It is quite impossible that any animal, except perhaps in the lowest zoological groups, should repeat all the ancestral stages in the history of the race; the limits of time available for individual development will not permit this. There is a tendency in all animals towards condensation of the ancestral history, towards striking a direct path from the egg to the adult. This tendency is best marked in the

---

252   5 mm.

253   As with some of AMM's other analogies, this is interesting and arresting but in the end one wonders if it contributes anything to biological understanding.

higher, the more complicated members of a group—*i.e.*, in those which have a longer and more tortuous pedigree; and, although greatly strengthened by the presence of food yolk in the egg, is apparently not due to this in the first instance.

Thus the simpler forms of Orbitolites, such as *O. tenuissima*, repeat in their development all the stages leading from a spiral to a discoidal shell; but in the more complicated species, as Dr. Carpenter has pointed out, there is a tendency towards precocious development of the adult characters, the earlier stages being hurried over in a modified form; while in the most complex examples, as in *O. complanata*, the earlier spiral stages may be entirely omitted, the shell acquiring almost from its earliest commencement the discoidal mode of growth. There is no question here of relative abundance of food yolk, but merely of early or precocious appearance of adult characters.

Of causes other than food yolk, or only indirectly connected with it, which tend to falsify the ancestral history, many are now known, but time will only permit me to notice the more important. These are distortion, whether in time or space; sudden or violent metamorphosis; a series of modifications, due chiefly to mechanical causes, and which may be spoken of as developmental conveniences; the important question of variability in development; and finally the great problem of degeneration.

Concerning distortion in time, all embryologists have noticed the tendency to anticipation or precocious development of characters which really belong to a later stage in the pedigree. The early attainment of the discoidal form in the shell of *Orbitolites complanata* is a case in point; and Würtenberger has specially noticed this tendency in Ammonites. Many early larvae show it markedly, the explanation in this case being that it is essential for them to hatch in a condition capable of independent existence—i.e., capable at any rate of obtaining and digesting their own food. Anachronisms, or actual reversal of the historical order of development of organs or parts, occur frequently. Thus the joint surfaces of bones acquire their characteristic curvatures before movement of one part on another is effected, and before even the joint cavities are formed.

Another good example is afforded by the development of the mesenterial filaments in Alcyonarians.[254] Wilson has shown, in the case of Renilla,[255] that in the development of an embryo from the egg the six endodermal filaments appear first, and the two long ectodermal filaments at a later period; but that in the formation of a bud this order of development is reversed, the ectodermal filaments being the first formed. He suggests in explanation, that as the endodermal filaments are the digestive organs, it is of primary importance to the free embryo that they should be formed quickly. The long ectodermal filaments are chiefly concerned with maintaining currents of water through the colony; in bud-development they appear before the endodermal filaments, because they enable the bud during its early stages to draw nutrient matter from the body fluid of the parent; while the endodermal filaments cannot come into use until the bud has acquired its own mouth and tentacles.

The completion of the ventricular septum in the heart of higher vertebrates before the auricular septum is an often quoted anachronism, and every embryologist could readily furnish many other cases. A curious instance is afforded by the development

254   Octocorallia, a group of sedentary cnidaria.
255   Sea Pansy.

of the teeth in mammals, if recent suggestions as to the origin of the milk dentition are confirmed, and the milk dentition prove to be a more recent acquisition than the permanent one.[256,257]

Distortion of a curious kind is seen in cases of abrupt metamorphosis where, as in the case of many Echinoderms, of Phoronis,[258] and of the metabolic insects,[259] the larva and the adult differ greatly in form, habits, mode of life, and very usually in the nature of their food and the mode of obtaining it; and the transition from the one stage to the other is not a gradual but an abrupt one, at any rate so far as external characters are concerned. Sudden changes of this kind, as from the free swimming Pluteus to the creeping Echinus, or from the sluggish leaf-eating caterpillar to the dainty butterfly, cannot possibly be recapitulatory, for even if small jumps are permissible in nature, there is no room for bounds forward of this magnitude.[260]

---

256 {M} This has since been disproved.—Ed.

257 This statement is a reference to the conclusions of Oldfield Thomas (On the Homologies and Succession of the Teeth in the Dasyuridae, with an attempt to trace the History of the Evolution of Mammalian Teeth in general. *Philosphical Transactions* 1887; **CLXXVIII**: 443–462.) that milk teeth have evolved more recently than permanent teeth. (The source is given in footnote 358, Lecture 13 where it is erroneously attributed to "Thomas Oldfield"). In the course of trying to classify the marsupialia, Thomas studied the dentition of museum specimens of extant and extinct (fossilised) dasyurids, a large class of marsupials with carnivore-like teeth (and which at that time included the now extinct Thylacine or Tasmanian Tiger). It is evident from Thomas's paper that the dasyurids were then believed to be ancestral to the carnivora, which is why his finding was thought to have general significance for the Mammalia. Such a pedigree is now rejected: placental and non-placental mammals had a much earlier common ancestor, so the similar dentition must be the result of convergent evolution. Thomas's conclusion was the opposite of the prevailing view, fully referenced in his paper, that milk teeth preceded permanent teeth. The curious feature, as far as this Lecture is concerned, is CFM's footnote. He does not indicate a source for the contradictory information, and no such comment appears against the reference to (Oldfield) Thomas's work in Lecture 13. Whatever the source, the matter seems to have remained under debate well into the 20th century (see Tims HWM, On the succession and homologies of the molar and premolar teeth in the mammalia. *Journal of Anatomy and Physiology* 1902; **36**(4): 321–343; Simpson GG, Article VI. Metacheiromys and the Edentata. *Bulletin of the American Museum of Natural History*, 1931; **59**: 295–381).

258 Horseshoe Worm.

259 Context indicates that this refers to those insects, now called holometabolous, which pass through the full set of distinct developmental stages (egg, larva, pupa, imago) during metamorphosis. Hemimetabolous insects, such as true bugs, undergo an incomplete sequence (egg, nymph and imago) in which the second stage partially resembles the adult.

260 This phrasing raises several issues. Richard Dawkins, Stephen Jay Gould and others have debated at length the issue of small and large steps in evolution, and there is no reason to go into it here. However, the size of a *jump* in the evolution of phenotype must be seen as a subjective interpretation. This is clearly evidenced

Cases of abrupt metamorphosis may always be viewed as due to secondary modifications, and rarely if ever have any significance beyond the particular group of animals
concerned. For example, a Pluteus larva may be recognised as belonging to the group
of Echinoidea before the adult urchin has commenced to be formed within it, and the
Lepidopteran caterpillar is already an unmistakable insect.[261] Hence, for the explanation of the metamorphoses in these cases it is useless to look outside the groups of
Echinoidea and Insecta respectively.

Abrupt metamorphosis is always associated with great change in external form
and appearance, in manner of life, and very usually in mode of nutrition. A gradual
transition in such cases is inadmissible, because in the intermediate stages the animal
would be adapted to neither the larval nor the adult condition; a gradual conversion of
the biting mouth of the caterpillar to the sucking proboscis of a moth would inevitably
lead to starvation. This difficulty is evaded by retaining the external form and habits
of one particular stage for an unduly long period, so that the relations of the animal to
the surrounding environment remain unchanged, while internally preparations for
the later stages are in progress.

Cinderella and the princess are equally possible entities, each being well
adapted to her environment. The exigencies of the situation do not permit however
of a gradual change from one to the other: the transformation, at least as regards
external appearance, must be abrupt.

Embryology supplies us with many unsolved problems, and it is not to be
wondered at that this should be the case. Some of these may fairly be spoken of as
mere curiosities of development, while others are clearly of greater moment. I do
not propose to catalogue these, but will merely mention one which I happen to have
recently run my head against, and remember vividly.

The solid condition of the oesophagus in dogfish embryos, first noticed by
Balfour, is a very curious point. The oesophagus has at first a well-developed lumen,
like the rest of the alimentary canal; but at an early period, stage K of Balfour's nomenclature, the part of the oesophagus overlying the heart, and immediately behind the

---

by the teleological adjective *forward* later in this sentence, implying purpose or
direction to the development. Using modern genetics, we may in fact find that a
single mutation in a developmental gene can have surprisingly profound effects on
phenotype, making apparently *large* jumps more understandable. Current developments, especially in cellular reprogramming with its astonishing potential in regeneration and repair, demonstrate that relatively small molecular changes can completely
alter the path of tissue differentiation.

261   As with this paragraph as a whole, the meaning of *secondary modification* is unclear.
AMM does not really explain what it signifies and it could be that it had an accepted
meaning which is now lost. The rest of the paragraph seems to say that certain
metamorphic changes are characteristic of certain groups, and that is all there is too
it. This is hardly an explanation; it just creates a *chicken–egg* problem in the sense
that the chicken is the egg's way of reproducing itself. The butterfly may be the caterpillar's way of reproducing itself—for that is what insects do (similarly Cinderella
and the princess, later paragraph). There is no *a priori* reason to think that one is the
developmental servant of the other unless the final stage of development is the target
of evolution.

branchial region, becomes solid and remains solid for a long time, the exact date of reappearance of the lumen not being yet ascertained. A similar solidification of the oesophagus occurs in tadpoles of the common frog. In young free swimming tadpoles the oesophagus is perforate, but in tadpoles of about 1/3 inch length it becomes solid and remains so until a length of about 1/2 inch has been attained. The solidification occurs at a stage closely corresponding with that in which it first appears in the dogfish, and a curious point about it is that in the frog the oesophagus becomes solid just before the mouth opening is formed, and remains solid for some little time after this important event. This closing of the oesophagus clearly cannot be recapitulation, but the fact that it occurs at corresponding periods in the frog and the dogfish suggests that it may possibly, as Balfour hinted,[262] *"turn out to have some unsuspected morphological bearing."*

A matter which at present is attracting much attention is the question of degeneration. Natural selection, though consistent with and capable of leading to steady upward progress and improvement, by no means involves such progress as a necessary consequence. All it says is that those animals will in each generation have the best chance of survival which are most in harmony with their environment, and such animals will not necessarily be those which are ideally the best or most perfect.[263]

If you go into a shop to purchase an umbrella the one you select is by no means necessarily that which most nearly approaches ideal perfection, but the one which best hits off the mean between your idea of what an umbrella should be and the amount of money you are prepared to give for it; the one in fact that is on the whole best suited to the circumstances of the case, or the environment for the time being. It might well happen that you had a violent antipathy to a crooked handle, or else were determined to have a catch of a particular kind to secure the ribs, and this might lead to the selection—*i.e.*, the survival, of an article that in other and even in more important respects was manifestly inferior to the average.[264]

So is it also with animals: the survival of a form that is ideally inferior is very possible. To animals living in profound darkness the possession of eyes is of no advantage, and forms devoid of eyes would not merely lose nothing thereby, but would actually gain, inasmuch as they would escape the dangers that might arise from injury to a delicate and complicated organ. In extreme cases, as in animals leading a parasitic existence, the conditions of life may be such as to render locomotor, digestive, sensory, and other organs entirely useless; and in such cases those forms will be most in harmony with their surroundings which avoid the waste of energy resulting from the formation and maintenance of these organs.

An excellent illustration of this downhill progress is afforded by the Rhizocephala, a curious group of parasitic Crustaceans, of which the genus Sacculina is perhaps the

---

262   Balfour FM, *A Monograph on the Development of the Elasmobranch Fishes*, 4 vols (London: Macmillan, 1878), Chapter 10: "The Alimentary Canal".

263   This sentence is contradictory. Survival requires (implies) harmony with the environment. *Best* and *perfect* only have meaning if we stray into teleology, yet that is just what the previous sentence and the first part of this one seek to avoid.

264   Strangely, this analogy is contradictory in exactly the same manner as the previous example. (Likewise *inferior* in the next sentence, *downhill progress* a little further on, etc.)

best known member. The adult Sacculina is found as a soft shapeless bag, an inch or so in length, attached to the under surface of the tail of a crab by a fleshy stalk which, passing through the skin of the crab, spreads out within it into a complicated system of branching tubular roots by which the parasite sucks up the juices of the crab, on which it depends for food. As regards structure, the Sacculina in its fully developed form is little more than a bag of eggs enclosed in a loosely fitting outer skin, with a single orifice through which the young escape. A nervous system is present, but there are no traces of limbs, digestive system, heart, breathing organs, or sense organs. Indeed, examination however careful of an adult Sacculina would fail to afford any clue as to its real zoological affinities.

Development however in obedience to the potent law of recapitulation, shows us at once that Sacculina is a Crustacean, more closely allied to the Barnacles than to any other of the more familiar members of the group. From each of the exceedingly numerous eggs which a Sacculina produces, there emerges a minute, free swimming larva of the type known as a Nauplius, characterised by possessing a somewhat pyriform[265] body, a single median eye, and three pairs of swimming appendages or legs. Nauplius larvae are widely spread amongst Crustaceans. All Entomostraeae, except the Cladocera,[266] hatch in this form, and Nauplius larvae are found in individual members in nearly all the higher groups as well. The only special peculiarity about the Sacculina Nauplius is that it has neither mouth nor digestive organs of any kind, these being unnecessary by the presence of a considerable quantity of food yolk. The Sacculina larva continues its free existence for a time; it casts its skin, or rather its cuticle many times, emerging each time rather more complicated in structure though actually smaller in size, for it cannot yet take in food. Ultimately it reaches the condition spoken of as the pupa stage. The pupa is enclosed in a bivalved carapace, very similar to that of a Cypris.[267] It possesses a pair of well-developed antennules in front and six pairs of swimming-legs behind. The Nauplius eye is still present, a little way behind the basal joints of the antennules.

Now comes the great change. The pupa meeting with a crab fastens itself to the under surface of the crab's tail by its antennules, and then goes to the bad with startling rapidity. Within three hours of the time of fixing itself to the crab, the six pairs of swimming legs, with the muscles moving them, and the whole posterior part of the body disintegrate and are cast off. The antennules become modified into a tube, piercing the skin of the crab; the head of the Sacculina remains as a bottle-shaped mass in connection with the modified antennules, but the bivalved carapace, with all the other organs including the eye, are cast off and lost. The Sacculina now passes for a time completely into the interior of the crab: later on, after increasing in size, it comes once more to the surface and becomes the bag-like mass which we have found to be the adult condition,

This is a typical instance of degeneration or retrograde development, the animal being more highly organised, and standing at a higher morphological level in its early stages than when adult. Yet inasmuch as the organs that are lost, such as the limbs and eye, would be of no use to it in its changed conditions of life, there is nothing in

---

265 Pear-shaped.
266 Water Fleas.
267 Barnacle larva.

the whole history that is in any way inconsistent with natural selection. This principle of degeneration, recognised by Darwin as a possible, and under certain conditions a necessary consequence of his theory, has been since advocated strongly by Dohrn, and later by Lankester, in an evening discourse delivered before the British Association at the Sheffield meeting, in 1879.[268] Both Dohrn and Lankester have suggested that degeneration may occur much more widely than is commonly supposed.

In animals which are parasitic when adult, but free swimming in their early stages, as in the case of Sacculina, degeneration is clear enough; so also is it in the case of the solitary Ascidians, in which the larva is a free swimming animal with a notochord, an elongated tubular nervous system, and sense organs, while the adult is fixed, devoid of the swimming tail, with no notochord, and with a greatly reduced nervous system and aborted sense organs. In such cases the animal, when adult, is, as regards the totality of its organisation, at a distinctly lower morphological level, is less highly differentiated than it is when young, and during individual development there is actual retrograde development of important systems and organs.

About such cases there is no doubt; but we are asked to extend the idea of degeneration much more widely. It is urged that we ought not to demand direct embryological evidence before accepting a group as degenerate. We are reminded of the tendency to abbreviation or to complete omission of ancestral stages of which we have quoted examples above; and it is suggested that if such larval stages were omitted in all the members of a group, we should have no direct evidence of degeneration in a group that might really be in an extremely degenerate condition. Supposing for instance the free larval stages of the solitary Ascidians were suppressed, say through the acquisition of food yolk, then it is urged that the degenerate condition of the group might easily escape detection. The supposition is by no means extravagant. Food yolk varies greatly in amount in allied animals, and cases like Hylodes, or amongst Ascidians, Pyrosoma, show how readily a mere increase in the amount of food yolk in the egg may lead to the omission of important ancestral stages.

The question then arises whether it is not possible, or even probable, that animals which now show no indication of degeneration in their development are in reality highly degenerate, and whether it is not legitimate to suppose such degeneration to have occurred in the case of animals whose affinities are obscure or difficult to determine. It is more especially with regard to the lower vertebrates that this argument has been employed; and at the present day zoologists of authority, relying on it, do not hesitate to speak of such forms as Amphioxus and the Cyclostomes as degenerate animals, as wolves in sheep's clothing, animals whose simplicity is acquired and deceptive rather than real and ancestral.

I cannot but think that cases such as these should be regarded with some jealousy;[269] there is at present, a tendency to invoke degeneration rather freely as a talisman to extricate us from morphological difficulties; and an inclination to accept such suggestions, at any rate provisionally, without requiring satisfactory evidence in their support.

---

268   The lecture was entitled *Degeneration*.

269   In the (now archaic) sense of antipathy. As amplified in the next paragraph, AMM is warning against appealing lazily to a particular explanation for want of a better, or better informed, one.

Degeneration of which there is direct embryological evidence stands on a very different footing from suspected degeneration, for which no direct evidence is forthcoming; and in the latter case the burden of proof undoubtedly rests with those who assume its existence. The alleged instances among the lower vertebrates must be regarded particularly closely, because in their case the suggestion of degeneration is admittedly put forward as a means of escape from difficulties arising through theoretical views concerning the relation between vertebrates and invertebrates.

Amphioxus itself, so far as I can see, shows in its development no sign of degeneration, except possibly with regard to the anterior gut diverticula, whose ultimate fate is not altogether clear. With regard to the earlier stages of development, concerning which, thanks to the patient investigations of Kowalevsky and Hatschek, our knowledge is precise, there is no animal known to us in which the sequence of events is simpler or more straightforward. Its various organs and systems are formed in what is recognised as a primitive manner; and the development of each is a steady upward progress towards the adult condition. Food yolk, the great cause of distortion in development, is almost absent, and there is not the slightest indication of the former possession of a larger quantity. Concerning the later stages our knowledge is incomplete, but so much as has been ascertained gives no support to the suggestion of general degeneration.

Our knowledge of the conditions leading to degeneration is undoubtedly incomplete, but it must be noticed that the conditions usually associated with degeneration do not occur. Amphioxus is not parasitic, is not attached when adult, and shows no evidence of having formerly possessed food yolk in quantity sufficient to have led to the omission of important ancestral stages. Its small size, as compared with other vertebrates, is one of the very few points that can be referred to as possibly indicating degeneration, but by no means proving its occurrence.

A consideration of much less importance, but deserving of mention, is that in its mode of life Amphioxus not merely differs, as already noticed, from those groups of animals which we know to be degenerate, but agrees with some, at any rate, of those which there is reason to regard as primitive or persistent types. Amphioxus, like Balanoglossus,[270] Lingula,[271] Dentalium[272] and Limulus,[273] is marine, and occurs in shallow water, usually with a sandy bottom, and like the three smaller of these genera it lives habitually buried almost completely in the sand, into which it burrows with great rapidity.

I do not wish to speak dogmatically. I merely wish to protest against a too ready assumption of degeneration; and to repeat that, so far as I can see, Amphioxus has not yet, either in its development, in its structure, or in its habits, been shown to present characters that suggest, still less that prove, the occurrence in it of general or extensive degeneration. In a sense all the higher animals are degenerate; that is, they can be shown to possess certain organs in a less highly developed condition than their ancestors, or even in a rudimentary state. Thus a crab as compared with a lobster is

---

270   Acorn Worm.
271   A brachiopod.
272   The Tooth Shell or Tusk Shell mollusc.
273   Horseshoe Crab.

degenerate in the matter of its tail, a horse as compared with Hipparion[274] in regard to its outer toes; but it is neither customary nor advisable to speak of a crab as a degenerate animal compared to a lobster; to do so would be misleading. An animal should only be spoken of as degenerate when the retrograde development is well marked, and has affected not one or two organs only, but the totality of its organisation.

It is impossible to draw a sharp line in such cases, and to limit precisely the use of the term degeneration. It must be borne in mind that no animal is at the top of the tree in all respects. Man himself is primitive as regards the number of his toes, and degenerate in respect to his ear muscles; and between two animals even of the same group it may be impossible to decide which of the two is to be called the higher and which the lower form.[275] Thus, to compare an oyster with a mussel. The oyster is more primitive than the mussel as regards the position of the ventricle of the heart and its relations to the alimentary canal; but is more modified in having but a single adductor muscle; and almost certainly degenerate in being devoid of a foot.

Care must also be taken to avoid speaking of an animal as degenerate in regard to a particular organ merely because that organ is less fully developed than in allied animals. An organ is not degenerate unless its present possessor has it in a less perfect condition than its ancestors had. A man is not degenerate in the matter of the length of his neck as compared with a giraffe, nor as compared with an elephant in respect of the size of his front teeth, for neither elephant nor giraffe enters into the pedigree of man. A man is however degenerate, whoever his ancestors may have been, in regard to his ear muscles; for he possesses these in a rudimentary and functionless condition, which can only be explained by descent from some better equipped progenitor.

We have now considered some of the more important of the influences which are recognised as affecting developmental history in such a way as to render the recapitulation of ancestral stages less complete than it might otherwise be; which tend to prevent ontogeny from correctly repeating the phylogenetic history.[276] It may at this point reasonably be asked whether there is any test by which we can determine whether a given larval character is or is not ancestral. Most assuredly there is no one rule, no single test, that will apply in all cases; but there are certain considerations which will help us, and which should be kept in view. A character that is of general occurrence among the members of a group, both high and low, may reasonably be

---

274   An extinct genus of horse.

275   These thoughts reveal an unexpected modesty, considering how frequently *high* and *low* and related terms are used throughout the Lectures (including later in this one) and in other writings of the time. It accords more closely with our modern view and strongly suggests that AMM was aware of the inadequacies of the language available to him.

276   The phrase *ontogeny recapitulates phylogeny* is usually attributed to Ernst Haeckel as a paraphrase of his *fundamental biogenetic law*, first formulated around 1866. The recapitulation theory it encapsulates, which greatly influenced leading biologists such as Weismann and which is supported by AMM in this Lecture and several others, is now largely discredited in favour of explanations based on the repositioning and retiming of morphological events during embryo development. The modern understanding is part of the so-called evolutionary-development (evo-devo) hypothesis.

regarded as having strong claims to ancestral rank; claims that are greatly strengthened if it occurs at corresponding developmental periods in all cases; and still more if
it occurs equally in forms that hatch early as free larva, and in forms with large eggs,
which develop directly into the adult. As examples of such characters may be cited the
mode of formation and relations of the notochord, and of the gill-clefts of vertebrates,
which satisfy all the conditions mentioned. Characters that are transitory in certain
groups, but retained throughout life in allied groups, may with tolerable certainty be
regarded as ancestral for the former; for instance, the symmetrical position of the
eyes in young flat-fish, the spiral shell of the young limpet, the superficial position of
the madreporite[277] in Elasipodous Holothurians,[278] or the suckerless condition of the
ambulacral feet in many Echinoderms.

A more important consideration is that if the developmental changes are to be
interpreted as a correct record of ancestral history, then the several stages must be
possible ones, the history must be one that could actually have occurred—i.e., the
several steps of the history as reconstructed must form a series, all the stages of which
are practicable ones. Natural selection explains the actual structure of a complex
organ as having been acquired by the preservation of a series of stages, each a distinct,
if slight, advance on the stage immediately preceding it, an advance so distinct as
to confer on its possessor an appreciable advantage in the struggle for existence. It
is not enough that the ultimate stage should be more advantageous than the initial
or earlier condition, but each intermediate stage must also be a distinct advance. If
then the development of an organ is strictly recapitulatory, it should present to us
a series of stages, each of which is not merely functional, but a distinct advance on
the stage immediately preceding it. Intermediate stages—e.g., the solid oesophagus
of the tadpole, which are not and could not be functional—can form no part of an
ancestral series; a consideration well expressed by Sedgwick thus: "Any phylogenetic
hypothesis which presents difficulties from a physiological standpoint must be regarded as
very provisional indeed."[279]

A good example of an embryological series fulfilling these conditions is afforded
by the development of the eye in the higher Cephalopoda. The earliest stage consists
in the depression of a slightly modified patch of skin; round the edge of the patch the
epidermis becomes raised up as a rim; this gradually grows inwards from all sides, so
that the depressed patch now forms a pit, communicating with the exterior through
a small hole or mouth. By further growth the mouth of the pit becomes still more
narrowed, and ultimately completely closed, so that the pit becomes converted into
a closed sac or vesicle; at the point at which final closure occurs formation of cuticle
takes place, which projects as a small transparent drop into the cavity of the sac; by the
formation of concentric layers of cuticle this drop becomes enlarged into the spherical
transparent lens of the eye; and the development is completed by histological changes
in the inner wall of the vesicle, which convert it into the retina, and by the formation
of folds of skin around the eye, which become the iris and the eyelids respectively.
Each stage of this developmental history is a distinct advance, physiologically, on
the preceding stage; and furthermore, each stage is retained at the present day as

---

277   Perforated plate at the site of water intake.
278   Sea Cucumbers.
279   Source not located.

the permanent condition of the eye in some member of the group Mollusca. The earliest stage, in which the eye is merely a slightly depressed and slightly modified patch of skin, represents the simplest condition of the Molluscan eye, and is retained throughout life in Solen.[280] The stage in which the eye is a pit with widely open mouth, is retained in the limpet; it is a distinct advance on the former, as through the greater depression the sensory cells are less exposed to accidental injury. The narrowing of the mouth of the pit in the next stage is a simple change, but a very important step forward. Up to this point the eye has served to distinguish light from darkness, but the formation of an image has been impossible. Now, owing to the smallness of the aperture, and the pigmentation of the walls of the pit which accompanies the change, light from any one part of an object can only fall on one particular part of the inner wall of the pit or retina, and so an image, though a dim one, is formed. This type of eye is permanently retained in the Nautilus. The closing of the mouth of the pit by a transparent membrane will not affect the optical properties of the eye, and will be a gain, as it will prevent the entrance of foreign bodies into the cavity of the eye.[281] The formation of the lens by deposit of cuticle is the next step. The gain here is increased distinctness and increased brightness of the image, for the lens will focus the rays of light more sharply on the retina, and will allow a greater quantity of light, a larger pencil of rays from each part of the object, to reach the corresponding part of the retina. The eye is now in the condition in which it remains throughout life in the snail and other gastropods. Finally the formation of the folds of skin known as iris and eyelids provides for the better protection of the eye, and is a clear advance on the somewhat clumsy method of withdrawal seen in the snail.

It is not always possible to point out so clearly as in the above instance the particular advantage gained at each step, even when a complete developmental series is known to us; but in such cases, as for instance in Orbitolites, our difficulties may be largely ascribed to ignorance of the particular conditions that confer advantage in the struggle for existence, in the case of the forms we are dealing with. That ontogeny really is a repetition of phylogeny must I think be admitted, in spite of the numerous and various ways in which the ancestral history may be distorted during the actual development.

Before leaving the subject, it is worth while inquiring whether any explanation can be found of recapitulation. A complete answer can certainly not be given at present, but a partial one may perhaps be obtained. Darwin himself suggested that the clue might be found in the consideration that at whatever age a variation first appears in the parent, it tends to reappear at a corresponding age in the offspring; but this must be regarded rather as a statement of the fundamental fact of embryology than as an explanation of it. It is probably safe to assume that animals would not recapitulate unless they were compelled to do so: that there must be some constraining influence at work, forcing them to repeat more or less closely the ancestral stages. It is impossible, for instance, to conceive what advantage it can be to a reptilian or mammalian embryo to develop gill-clefts which are never used, and which disappear at a slightly later stage; or how it can benefit a whale, that in its embryonic condition

---

280   Bivalve molluscs, including Razor Shell.

281   In fact, the refractive properties of the membrane, especially in water, may enhance the image.

it should possess teeth which never cut the gum, and which are lost before birth. Moreover, the history of development in different animals or groups of animals offers to us, as we have seen, a series of ingenious, determined, varied, but more or less unsuccessful efforts to escape from the necessity of recapitulating, and to substitute for the ancestral process a more direct method.

A further consideration of importance is that recapitulation is not seen in all forms of development, but only in sexual development; or at least only in development from the egg. In the several forms of asexual development, of which budding is the most frequent and most familiar, there is no repetition of ancestral phases; neither is there in cases of regeneration of lost parts, such as the tentacle of a snail, the arm of a starfish, or the tail of a lizard. In such regeneration it is not a larval tentacle, or arm, or tail, that is produced, but an adult one.

The most striking point about the development of the higher animals is that they all alike commence as eggs. Looking more closely at the egg and the conditions of its development, two facts impress us as of special importance. First, the egg is a single cell, and therefore represents morphologically the Protozoon, or earliest ancestral phase; secondly, the egg, before it can develop, must be fertilised by a spermatozoon, just as the stimulus of fertilisation by the pollen grain is necessary before the ovum of a plant will commence to develop into the plant-embryo,

The advantage of cross-fertilisation in increasing the vigour of the offspring is well known, and in plants devices of the most varied and even extra-ordinary kind are adopted to ensure that such cross-fertilisation occurs. The essence of the act of cross-fertilisation, which is already established among Protozoa, consists in combination of the nuclei of two cells, male and female, derived from different individuals. The nature of the process is of such a kind that two individual cells are alone concerned in it; and it may I think be reasonably argued that the reason why animals commence their existence as eggs—i.e., as single cells—is because it is in this way only that the advantage of cross-fertilisation can be secured, an advantage admittedly of the greatest importance, and to secure which natural selection would operate powerfully.

The occurrence of parthenogenesis, either occasionally or normally, in certain groups is not I think a serious objection to this view. There are very strong reasons for holding that parthenogenetic development is a modified form, derived from the sexual method. Moreover, the view advanced above does not require that cross-fertilisation should be essential to individual development, but merely that it should be in the highest degree advantageous to the species; and hence leaves room for the occurrence, exceptionally, of parthenogenetic development.

If it be objected that this is laying too much stress on sexual reproduction, and on the advantage of cross-fertilisation, then it may be pointed out in reply that sexual reproduction is the characteristic and essential mode of multiplication among Metazoa: that it occurs in all Metazoa, and that when asexual reproduction, as by budding, occurs, this merely alternates with the sexual process which, sooner or later, becomes essential.

If the fundamental importance of sexual reproduction to the welfare of the species be granted, and if it be further admitted that Metazoa are descended from Protozoa, then we see that there is really a constraining force of a most powerful nature compelling every animal to commence its life-history in the unicellular condition, the only condition in which the advantage of cross-fertilisation can be obtained—i.e., constraining every animal to begin its development at its earliest ancestral stage, at the very bottom of its genealogical tree.

On this view the actual development of any animal is strictly limited at both ends: it must commence as an egg, and it must end in the likeness of the parent. The problem of recapitulation becomes thereby greatly narrowed; all that remains being to explain why the intermediate stages in the actual development should repeat the intermediate stages of the ancestral history.[282]

Although narrowed in this way, the problem still remains one of extreme difficulty. It is a consequence of the theory of Natural Selection that identity of structure involves community of descent; a given result can only be arrived at through a given sequence of events: the same morphological goal cannot be reached by two independent paths. A negro and a white man have had common ancestors in the past; and it is through the long-continued action of selection and environment that the two types have been gradually evolved. You cannot turn a white man into a negro merely by sending him to live in Africa: to create a negro the whole ancestral history would have to be repeated; and it may be that it is for the same reason that the embryo must repeat or recapitulate its ancestral history in order to reach the adult goal.

I am not sure that we can get much further than this at present. However, be the explanation what it may, there can I think be no doubt as to the general truth of the Recapitulation Theory, and the wonderful assistance which it gives us in reconstructing the pedigrees of animals. Yet it must not be supposed that all we have to do in order to determine the past history of a species is to study the actual development of the existing members of that species.

Embryology is not to be regarded as a master key that will open the gates of knowledge, and remove all obstacles from our path without further trouble on our part; it is rather to be viewed and treated as a delicate and complicated instrument, the proper handling of which requires the utmost nicety of balance and adjustment, and which unless employed with the greatest skill and judgment may yield false instead of true results. We are indeed only just beginning to understand the real power of our weapons and the right way of employing them; and in the future embryology, especially when studied as it should be in conjunction with palaeontology, may be confidently relied on to afford a far clearer insight than we have yet obtained into the history of life on the earth.

---

282   AMM is hinting here, and in the following paragraphs, at a problem with the recapitulation theory—the *why* question. The modern (evo-devo) view resolves this by changing the perspective. Patterns of embryological development are not goal-orientated; rather they survive because the resulting phenotypes are successful.

LECTURE 11

SOME RECENT EMBRYOLOGICAL INVESTIGATIONS

The date of this lecture to the Manchester Microscopical Society, 21 January 1893, turns out to be crucial in reviewing AMM's unexpected reading of the Krause embryo controversy, discussed in the first few paragraphs. (For a detailed analysis of this controversy and its importance in the depiction and interpretation of the stages of embryological development, see Hopwood N, Producing development: the anatomy of human embryos and the norms of Wilhelm His. *Bulletin of the History of Medicine* 2000; **74:** 29–79; and Hopwood N, *Haeckel's Embryos: Images, Evolution and Fraud* (University of Chicago Press, 2015).

Briefly, Wilhelm His set out to describe, as objectively as possible, the developmental stages of the human embryo. He took a different approach from that of many contemporaries by avoiding the assumption that superficial morphological resemblances between early embryos of different vertebrates meant that the development of "advanced" animals built on that of less complex ones. Instead, he obtained as many embryo specimens as he could from medical and midwifery contacts. He fixed and sliced them, using his own techniques, and eventually created a *Normentafel*, a pictorial reference chart of carefully produced sketches representing accurately timed stages of pregnancy. Crucially, he refused to idealise his drawings or to fill gaps between stages by interpolation.

In the meantime, as AMM records, Johann Krause described a four-week human embryo with an allantoic structure different from those of later stages. This seemed to show that the human embryo passed through a bird embryo-like stage. Ernst Haeckel used Krause's specimen to try to discredit His and his approach, and others were (disgracefully) easily persuaded. His countered by decrying the specimen as avian, not human, and questioning Haeckel's scientific legitimacy. The dispute was public, personal and vitriolic.

Krause was initially reluctant to let others examine his specimen but when it was eventually shown at gatherings of developmental biologists in Gottingen and Marburg in 1882, there was final agreement that it was in fact the embryo of a bird.

The curiosity in this Lecture, presented a decade after the resolution of the controversy, is AMM's reference to the issue being put to rest *within the last twelve months* and that *Krause was right* to claim that the disputed embryo was human. Nick Hopwood has kindly confirmed that, to his knowledge, no further evidence on the matter appeared after 1882 (and the fate of Krause's specimen is unknown).

It seems most likely that AMM's conclusion was influenced by his friend and mentor Francis Balfour. According to Hopwood, Balfour became involved in the debate in 1880 when he and others tried to reconcile opposing views by suggesting that the embryo was human but abnormal. There is no evidence of direct contact between AMM and Krause, and their visits to the Naples Zoological Station (5–26 April 1884 and 24 October 1885–6 March 1886, respectively) did not coincide.

The marginal note by Duckworth provides an intriguing coda to the story. Whether Wilson's specimen was abnormal or not, Duckworth's final "after all" shows that he too would have expected AMM to report Krause's embryo as avian. Indeed, in the fourth paragraph of the lecture AMM himself warns against over-interpretation of species similarities. AMM evidently had a high opinion of His, recommending him for foreign membership of the Physiological Society.

A little later in the lecture, AMM resumes his emphasis on the superiority of evidence over prejudice ("facts must always be accepted"), with which His would surely have concurred. Much of the rest of the material is an attempt to demonstrate the diversity of ways in which animal embryos develop. Although the final list has more possibilities in it than we would probably now accept, AMM's sense of wonder at the natural world shines through.

LECTURE II

SOME RECENT EMBRYOLOGICAL INVESTIGATIONS

THE close resemblance between the embryos of animals which, when adult, are widely different in form, in size, and even in structure, greatly impressed the earlier embryologists, and was often insisted on by them. A reptile, a bird, and a mammal are in their early stages of development so closely similar that v. Baer himself was unable to decide to which of these groups three unnamed embryos in his collection were to be referred.

A still more striking illustration is afforded by the controversy which raged for many years over Krause's famous embryo. In 1875 Krause described[283] an early human embryo which appeared to differ from all known human embryos in having a large vesicular allantois like that of a chick or reptile, instead of the thick allantoid stalk by which the human embryo is normally connected with the chorion. This peculiarity with regard to the allantois was so marked that doubts were at once raised as to the embryo being really a human one; and Professor His, one of the most expert of embryologists, asserted roundly that Krause must have made a mistake, and that his specimen was a chick embryo and not a human one at all. An ardent, almost furious, discussion arose, and continued for many years: it is indeed only within the last twelve months that the points at issue have been finally put at rest, and it has been shown that while Krause was right in describing his embryo as a human one, he was mistaken in regard to the supposed peculiarity in the allantois, the bladder-like vesicle which he took for the allantois being merely a pathological dilatation of the allantoic stalk.[284]

---

283 Ueber die Allantois des Menschen, *Archiv für Anatomie, Physiologie und wissenschaftliche Medicin* (1875).

284 A marginal note in the Hammond Library copy of Biological Lectures and Addresses, evidently by Duckworth and dated 21/x/1923, says: *"However Professor Wilson has seen a human embryo with such an Allantois which he considers was not a path dilatation, so Krause may have been right after all."* The communicant is presumably James Thomas Wilson MB FRS (1861–1945), Professor of Anatomy at the University of Cambridge from 1920 to 1935, and fellow of St John's College. During a period as Challis Professor of Anatomy, University of Sydney, Wilson wrote a paper "Observations on young human embryos", *Journal of Anatomy and Physiology* 1914; **48**(3): 315–351. [As further evidence pointing to Wilson's identity, other books in the HL contain a bookplate bearing the St John's College crest and "Ex Libris Prof. J. T. Wilson". One of these is Bradley M Patten's *The Early Embryology of the Chick*, published in 1920, which cites AMM's book *Vertebrate Embryology* (1893) and his paper "The development of the cranial nerves in the chick", *Quarterly Journal of Microscopical Science* 1878; **XVIII**.]

Among Invertebrates the resemblances between the early larval forms of allied groups are equally striking: the veliger[285] and trochophore[286] larva, for example, or the nauplius larva[287] of Crustacea, having not merely very wide zoological distribution, but presenting marked constancy in essential, or even in minor characters in the groups in which they occur.

That the early larval stages of a prawn, a cyclops,[288] and a barnacle; or of a reptile, a chick, and a man, should be so closely similar that it is possible to mistake one for the other, is undoubtedly a very remarkable thing; and it was perhaps inevitable that there should be at first a tendency to overestimate the exactness of this resemblance.[289] Embryologists have indeed too often overrun their facts; and, misled by the undoubtedly striking similarity between embryos of forms zoologically akin, have not hesitated to fill the gaps in their knowledge of the developmental history of certain animals, by reference to the known processes in allied forms. More especially is this the case with regard to man himself: material for direct observation is difficult to obtain, and it is only too common to find that descriptions purporting to be of human embryos are really founded on observations derived from the study of pigs and rabbits, or even of animals so remote zoologically as chicken, lizards, or dog-fish.

Of late years however there has been a marked reaction in this respect, and the pendulum has shown a tendency to swing over to the opposite side. More exact observations on many groups of animals have proved that even in allied forms the course of development may be markedly different; in some cases indeed not only genera and species, but even the eggs of the same brood, may develop in ways curiously unlike one another.

This inconstancy in the mode of development, more especially in the earliest stages, is one of the most striking results of recent embryological investigation, and may well claim attention on the present occasion. If at first sight it appear bewildering or even unwelcome to those who, from the study of development of existing animals, would seek to unravel the past history of the race, it is to be remembered that facts must always be accepted; and it may even be that in this variability, by which his labours are so greatly extended, the embryologist will find the clue he has so long waited for, which will enable him to distinguish the real from the spurious, the ancestral from the acquired characters.[290]

The earliest stage of development in every Metazoon is the act of cell-division, or segmentation as it is commonly termed, by which the single cell, or egg, becomes

---

285   Planktonic larvae of marine gastropod and bivalve molluscs.

286   Planktonic larvae of marine annelids and molluscs.

287   Larvae of copepods, which use head appendages for swimming.

288   A fresh-water copepod.

289   This phrase and the next two sentences go to the heart of the His–Haeckel debate. The two subsequent paragraphs take His's stance on the matter. They can also be taken as recognizing the inadequacies of the *ontogeny recapitulates phylogeny* concept that AMM makes use of in Lectures 10 and 13.

290   This is a somewhat curious additional idea. Presumably "acquired" refers not to characters resulting from the direct effects of the environment (in a Lamarckian sense) but to those appearing during evolution which distinguish an organism from its ancestors.

divided into a number of cells or blastomeres. The details of the process vary greatly in the eggs of different animals, and are affected more particularly by the amount of food yolk, or deutoplasm, present in the egg; this food yolk offering mechanical hindrance to the division of the egg.[291] In each species or genus, or even in larger groups, the mode of segmentation was formerly assumed to be constant, but the more recent researches have shown that this is by no means always the case.

In 1878, Kleinenberg noticed that the eggs of an earth-worm, *Lumbricus trapezoids*,[292] which he was investigating,[293] showed considerable individual differences in the mode and order in which the successive cell-divisions were effected; an observation which was afterwards confirmed by Wilson, and extended to other species of earthworms.[294] In 1884 while studying the development of Renilla, one of the Pennatulida or Sea-Pens, Wilson found an extraordinary range of variation in the mode of segmentation of eggs, even of the same brood. In some cases the egg divided into two blastomeres in the normal manner, each of these in its turn again dividing; in other cases however the egg divided at once into eight, sixteen, or even thirty-two blastomeres, which in different specimens were approximately equal or markedly unequal in size. Sometimes a preliminary change of form occurred without any further result, the egg returning to its spherical shape, and pausing for a time before recommencing the attempt to segment. Segmentation sometimes commenced at one pole, as in the telolecithal[295] eggs of birds or reptiles, with the formation of four or five small segments, the rest of the egg breaking up later, either simultaneously or progressively, into segments about equal in size to those first formed; while lastly, in some instances segmentation was very irregular, following no apparent law. Similar modifications in the segmentation of the egg have been described in the oyster by Brooks,[296] in Anodonta[297] and in other Mollusca, and in Hydra. In the different species of Peripatus[298] there appear also to be considerable variations in the details of segmentation.

It has long been known that the eggs of Amphibia may vary greatly in the mode in which the early stages of segmentation are effected; and recently Jordan and Eycleshymer have described these variations in detail.[299] The first cleft usually corresponds with the median sagittal plane, dividing the egg into two blastomeres, which give rise respectively to the right and left halves of the embryo. The cleft may however

---

291   See Lecture 10 for a discussion of this concept.

292   Genus name is now *Aporrectodea* or *Allolobophora*.

293   Sull'origine del Sistema nervosa centrale degli Anellidi. *Memorie della Reale Accademia dei Lincei* 1881–2, Ser. 3, Annata 279, Vol 6.

294   The embryology of the earthworm. *Journal of Morphology* 1889; **3**: 338–463.

295   Having a large, non-dividing yolk.

296   Development of the American oyster, in *Report of the Commissioners of Fisheries of Maryland* 1880; **39**: 3–4 (see Keiner C, *Journal of the History of Biology* 1998; **31**: 383. doi:10.1023/A:1004393608972).

297   Freshwater or river mussels.

298   Velvet worms.

299   Reported in Jordan EO and Eycleshymer AC, On the cleavage of amphibian ova. *Journal of Morphology* 1894; **9**(3): 407–416.

be oblique, or even transverse to this plane. The two first blastomeres are usually equal, but may be very unequal, one being sometimes twice the size of the other.

The second cleft is usually at right angles to the first, and divides the egg into anterior and posterior halves; but it may cut the egg obliquely. The third and fourth clefts also present considerable variability in their position and relations to the earlier clefts. In all the cases described above the variability was confined to the earliest phases of segmentation: the earlier stages of development were the same, whatever the mode in which segmentation was effected; and apparently identical, and certainly normal embryos, resulted in all cases.

Some recent observations of Loeb give a possible clue to these phenomena. He found by experimenting on the eggs of sea urchins that the process of segmentation could be retarded, or modified, by varying the proportion of sodium chloride present in the sea-water in which the eggs were laid. A slight increase in the normal proportion of sodium chloride delayed the occurrence of segmentation; but on the return of the eggs to normal sea-water they very quickly divided, often simultaneously, into a number of blastomeres. He states his results thus: *"If we bring impregnated eggs into seawater of a certain higher concentration, no segmentation takes place; but if we bring them back into normal sea-water, they divide in about twenty minutes directly into nearly, but not quite so many, cleavage spheres as they would contain by that time if they had remained in normal sea-water all the time."*[300] Further investigation showed that, although when placed in water containing more than the normal proportion of sodium chloride, the eggs did not segment, yet that division of the nucleus occurred: unsegmented eggs were in this way obtained with from four to as many as thirty nuclei.

It is at least possible that some of the variations in the normal process of segmentation mentioned above as observed in other animals may be due to changes in the composition, or perhaps of the temperature of the water in which the eggs were developing. Herbst's interesting experiments on the modifications in the development of Echinoderm larvae produced by the addition of potassium or lithium salts to the sea-water in which they were contained, led him to conclude that the influence on the developing ova was not directly chemical, but was due to the altered osmotic pressure of the sea-water.[301] The field of enquiry thus opened up is a most promising one, and is certain to receive the attention of embryologists in the immediate future. We are at present ignorant of the causes which determine the division of a cell, or the segmentation of an egg; but we shall have made an appreciable step towards a right understanding of the phenomena when we have determined with precision in what way they can be modified by slight, but known, alterations in the environment.

Variations occur in the later as well as in the earlier stages of development; and the differences between allied genera or species, or even between closely related individuals, may be very marked, Among Coelenterates, for example, the mode of formation of the inner germinal layer, or hypoblast, presents most perplexing

300   Exact source not located. Loeb's embryological work is described in his *Artificial Parthenogenesis and Fertilization* (Chicago: University of Chicago Press, 1913).

301   Experiments described in Experimentelle Untersuchungen über den Einfluss der veränderten chemischen Zusammensetzung des umgebenden Mediums auf die Entwickelung der Thiere. *Archiv für Entwicklungsmechanik der Organismen* 1896; **2**: 455, doi:10.1007/BF02084503.

modifications: it may arise as a true gastrula invagination; as cells budded off from one pole of the blastula into its cavity; as cells budded off from various parts of the wall of the blastula; by delamination, or actual division of each cell of the blastula wall into outer or epiblastic, and inner or hypoblastic elements; or it may be present from the first as a solid mass of cells enclosed by the epiblast cells. Another good illustration is afforded by the extraordinary modifications in the position, and in every detail of formation of the middle germinal layer, or mesoblast, in different and often in closely allied forms of the higher Metazoa: differences which have given rise to ardent discussion, and have led to the proposal of theory after theory, each rejected in its turn as affording only a partial explanation, and finally culminating in KIeinenberg's protest against the use of the term mesoblast at all, at any rate in a sense implying any possibility of comparison with the primary cellular layers, epiblast and hypoblast, of Coelenterata. Amongst Vertebrates the frog and the newt differ greatly in important developmental points. The stages immediately following segmentation of the egg are very unlike in the two cases. The epiblast is but one cell thick in the newt, while in the frog it consists of two distinct layers from the first; the ear of the newt develops as a pit-like depression of the skin, while in the frog it is from the first a closed sac; and many other differences will readily occur to embryologists. In the common English frog, *Rana temporia* and in the closely allied *Rana esculenta,* the branchial blood-vessels develop in very different manner, although ultimately reaching the same condition.

But the most remarkable series of facts with regard to variation in the later stages of development is afforded by the recently published observations of Professor Brooks and Mr. Herrick on the metamorphoses of certain small decapod Crustacea of the genus Alpheus.[302] These are essentially tropical forms, not unlike small crayfish in appearance, and about an inch or so in length. They are brilliantly coloured and occur most abundantly in shallow water, more especially amongst coral reefs. A few species live freely, but the majority are found inhabiting the tubes of sponges, or dwelling in holes and crannies in the porous coral limestone. The development of thirteen species was carefully studied by Messrs. Brooks and Herrick, and the remarkable conclusion was arrived at that *"individuals of a single species sometimes differ more from each other as regards their metamorphoses than do the individuals of two very distinct species."* It is not merely that different individuals hatch at different stages of development: one individual may appear with characters, *e.g.,* the number and form of the legs, which occur at no stage in the development of other individuals of the same species.[303] In some cases the differences in developmental history were associated with differences in the locality from which the specimens were obtained. Thus individuals of a given species observed at Key West, in Florida, differed constantly from individuals of the same species from N. Carolina, and these again from individuals

---

302   Brooks WK and Herrick FH, The embryology and metamorphosis of the macroura, *Memoirs of National Academy of Science Washington* 1891; **5**: 319–576. *Alpheus* is a Snapping Shrimp.

303   Subsequent research on Alpheids has demonstrated a very large number of species (>250), some of them distinguished by differences in their rate and stages of development. Thus it is likely that *individuals of the same species* do not in fact differ in the manner described here. There may also be morphologically distinguishable subspecies separated by location, as indicated in the rest of this paragraph.

living at the Bahamas. In other instances the differences in development appeared to be related to the conditions of life. At the Bahamas a species of Alpheus was observed dwelling in the chambers of sponges. Two kinds of sponge were employed by the prawn, one sponge being green, the other brown. The prawns living in the green sponge produced considerable numbers of small eggs; while the individuals of the same species which dwelt in the brown sponge gave rise to a few eggs of large size, from which the young prawns were hatched at a stage corresponding to that reached at the third moult by the young developed from the smaller eggs in the green sponge.

Another very curious series of facts has been made known to us by the careful researches of the Russian embryologist, Salensky, on the early development of certain Ascidians.[304] The normal course of development in a Metazoon, as is well known, is that the fertilised egg segments, i.e., divides into a number of nucleated cells from which, by further division, and accompanying differentiation, all the various parts and tissues of the embryo, and finally of the adult animal, are produced. Every individual cell of the adult, whether an epithelial cell, a nerve cell, a muscle cell, or a bone cell, owes its origin to direct descent from the original egg-cell or ovum. The most striking fact in the whole range of embryology is that all animals above Protozoa commence their existence as eggs—i.e., as single cells—from which all parts of the adult body are ultimately derived. The purpose of this arrangement appears to be to secure the advantage to be derived from cross fertilisation, a process which concerns two cells only, male and female respectively; and which necessitates a unicellular stage as the initial one in individual development.[305] The eggs of Metazoa, while still in the ovary, are very commonly enclosed in follicles. These consist of one or more layers of cells which are of epithelial origin,[306] and which are of the same order as the ova themselves, and in their earlier stages indistinguishable from these. The follicle cells play an important part in the nourishment of the egg which they surround; but they are left behind in the ovary, or are rubbed off after the egg has ripened and discharged from the ovary, and they have nothing to do with the formation of the embryo.

In the genus Pyrosoma however, a well-known colonial and pelagic Ascidian, Salensky discovered a very peculiar condition of things.[307] Each Pyrosoma individual produces one very large egg, which is closely invested by a follicle or capsule. The egg is meroblastic; and as in the hen's egg, segmentation is confined to one pole, and results in the formation of a small cap of cells or blastoderm, lying on the top of the mass of unsegmented yolk formed by the rest of the egg. As segmentation proceeds, certain of the follicle cells investing the egg grow in between the segmentation cells or

---

304 Salensky W, Beiträge zur Entwicklungsgeschichte der Synascidian. 2. Über die Entwicklung von Didemnum niveum. *Mittheilungen aus der Zoologischen Station zu Neapel* 1895; **11**: 488–552. Ascidians (sea squirts) are non-vertebrate marine chordates (see Lecture 20).

305 For a more recent, theoretical discussion of this question, including the evolution of multicellularity and mechanisms of sexual reproduction in Metazoa, see Maynard Smith J and Szathmáry E, *The Origins of Life* (Oxford: Oxford University Press, 1999).

306 Granulosa, theca and interstitial cells, depending on the type of animal.

307 On the development of Pyrosoma, *Annals and Magazine of Natural History* Series 6, **6** (33): 236–244 (trans. from *Biologisches Centralblatt* 1890; Band x, Heft 8: 225 *et seq.*).

blastomeres. These follicle cells, or kalymmocytes[308] as they are called by Salensky, are at first very distinct in appearance from the blastomeres, and easily recognised from these; but as development proceeds the differences become less and less marked, and finally cease to be evident. These ingrowing follicle cells, or kalymmocytes, enter directly into the formation of the embryo, which is thus built up partly from cells derived, as in other Metazoa, from division of the fertilised egg or ovum; and partly from cells of independent origin, the kalymmocytes, which are unfertilised, and have grown into the egg from the surrounding follicle.[309] In the genus *Salpa* a similar condition of things has been described by Salensky; the only difference, as compared with Pyrosoma, being that in Salpa the kalymmocytes are actually more numerous and more bulky than the blastomeres formed by segmentation of the ovum, so that the major part of the embryo is composed of unfertilised cells.

These observations are of the greatest possible interest, and we wait anxiously to learn whether this curious mode of development is confined to the groups of Ascidians, or whether it occurs in other animals as well. It is perhaps worth while suggesting that if the process were carried one stage further than it actually is in Salpa, we should arrive at a condition of things curiously resembling the mode of formation of gemmules[310] in Sponges, or of statoblasts in Polyzoa. In Pyrosoma, the follicle cells or kalymmocytes form part, but the smaller part, of the embryo. In Salpa, the kalymmocytes are more abundant than the blastomeres, so that the greater part of the embryo is formed from the unfertilised kalymmocytes or follicle cells. If now we imagine this carried one step further, if we suppose the blastomeres, or fertilised elements, to be completely absent, then the whole embryo would be formed by an aggregation of unfertilised cells, and the process would become most suggestively similar to that by which a Sponge gemmule is formed.

The last series of phenomena to which I wish to refer are those resulting from the natural or artificial division of an egg, in the early stages of its development, into two or more fragments. In 1869, Haeckel, while studying the development of *Crystalloides*,[311] a genus of Siphonophora or pelagic Hydroids, was struck with the fact that the polyhedral cells of which the egg consisted at the close of segmentation exhibited active amoeboid changes of shape, and appeared to possess a certain amount of independence. The idea occurred to him to test this power of independent existence by breaking up the embryo into fragments, and following their fate. By means of needles, eggs at the close of segmentation on the second day of development were broken into two, three, or four pieces, and it was found that these not only lived for eight or ten days, but developed and gave rise, in almost normal manner,

---

308   The translation of Selensky's paper (above) spells this "kalymocyte"; other sources give it as "calymnocyte".

309   The current view is that Pyrosoma has both sexual and asexual phases in its life cycle, sometimes alternating. The yolk-containing egg, or oozooid, undergoes budding to produce blastozooids. These form the subsequent colony of new individuals whilst the original oozooid degenerates. This interpretation makes AMM's drawing of parallels with sponge development, in the next paragraph, redundant.

310   For use of this word in a different sense by Darwin, see Lecture 7 and (briefly) Lecture 12.

311   See Lecture 7.

to rudimentary Siphonophoran colonies. These observations of Haeckel's are the earliest experiments on lines which have recently led to remarkable results.

Kleinenberg, in 1878, published an account of the early stages of development of an earthworm, *Lumbricus trapezoides,* which he obtained abundantly in the gardens of the island of Ischia.[312] The worms lay their eggs in capsules: each capsule is from one to eight millimetres in length, and is filled with an albuminous mass in which are contained bundles of ripe spermatozoa, and from three to eight eggs. Of the eggs only one develops as a rule, the remainder gradually disappearing. At an early stage in development, as soon as the germinal layers are established, and before the appearance of any of the organs or parts of the embryo, the ovum divides into two parts which are usually equal and similar. Each of these parts develops into an embryo, and subsequently into an adult worm; the two remaining for a time connected together like Siamese twins, but ultimately separating.

The upshot of the process is that what Haeckel effected artificially for *Crystalloides*—i.e., the division of the segmented egg into two parts each of which develops into an embryo—is found to be a natural or even normal occurrence in *Lumbricus trapezoides.* Kleinenberg supposed that the tendency to form twins, which is so marked a feature in *Lumbricus trapezoides,* was due to a process of double fertilisation, two spermatozoa instead of one being concerned in the act. Vejdovsky[313] however suggested that the twinning was perhaps influenced by warmth, for it occurred most frequently in warm weather; a suggestion which receives much support from some experiments made by Driesch on Echinoid eggs, which when artificially warmed were found to have a marked tendency to develop twin-embryos.[314]

Another line of research was initiated independently by Chabry and by Roux,[315] who investigated the effects of injuring or destroying one of the two or four blastomeres

---

312   The development of the earth-worm Lumbricus trapezoids, Dugés. *Journal of Cell Science* 1879; s2-19: 206–244.

313   Vejdovsky Fr, *Ueber Zwillingsbildungen der Lumbriciden. Entwicklungsgeschichtliche Untersuchung,* 2 (1888) (see Newman HH, *The Physiology of Twinning,* Chicago: University of Chicago Press, 1923).

314   Entwicklungsmechanische Studien: I. Der Werthe der beiden ersten Furchungszellen in der Echinogdermenentwicklung. Experimentelle Erzeugung von Theil- und Doppelbildungen. II. Über die Beziehungen des Lichtez zur ersten Etappe der thierischen Form-bildung. *Zeitschrift für wissenschaftliche Zoologie* 1891; **53**: 160–184. Translated as: The Potency of the First Two Cleavage Cells in Echinoderm Development. Experimental Production of Partial and Double Formations. In Benjamin H Willier and Jane M Oppenheimer (eds), *Foundations of Experimental Embryology* (New York: Hafner Press, 1964), pp. 38–50.

315   Roux, Wilhelm, Beiträge zur Entwickelungsmechanik des Embryo. Über die künstliche Hervorbringung halber Embryonen durch Zerstörung einer der beiden ersten Furchungskugeln, sowie über die Nachentwickelung (Postgeneration) der fehlenden Körperhälfte. *Virchows Archiv für Pathologische Anatomie und Physiologie und für Klinische Medizin* 1888; 113–53. Translated as: Contributions to the Development of the Embryo. On the Artificial Production of One of the First Two Blastomeres, and the Later Development (Postgeneration) of the Missing Half of the Body. In Benjamin H Willier and Jane M Oppenheimer (eds), *Foundations of Experimental*

resulting from the first or second cleavage of the egg. Chabry, whose results were published in 1887, used for his experiments the egg of an Ascidian, *Ascidia aspersa*. Taking eggs in which segmentation had just commenced, and division into two cells had been effected, he destroyed one of the two cells by pricking with a needle. Under such circumstances he found that the surviving cell developed into a half-embryo. By experimenting in similar manner on eggs which had divided into four cells or blastomeres, he was able to produce either quarter, half, or three-quarter embryos. Chabry concluded from his results that each blastomere, at any rate in the early stages of development, has a determined destiny and represents a definite part of the larva. This view was first propounded by Professor His who, in 1874, maintained the existence of *"special regions in the germ, which give rise to special organs,"*[316] and held that each organ of the embryo is represented by a definite part of the body of the egg.

This view is closely similar to the old doctrine of "Evolution" or "Preformation," according to which it was held that all the organs and parts of the embryo were already present in the egg when laid; and that development consisted in an unfolding and perfecting of these parts, much as the flower is formed by the expansion of parts already present in the bud. This doctrine of Preformation, in its original form, was overthrown in 1759 by Wolff, who substituted for it the doctrine of Epigenesis-*i.e.*, that there is no trace of the embryo or of any of its parts in the egg, and that the formation of the embryo is an entirely new process. It is not a little curious to find the older doctrine, which embryologists thought disposed of for ever, coming up again in somewhat modified form, as a result of more recent investigations.[317]

Almost simultaneously with Chabry, Roux investigated in similar manner, but in more detail, the results of localised injuries to frog embryos at early stages of development. His method consisted in destroying one or more of the cells of a segmenting egg by puncturing them with a hot needle. He took a frog's egg at the completion of the first cleft—an egg which had just divided into the first two cells or blastomeres—and destroyed one of the two cells. The surviving cell developed into a half-embryo, from which by a process of regeneration the missing half was gradually formed, a whole embryo ultimately resulting. A more satisfactory method of experimenting, inasmuch as it does not involve the destruction of any of the cells, consists in shaking

---

Embryology (New York: Hafner Press, 196), pp. 42–37. For discussion of these embryological studies, see: Sunderland ME, "Hans Adolf Eduard Driesch (1867–1941)", *Embryo Project Encyclopedia* (2007-11-01). ISSN: 1940-5030 http://embryo.asu.edu/handle/10776/1679 and Kearl M, "Contributions to the Development of the Embryo. On the Artificial Production of One of the First Two Blastomeres, and the Later Development (Postgeneration) of the Missing Half of the Body (1888), by Wilhelm Roux", *Embryo Project Encyclopedia* (2009-07-20). ISSN: 1940-5030 http://embryo.asu.edu/handle/10776/2000.

316   Apparently Weismann attributes this quote to Driesch (1891); see: Stanford PK, *Exceeding Our Grasp: Science, History, and the Problem of Unconceived Alternatives* (Oxford: Oxford University Press, 2010).

317   AMM seems to be pointing here to mosaicism (see below) as being partially consistent with preformation. Wolff's use of *epigenesis* is quite different from the modern usage of the word, where the *gen* root takes on its 20th-century molecular meaning. See footnote 22, Lecture 1.

apart the blastomeres at an early stage, and then following their subsequent fate. This was first done, in 1877, by Chun,[318] and has since been repeated by other observers. Chun experimented with the eggs of Ctenophora, and found that if the two first segmentation cells, or blastomeres, were separated by shaking, each developed into a half-larva, which actually became sexual, and which ultimately regenerated the missing half by a process of budding.[319]

Driesch, in 1891, carried out a similar but more complete series of experiments at Trieste, employing for the purpose the eggs of sea-urchins, chiefly *Echinus microtuberculatus* and *Sphaerechinus granularis*. Selecting eggs which had just completed the first division, into two segmentation cells or blastomeres, he put from fifty to a hundred in a small quantity of sea-water in a glass tube about an inch and a half long and a quarter of an inch in diameter. By vigorously shaking the tube for five minutes or longer he succeeded in rupturing the egg membranes, and isolating the blastomeres. The contents of the tube were then poured into a shallow vessel, and the isolated blastomeres picked out with a fine pipette, and placed in separate watch-glasses, in which their further development could be followed. Each blastomere so treated developed at first into a half-embryo. By the evening of the first day a hemi-blastula was formed. During the night a change took place, and by the following morning each hemi-blastula had become a complete blastula or sphere, but of half the normal size. By the end of the second day invagination commenced. Each blastula became a gastrula in the normal manner; and the formation of the various organs, mouth, arms, coeIom, and ambulacral system was effected in the typical fashion. In some cases the shaking had effected imperfect separation of the two blastomeres, without rupturing the egg membrane. From these eggs either twins or double embryos were produced, a result which suggests that the twinning which occurs so commonly in *Lumbricus trapezoides* may not after all be a result of double fertilisation as was supposed by KIeinenberg.

The most recent and the most complete of the experiments on the results of shaking apart the blastomeres of segmenting eggs are those performed by Wilson on Amphioxus eggs in 1892.[320] The eggs of Amphioxus are very minute, about 0.1 mm. in diameter; and their early stages of development are extremely simple. The egg divides by a vertical cleft into two equal blastomeres: by a second vertical cleft, at right angles to the first, each of the two blastomeres is bisected, and four blastomeres of equal size result. The third cleft is a horizontal one, and is rather nearer the upper pole than the lower: by it each of the four blastomeres is divided into an upper and rather smaller cell, and a lower and rather larger one. In the later stages the number of cells is rapidly increased; a blastula, and then by invagination a gastrula is formed, and the several organs of the embryo quickly appear. In the early stages the blastomeres hang

---

318   Original source not located. C Chun and K Chun are credited with describing several species of siphonophora and other plankton. See Oliveira OMP et al, *Zootaxa* 2016; **4194**(1): 1–256.

319   In a current context, the differences in the developmental potentials of cells separated at different stages of development (discussed here and throughout the rest of the Lecture) are encapsulated in the distinction between pluripotency and multipotency. The genes that determine these potentials are matters of much current research.

320   Edmund Beecher Wilson, Amphioxus and the mosaic theory of development. *Journal of Morphology* 1893; **VIII**: 579–638.

together very loosely, and are easily separated by shaking. Commencing with the stage at which two blastomeres are present, Wilson isolated them by shaking, and found that each developed as though it were a complete egg, segmenting in the normal fashion, giving rise to a blastula, and then a gastrula, and ultimately becoming a larva, which differed from a normal one merely in being half the usual size.

Wilson next took the stage with four blastomeres, and found that each of the four, if isolated, developed into a larva of typical form, but one-fourth the usual size: if two or three of the blastomeres held together a larva resulted which was half or three-quarters the normal size. It appears therefore that at the stage with four blasto-meres, either one, two, three, or all four of the blastomeres have the power of devel-oping into an embryo which will be normal in all respects, saving only as regards size. Anxious to determine the limits to which this extraordinary process could be continued, Wilson tried the next stage, that in which eight blastomeres are present, of which four are rather smaller, and four rather larger in size. Each blastomere when isolated commenced to develop, but never became a complete embryo. Flat plates, curved plates, even blastulas one-eighth the normal size were formed, which swam about freely by means of cilia, but which underwent no further development.

These results are of very great interest. They must I think be regarded as fatal to the doctrine of "Evolution" in its new form, as maintained by His and Weismann; for if any one of the first four blastomeres is able by itself to develop directly into an embryo, it seems impossible to hold that each blastomere has a predestined part to play in the formation of the embryo.

Another very interesting result is the precise point at which the power of devel-oping an embryo from a single blastomere ceases. Discussion has taken place as to whether the failure is due to quantitative or to qualitative considerations; to an insufficiency in amount of living matter, or to incompleteness in its structure or composition. Wilson is of opinion that the difficulty is mainly a qualitative one, and has advanced arguments in support of his view. Perhaps the most important consid-eration in its favour is that so long as the blastomeres are precisely alike—i.e., up to the stage with four blastomeres—each one possesses the power when isolated of developing into an embryo; while as soon as a distinction between smaller and larger blastomeres appears—i.e., at the third cleavage—this power ceases.

This last consideration becomes still more significant if we regard a Metazoan as comparable to a colony of Protozoa, in which differentiation between the component units, and consequent mutual dependence, and the necessity for holding together have become established; and if further we view the early development of a Metazoan as a shadowing of the process by which this condition was originally attained. So long as the cell units or blastomeres remain identical, so long does each one retain the power of independent existence and development; but as soon as differentiation is established between one and another this power of separate life ceases. This view, though extremely suggestive, must however be regarded as entirely provisional, until we gain more complete and exact knowledge concerning the conditions under which the power of independent development can be exerted, not only in Amphioxus, but in other animals as well.

The present paper is merely a record of some recent investigations, and has no pretence to completeness. These researches however all tend in one direction, namely to emphasise the fact that the development of animals takes place in far more varied manner than is generally supposed, and that marked differences may occur both in the earlier and the later stages of the embryological history, not merely in allied genera and species, but even between individual members of the same brood. I will conclude with a brief summary of the principal modes in which the development of a Metazoan may be effected. It should be noted that the adult structure as a rule gives no indication of the particular mode in which development has occurred: a sponge developed from a gemmule is, so far as we know, indistinguishable in its adult form from one developed from a fertilised egg.

AN ADULT METAZOON MAY BE DEVELOPED,

1. From a group of (mesoblast?) cells, which are originally independent of one another, and of which none are fertilised. *Examples:* The gemmules of Sponges, or the statoblasts of Polyzoa.

2. Partly from a fertilised ovum, and partly from unfertilised follicle cells. *Examples:* Salpa and Pyrosoma.

3. From a single fertilised ovum, together with a greater or less number of yolk cells, which are at first independent cells, but which ultimately become absorbed into the ovum. *Examples:* Fasciola, Pisidium.

4. From a single fertilised ovum, aided by nutrient matter derived from the surrounding follicle cells. *Examples:* Cephalopoda, and most Vertebrates.

5. From a single fertilised ovum, without the aid of follicle cells. *Example:* Amphioxus.

6. From an unfertilised cell, or ovum. *Examples:* The parthenogenetic ova of Entomostraca or of Rotifera.

7. From one or more of the blastomeres resulting from the segmentation of a fertilised ovum. *Examples:* Echinus, Amphioxus.

8. From a portion, large or small, of the body of the parent. *Examples:* Hydra, Sponge.

9. By local proliferation of the cellular elements of the parent, forming an external or internal bud. *Examples:* Sporocyst, Hydra.

LECTURE 12

DEATH

In this lecture to the Manchester Microscopical Society, as in other presentations, AMM engages his audience from the outset with some gentle humour but then quickly moves to matters they may not have thought about—the essential difference between unicellular fission and multicellular reproduction as regards the involvement of birth and death in the perpetuation of life.

The topic of death and its significance greatly interested leading 19th-century biologists including August Weismann, and before him E Ray Lankester, both of whom AMM evidently admired. Here he presents an analysis and discussion of Weismann's theories, including several lengthy quotations. Weismann is interesting for the manner in which some of his ideas anticipated 20th-century cell biology, even if it took many technical and theoretical developments to discover the mechanisms involved. As indicated in AMM's quotes, Weismann seems to have accepted the inadequacy of late 19th-century technology in testing his hypotheses and is wistful of future developments. (Further discussion of this is beyond present scope; for a comprehensive guide to Weismann's life and ideas, including a bibliography, see Churchill FB, *August Weismann: Development, Heredity and Evolution*, Cambrige, MA, Harvard University Press, 2015.) AMM points to the value of Maupas' lengthy and painstaking experiments on infusoria (fresh water protozoa) in testing some of Weismann's ideas.

Other parts of the lecture are less informative. The *comparison* of reproductive events in protozoa and metazoa appears somewhat contrived to modern eyes. It is unclear whether AMM is trying to imply common ancestry in the existence of particular features (alternating sexual and asexual phases, the stages at which cell death occur, etc.) or whether we should just take them as mechanisms with survival/reproductive value in different groups of animals. The summary list of relationships between death, sexual reproduction and metazoan development suggest the former. The final conclusion in the list (about metazoa) has the same intrinsic truth as the one it extends (about protozoa) and one feels that its very inevitability, echoed by the rather obvious statement about the continuity of living matter, rather detracts from its heuristic value.

LECTURE 12

DEATH

THE subject of my address is I admit open to objection from more than one standpoint. It is commonly regarded as dull and uninviting; and although in the long run it must concern us all, there would seem to be little reason for urging its consideration prematurely. Then again it may be objected that having, from the nature of the case, no personal experience of the phenomenon, I am not in a position to speak about it with authority.

I can only reply that a matter which concerns vitally all living things, animal or vegetable, cannot be devoid of interest. The problems of life are the most fascinating of all scientific enquiries; and surely death, the cessation of life, must have something to teach us, must throw some light on the nature of life itself. The beginnings of life are at present hidden from us; the other end of the series, the termination of life, we have daily opportunities of studying. With regard to the second objection, it is true that from personal experience I can say nothing on the subject of death; but this is a disqualification which I share with all members of our Society, and with all living men. If nobody is to be allowed to talk about death until he has qualified by personal acquaintance with the "fell sergeant," it is clear that any knowledge to be derived, any lessons to be learnt from its study, are lost to us for ever.

My choice of the subject has been determined mainly by the consideration that of recent years it has attracted considerable attention, and has given rise to interesting speculations, many of which are based on facts made known to us by those extraordinarily minute investigations into the structure of the lower animals, which the modern improvements in microscopical methods and appliances have rendered possible. My purpose this evening is to give a summary of these more recent contributions to our knowledge of the nature and causation of death, and an indication of the paths along which it seems probable that advances will be made in the future.

In a scientific enquiry it is above all things desirable to have clear ideas as to the nature of the subject we are dealing with. Unfortunately it is not easy to define with any degree of precision what it is that we understand by death. Götte regards death as something inherent in life itself; a view which is held more or less explicitly by the majority of mankind. This is however disputed by Weismann, whose contributions to the subject are of importance. Weismann defines death as *"an arrest of life, from which no lengthened revival, either of the whole or any of its parts, can take place; or to put it concisely, as a* definite arrest of life." *"The real proof of death,"* according to Weismann, *"is that the organised substance which previously gave rise to the phenomena of life, for ever ceases to originate such phenomena."* He adds to this the corollary that death involves the presence of something dead—i.e., a corpse. Weismann next challenges the statement that death is a necessity, inseparable from the idea or existence of life. He calls attention to the conditions which obtain among animals such as Protozoa, in which reproduction is normally effected by fission; and points out that in the life-history of such animals natural death does not occur. An Amoeba for example reproduces by simply dividing into two. In such an act of fission the parent generation disappears, but nothing has died. If the original Amoeba be called Tom, and the products of fission Dick and Harry, the upshot of the process may be expressed by saying that Tom has disappeared without having died, while Dick and Harry have come into existence without having been born. Nothing has died, there is no corpse to bury, and our ordinary ideas with regard to individuality and identity fail altogether to afford answer to the question—Where is Tom at the end of the process?

Hence arises the idea of the immortality of the Protozoa. An Amoeba or other Protozoon reproducing by simple fission can indeed be killed, as by boiling the water in which it is contained; but it does not, in the ordinary course of events, die. The production of one generation involves the disappearance but not the death of the parent generation. From the first Amoeba to the present day there has been direct continuity of living matter. Death may occur through violence, but it is not a necessary accompaniment or consequence of life. Moreover death, when it does happen in the case of an Amoeba, causes a final interruption, an absolute break in the chain. No Ameoba that has died has left offspring, for such offspring can only arise by the division of the living body of the parent. "No Amoeba," it has been well expressed, "has ever lost an ancestor by death."

The above considerations, which clearly apply not only to Protozoa but to any other animals which reproduce by fission, form the basis on which Weismann has built up his theory of the origin of death. This theory may be briefly summarised as follows: Protozoa, reproducing by fission, are immortal in the sense that death does not occur as a necessary or natural termination of the life-cycle. Natural death must therefore be limited to Metazoa, or multicellular animals. In Metazoa a distinction is always found between somatic cells and reproductive cells; the former being the component elements of which the body of the individual is constructed, while the latter are the units from which the individuals of the next succeeding generation will be developed. The somatic cells are concerned with the existence and welfare of the individual; the reproductive cells with the perpetuation of the species.[321]

The normal life-cycle of a Metazoon is as follows: The fertilised egg, or reproductive cell, by repeated division gives rise to a number of cells, of which some become the somatic cells—i.e., the body of the individual animal—while others remain as its reproductive cells; from these latter, in due course and in similar manner, the individuals of the next generation are formed. Of the two kinds of cells, the somatic cells alone are liable to natural death: the reproductive cells survive as the individuals of

---

321   It was Weismann who first proposed a distinction between somatic and reproductive cells in metazoans. His *germ-plasm* theory began to take shape in the late 1870s and was first stated formally in 1885. His major work, *Das Keimplasma: Eine Theorie der Vererbung* was first published in 1892 and appeared in English translation (dedicated to Darwin) the following year. The germ-plasm theory attracted criticism from others, including his erstwhile friend and colleague Haeckel, but he continued to defend and develop it virtually until his death in 1914. It took on greater significance at the start of the 20th century when Mendel's laws were rediscovered, but it was overtaken by events as by this time chromosomes were well known and the nuclear events of mitosis and meiosis, as well as polyploid conditions, were becoming clear. Weismann's historical importance thus lies more in the concepts he developed than in the elucidation of mechanisms.

the succeeding generation. The reproductive cells of Metazoa are therefore immortal, in exactly the same sense as are Amoebae. The reproductive cells, like the Amoebae, can be destroyed, but they do not die naturally. Each reproductive cell is derived by fission from a corresponding cell of the preceding generation; and in Metazoa, as in Protozoa, there has been from generation to generation direct continuity of living matter.[322] We are apt to think of the somatic cells—i.e., the body of the individual animal—as the essential part, by reason of its greater bulk and impressiveness; and to regard the reproductive cells as structures whose purpose it is to give rise to the somatic cells—i.e., the individuals of the next generation. In doing so however we lose sight of the true relation between the two groups or cells. The reproductive cells are the really essential elements, and the part of the somatic cells is a subordinate one; their purpose being to nourish and protect the reproductive cells in such way as to afford them the best chance of completing their special duty, the perpetuation of the species.[323]

The above considerations bring us to the final point in Weismann's argument. If natural death affects the somatic cells only, and is a character acquired by them, it remains to inquire why such natural death should occur, and what determines the time of its occurrence. Weismann answers these questions by saying that death occurs because it is advantageous to the species that it should do so; and that the normal time for such death to occur is the end of the reproductive period of the individual. Both these points require further consideration. With regard to the former it is of great importance to distinguish clearly between what is for the good of the individual on the one hand, and on the other hand what is advantageous for the species.[324] A good illustration is afforded by the elaborate provision which insects

---

322 The logic of this seems inescapable but only if AMM's distinction between the two types of cell is taken at face value. In fact, the first (diploid) somatic cell of a multi-cellular organism is formed from two (haploid) germ cells and the next generation primordial germ cells arise from it by asymmetric division in the very early embryo (4- to 32-cell stage, and by one of several different mechanisms, depending on the type of organism). The primordial germ cells migrate to the gonads or equivalent structures and are subsequently characterized by meiotic rather than mitotic division. In other words, reproductive cells do not regenerate themselves directly. Rather the genes occupy a brief intervening period of somatic existence, until the early embryo cells begin to differentiate. It nevertheless remains true that germ cells outlive the bodies that produce them.

323 Substitute *genes* for *species* and we have a Richard Dawkins-style view of the perpetuation of life (and see next footnote).

324 It has become unfashionable (since the 1970s at least) to speak of biological developments as being for the good of the species. The concept that genes are inherently selfish, that cells and bodies are a gene's way of ensuring its survival and spread, and that species are just groups of related carriers of replicated genes, has become all pervasive. Group selection, kin selection, mate choice and other phenomena are all interpreted using the mathematical precision of genetic distribution and its optimization. Although gene distributions can be measured at population level, there is no mechanism for species-level advantage and we now manage without it. AMM knew nothing of this, so we should avoid being critical of his choice of words. The

make for their offspring, which they will never see. Certain wasps have the habit of stinging the larvae of beetles in their nerve centres in such manner as to paralyse their victims without killing them. On the body of the paralysed larva a single egg is laid by the wasp, and then left to its fate. From the egg a grub is hatched in due time, which at once begins to suck the juices of the larva; the victim supplying it with food sufficient for the whole period of its development. The grub changes to a pupa on the skin of its victim, and passing through the winter in the pupa state, emerges in the spring as a wasp with the same instincts and habits as its parent. Difficulty is sometimes felt in accounting for such instincts. The individual wasp, it is true, derives no advantage whatever from its ingenuity; but the gain to the species is enormous; and the preservation of the habit is due to the fact that those individuals which took the greatest care to make provision for their young would be most likely to give rise to offspring which would survive in the struggle for existence. Natural selection would tend to preserve the instinct because it is advantageous to the species, though of no benefit whatever to the individual.

So with regard to death; Weismann argues that the origin thereof is to be found in the consideration that it is advantageous to the species that individuals should die. The argument is perhaps best stated in his own words. *"Let us imagine,"* he says,[325] *"that one of the higher animals became immortal; it then becomes perfectly obvious that it would cease to be of value to the species to which it belonged. Suppose that such an immortal individual could escape all fatal accidents through infinite time,—a supposition which is of course hardly conceivable. The individual would nevertheless be unable to avoid from time to time slight injuries to one or another part of its body. The injured parts could not regain their former integrity, and thus the longer the individual lived the more defective and crippled it would become, and the less perfectly would it fulfil the purpose of its species. Individuals are injured by the operation of external forces, and for this reason alone it is necessary that new and perfect individuals should continually arise and take their place, and this necessity would remain even if the individuals possessed the power of living eternally."* *"From this follows, on the one hand the necessity of reproduction, and on the other the utility of death. Worn-out individuals are not only valueless to the species, but they are even harmful, for they take the place of those which are sound. Hence by the operation of natural selection the life of our hypothetically immortal individual would be shortened by the amount which was useless to the species. It would be reduced to a length which would afford the most favourable conditions for the existence of as large a number as possible of vigorous individuals at the same time."*

The passage just quoted is extremely ingenious, but is hardly convincing, for it does not attempt to explain the real nature of death, nor how it came about in the first instance. The distinction between somatic and reproductive cells is a real one in Metazoa. The actual steps by which it was established have yet to be traced: but it would seem probable that in the earliest Metazoa all the cells originally retained

---

wasp example which follows (and the other insect and vertebrate examples a few paragraphs further on) is easy to read with modern eyes whilst still appreciating AMM's meaning. The summary statement (item v in the list at the end of the lecture) also clarifies the matter. The only adjustment we might want to make is deletion of this paragraph's final sentence. (See also the very end of Lecture 19.)

325   *The Duration of Life* (1881).

their reproductive power, and that the process by which this power became restricted to particular groups of cells was a gradual one. The gemmules of sponges, or the statoblasts of Polyzoa,[326] are perhaps to be interpreted as examples of the retention of reproductive power by groups of cells which in allied animals have become exclusively somatic in character; and similar instances could be quoted from the vegetable kingdom.

Weismann's contention that the reproductive cells of Metazoa are immortal in the same sense as an Amoeba, must also be admitted to be established. He however leaves altogether undecided the question of the way in which death of the somatic cells arose in the first instance; and we shall find later on that there are strong grounds for holding that natural death appeared first, not as Weismann supposed among Metazoa, but in the Protozoa themselves.

Before considering these more recent aspects of the problem it will be well to refer briefly to Weismann's views, which have already been mentioned, in regard to the causes determining the duration of life in different animals. According to Weismann the duration of life in a given case is that which is most advantageous for the species. In the simpler cases death occurs at the close of the reproductive period. In the silkworm moth, or the May fly, the adult existence is only of a few hours duration; the insect laying all its eggs simultaneously, and then dying. In other insects, as in many of the hawkmoths, and in most butterflies, the eggs are laid at intervals and in different places; in such cases the life of the adult is prolonged until a sufficient number of eggs have been laid to ensure the perpetuation of the species, and then the insect dies. In birds, owing to the small number of the eggs that are produced at any one time, and the great destruction to which the eggs are liable from the attacks of enemies, several years may elapse before enough eggs are produced to ensure survival of the species; and the life of the adult is consequently prolonged for many years. A further lengthening of life takes place in mammals and other animals which tend or rear their young, either by retaining the eggs or embryos within their bodies for a longer or shorter portion of their development, or by protecting and feeding the young after birth.

The whole subject thus opened up is of extreme interest, but it would be impossible to treat it adequately on the present occasion. Weismann himself has accumulated a large number of statistics with regard to different groups of animals, which lead him to the conclusion that "*the end of the reproductive period is usually more or less coincident with death;* " while in cases in which the duration of life is prolonged, owing to the parent tending or nursing the young after birth, he concludes that "*as a general rule the increase in length of life is exactly proportional to the time which is demanded by the care of the young.*"[327]

---

326   See Lectures 1 and 6.

327   The source of the quotes in this paragraph is unclear but is likely to be *The Duration of Life* (1881). It is now common to add to this the role played by non-reproductive individuals (e.g., post-menopausal females in the case of humans and a few other species, drones in social insects) in raising young and in providing a repository of accumulated social knowledge.

With regard to the actual cause of death in Metazoa, Weismann suggested in his earlier essay published in 1881,[328] that death may be due to the somatic cells losing the power of reproduction by cell-division after a certain number of generations.[329] In 1883 in a further essay on life and death,[330] he repeated this suggestion and developed it more fully. After referring to his former view that the varying duration of life in the animal kingdom *"is determined in different species by the varying number of somatic cell-generations,"* he frankly admits that he is quite unable to indicate the changes in the physical constitution of protoplasm upon which the variations in the capacity for cell-division depend, or the causes which determine the greater or smaller number of cell-generations; but he urges that if we must wait until we understand the molecular structure of cells before advancing views concerning the nature and limits of their activities we shall probably never solve the problem. *"Therefore,"* he continues, *"it is in my opinion an advance if we may assume that length of life is dependent upon the number of generations of somatic cells which can succeed one another in the course of a single life; and furthermore that this number, as well as the duration of each single cell-generation, is predestined in the germ itself. This view seems to me to derive support from the obvious fact that the duration of each cell-generation, and also the number of generations, undergo considerable increase as we pass from the lowest to the highest Metazoa."*

This bold suggestion, based entirely on theoretical considerations, received striking confirmation a couple of years later, from the results of Maupas' famous researches into the reproduction of Infusoria.[331] The normal mode of reproduction among the ciliated Infusoria is by means of fission, essentially similar to the fission of Amoebae. In addition to this a process of conjugation has long been known to occur in Infusoria, though its real nature has been much disputed. Balbiani described the process in great detail as long ago as 1858,[332] and was led to the conclusion that it was a true sexual act, comparable to fertilisation in the higher animals. Balbiani's views were for a long time discredited, mainly through the unwillingness of zoologists to admit the possibility of sexual reproduction occurring in unicellular animals. More recent researches however and in particular those of Maupas, have shown that while Balbiani was not absolutely right with regard to all the details of the process, yet that he was correct in his interpretation of the conjunction of Infusoria as a true sexual act.

The problem which Maupas set himself to solve was to determine the relation between the two modes of reproduction in Infusoria, to ascertain the conditions under which asexual and sexual reproduction respectively occur, and to find out what causes lead to the substitution at particular times of one process for the other. To do this it

---

328   *Über die Dauer des Lebens, Tageblatt der 54. Versammlung deutscher Naturforscher und Aertz in Salzburg vom 18.-24. September 1881* (Salzburg).

329   This anticipates our current understanding of the role of telomeres in limiting the number of cell divisions.

330   *Über Leben und Tod* (Jena: Gustav Fischer, 1884).

331   For a contemporaneous account, see Hartog MM, Abstract of Maupas's researches on multiplication and fertilisation in ciliate Infusorians. *Journal of Cell Science* 1891; **32** (series 2): 599–614. See also Binet A, The immortality of Infusoria. *The Monist* 1890; **1**(1): 21–37.

332   Note relative l'existence d'une generation sexualle chez les infusoires. *Journal of Physiology* **1**: 347–352 and *Comptes rendus de l'Académie des Sciences* **46**: 628–632.

was necessary to isolate an individual Infusorian, to place it under known conditions as to temperature and food supply, and then to follow the fate of the successive generations of offspring to which it gave rise by fission.

The investigation was an extraordinarily laborious one. Continuous observations for over five months were necessary; the production and fate of from 200 to 300 successive generations had to be followed accurately, and in some cases the observations extended over more than 600 generations. The most suitable temperature at which to conduct the experiments, and the kind of food best suited to the particular Infusorian under observation, had in each case to be determined. No less than twenty different species were experimented upon, and the whole research, both as regards the extreme patience necessary for its proper conduct, and the great importance of the results obtained, justly ranks among the most famous of its kind.

The general results of Maupas' investigation, which deserve to be followed in detail by all microscopists, were to this effect: In conjugation the paranucleus of each of the conjugating individuals acts as a hermaphrodite sexual element: it undergoes successive divisions, and parts are extruded and lost completely. The remaining parts of each paranucleus become differentiated into male and female pronuclei; interchange of the male pronuclei takes place between the conjugating individuals, and the male pronuclei then fuse with the female pronuclei of the individuals to which they have been transferred. The entire nuclear apparatus of each of the conjugating individuals is now reconstituted, and the two Infusoria separate and become independent once more. The purpose of conjugation appears to be to stimulate the asexual act of fission. Individuals which previous to conjugation showed no tendency to divide, begin to do so actively as soon as conjugation has been completed. In order however for conjugation to be effective, Maupas finds that it must take place between individuals which are not closely related to each other. Members of the same family show no tendency to conjugate with each other, and appear indeed to be incapable of doing so, for attempts at such conjugation have been found to prove abortive.

By his method of isolating Infusoria and following the fate of the successive generations produced by fission, Maupas was enabled to prove that there are natural and definite limits to the continuance of asexual reproduction. An Infusorian, *Stylonychia pustulata*,[333] was isolated in November 1885, and the successive generations followed until March 1886. By that time there had been 215 generations produced by successive acts of fission. Conjugation had not occurred, since nearly related individuals will not conjugate, and unrelated forms were excluded by the conditions of the experiment. Towards the close of the experiment the tendency to multiply by fission became less manifest, and the individuals themselves were in a condition well described as senescent; they were of reduced size, often distorted in shape, or actually malformed; their nuclear apparatus was in a degenerate condition, and they no longer had the power, which all the younger generations possessed, of conjugating effectively with unrelated individuals when transferred to water containing such. Finally the nutritive powers of these senescent members failed and death of these exhausted forms brought the experiment to an end. The same results were obtained from experiments conducted in a similar way with other Infusorians. The general conclusions

333   An ovoid hypotrich, representing a large genus of protozoa found in soil and fresh water.

arrived at were: (i.) That in those Protozoa which reproduce both by fission and by a sexual process of conjugation there are definite limits to the number of generations which can be produced asexually; (ii.) that conjugation only occurs between individuals which are not nearly related to each other; (iii.) that if conjugation with unrelated forms be prevented, senescence, and finally death occur.

These results are of the utmost interest in connection with Weismann's views regarding the nature and origin of death. They show that Weismann was wrong in supposing that death occurred first amongst Metazoa. Natural death occurs among Protozoa; and the tendency to it and inability to escape from it, are probably inherited by Metazoa from their Protozoon ancestors. On the other hand, Maupas' results confirm in the fullest manner Weismann's bold suggestions, (i.) that the original occurrence of death is intimately connected with sexual reproduction, if not indeed an actual consequence of it; (ii.) that the number of generations of somatic cells which can succeed one another in the course of a single life may be strictly limited. Maupas' experiments seem to me to afford the very evidence of which Weismann was in search. They prove that amongst Infusoria asexual reproduction by cell division cannot be continued indefinitely, but that it leads in time to senescence and ultimately to death.

If we apply these results to Metazoa the conclusions become very striking. In Metazoa, as in Infusoria, there is alternation of sexual and asexual modes of reproduction. The fusion of male and female pronuclei in the act of fertilisation of the egg is the sexual process, and is equivalent to the similar fusion of male and female pronuclei of unrelated cells, seen in the conjugation of Infusoria. On the other hand the successive acts of cell division, by which the fertilised egg gives rise to the embryo and the embryo becomes converted into the adult, are asexual processes, equivalent to the repeated acts of cell division by which the successive generations of the Infusorian are produced. In the Infusorian the number of such asexually produced generations that can succeed one another is limited; so also is it in the Metazoon; and the gradual failure of the power to divide further leads in both cases alike first to senescence or old age, and ultimately to death,

This comparison between Protozoa and Metazoa in regard to the modes in which reproduction is effected appears to be a just one. The striking difference, that in the Protozoon the products of the asexual process of cell division become independent and similar unicellular animals, while in the Metazoon they are component and differentiated units in the body of a multicellular animal, does not affect the comparison so far as concerns the essential point – i.e., the mode in which successive cell generations come into existence in the two cases alike. A further point of difference is found in the consideration that in the Infusoria all the asexually produced cells retain, at any rate for a number of generations, the power of conjugating with other cells; while in Metazoa the power appears to be lost very early by the majority of the cells, and retained only by the reproductive cells. This however does not in any way invalidate the comparison, and is merely an example of that structural and physiological differentiation which distinguishes Metazoa from colonial Protozoa, and which affords the key to all Metazoon structure.

Moreover, it is at present entirely unknown to us at what period or to what extent somatic cells of a Metazoon lose their power of conjugating. From this standpoint it would be of the greatest interest to know precisely what happens in cases of introduction of cells from without into a living Metazoon; for example, in vaccination or in other methods of inoculation; or in cases of transfusion of blood on a large scale. Theoretically it seems possible that rejuvenescence of the somatic, as of the reproductive cells of a Metazoon might be effected, and a new lease of life obtained for these cells and their descendants. This however is a matter of mere speculation, and not to be lightly entered upon. The general conclusions to which these recent investigations have led us may be briefly summarised as follows:—

i.   Death is not an intrinsic necessity, either of life or of organisation.

ii.   Natural death first appeared, so far as we know at present, among the higher Protozoa.

iii.   Death is closely associated with the occurrence of conjugation, and the consequent alternation of sexual and asexual modes of reproduction.

iv.   The asexual mode of reproduction, by fission, is the more primitive one. Conjugation, or sexual reproduction, gives an advantage in the struggle for existence; and at first a luxury, has through the action of natural selection become a necessity.

v.   The normal duration of life of a given species is that which is most advantageous to the species—i.e., an animal dies when it has produced sufficient young to ensure the perpetuation of the species under existing conditions.

vi.   In Metazoa, as in Protozoa, there is direct continuity of living matter by cell division from generation to generation.

vii.   The statement that *"no Protozoon has ever lost an ancestor by death"* may now be extended thus: *There is not a single component cell in the body of a Metazoon that has ever lost an ancestor by death.* For each component cell in the body of a Metazoon is descended by direct fission from the egg. The egg was a body cell of the parent, and in its turn was derived by cell division from the egg cell from which the parent was developed; and so on, generation behind generation, there has been unbroken continuity of living matter.

LECTURE 13

THE RECAPITULATION THEORY

This lecture was the 1890 President's address to the Biological Section of the British Association [for the Advancement of Science] delivered at its meeting in Leeds. The original BAAS meeting report shows that it was delivered on Thursday, 4 September. It appears there without a title so the one given here must have been added by CFM. The unusually large number of footnote references to cited work presumably reflects the scientific nature of the audience. {Curly parentheses contain text from the BA transcript where this is significantly different from the published lecture.}

It is the longest of all the essays and, as CFM notes in his Preface, it discusses matters dealt with in several other lectures, especially 10 but also 6, 8 and 11. An interesting feature is the way that AMM, although professing acceptance of the theory of recapitulation, cites many instances where it does not provide an unqualified explanation for what is observed during animal development. He finds ways around this problem, but one is left feeling that he is on the verge of finding the theory to be inadequate or incorrect, as indeed later biologists were to do. It is a classic example of the Kuhnian progress of science: working new evidence into the current paradigm until it no longer fits, at which point a paradigm shift, or even a revolution, is the only escape.

The other tension evident in this lecture, as much as or more than in others, is that of terminology. AMM sometimes uses the language of hierarchy (*high*, *low*, *advanced*) which we might find hard to accept; unless we make a careful reading, it often seems to imply goal-directed development. At other times, the phraseology is more clearly Darwinian and easier to live with. It is further evidence of a transition in discourse and understanding, running alongside that of scientific knowledge.

The Interlude which follows this essay provides a broader perspective on AMM's position on several of the matters addressed in this lecture. For detailed, critical accounts of the rise and fall of the Recapitulation Theory, of the scientists who accepted or rejected it, and of its influence on biology and sociobiology into the 20th century, see: Churchill FB, *August Weismann: a developmental evolutionist*. In Churchill FB and Risler H, *August Weismann: Selected Letters and Documents* (Freiburg: Universitätsbibliothek Freiburg,1999), Vol. 2, Section 6, pp. 749–798; Gould SJ, *Ontogeny and Phylogeny* (Cambridge, MA: Belknap Press, 1977); Hopwood N, *Haeckel's Embryos* (Chicago: Chicago University Press, 2015).

LECTURE 13

THE RECAPITULATION THEORY

A s my theme for this morning's address I have selected the Development of Animals. I have made this choice from no desire to extol one particular branch of biological study at the expense of others, nor through failure to appreciate or at least admire the work done and the results achieved in recent years by those who are attacking the great problems of life from other sides and with other weapons. My choice is determined by the necessity that is laid upon me, through the wide range of sciences whose encouragement and advancement are the peculiar privilege of this Section, to keep within reasonable limits the direction and scope of my remarks; and is confirmed by the thought that, in addressing those specially interested in and conversant with biological study, your President acts wisely in selecting as the subject-matter of his discourse some branch with which his own studies and inclinations have brought him into close relation.

Embryology, referred to by the greatest of naturalists as "one of the most important subjects in the whole round of Natural History," is still in its youth, but has of late years thriven so mightily that fear has been expressed lest it should absorb unduly the attention of zoologists, or even check the progress of science by diverting interest from other and equally important branches.[334] Nor is the reason of this phenomenal success hard to find. The actual study of the processes of development; the gradual building up of the embryo, and then of the young animal, within the egg; the fashioning of its various parts and organs; the devices for supplying it with food,

---

334   The development and importance embryology in 19th-century biology has been discussed at length by many authors. Churchill (2015) points out that it was principally a descriptive rather than experimental science. He identifies (as does AMM in Lectures 1 and 3) von Baer as the key historical figure in its early development, based on his discovery of the mammalian egg (in 1827) and his two-volume treatise on the subject, *Über Enwickelungsgeschichte der Thiere: Beobachtung und Reflexion* (Königsberg: Gebrüdern Borntrager), published in 1828 and 1837. Embryology took on additional significance after the publication of Darwin's theories in the middle of the century, by feeding many controversies, reinterpretations and extreme positions, including those associated with cell theory, reproductive mechanisms, the inheritance of acquired characteristics, recapitulation, degeneration and ancestral lineages. Embryology pervades AMM's approach to all these matters, as is clear from the lectures in this collection. Besides the many scientists referred to in the present lecture and others, the key educational influences on him in this regard were FM Balfour, E Ray Lankester and TH Huxley. He was undoubtedly further enlightened by his studies at the Naples Zoological Station, prior to establishing his own academic career in England (1875) and on a brief later visit (1884). Amongst the scientists he cites, those who worked in Naples at some point, besides its founder Dohrn, include Balfour, Boveri, Driesch, Eisig, Hertwig, Kleinenberg, Kowalevsky, Krause, Lankester, Loeb and Weismann.

and for ensuring that the respiratory and other interchanges are duly performed at all stages: all these are matters of absorbing interest. Add to these the extraordinary changes which may take place after leaving the egg, the conversion, for instance, of the aquatic gill-breathing tadpole—a true fish as regards all essential points of its anatomy—into a four-legged frog, devoid of tail, and breathing by lungs; or the history of the metamorphosis by which the sea-urchin is gradually built up within the body of its pelagic larva, or the butterfly derived from its grub. Add to these again the far wider interest aroused by comparing the life-histories of allied animals, or by tracing the mode of development of a complicated organ—e.g., the eye or the brain—in the various animal groups, from its simplest commencement, through gradually increasing grades of efficiency, up to its most perfect form as seen in the highest animals. Consider this, and it becomes easy to understand the fascination which embryology exercises over those who study it.

But all this is of trifling moment compared with the great generalisation which tells us that the development of animals has a far higher meaning; that the several embryological stages and the order of their occurrence are no mere accidents, but are forced on an animal in accordance with a law, the determination of which ranks as one of the greatest achievements of biological science. The doctrine of descent, or of Evolution, teaches us that as individual animals arise, not spontaneously, but by direct descent from pre-existing animals, so also is it with species, with families, and with larger groups of animals, and so also has it been for all time; that as the animals of succeeding generations are related together, so also are those of successive geologic periods; that all animals, living or that have lived, are united together by blood relationship of varying nearness or remoteness; and that every animal now in existence has a pedigree stretching back, not merely for ten or a hundred generations, but through all geologic time since the dawn of life on this globe.

The study of Development, in its turn, has revealed to us that each animal bears the mark of its ancestry, and is compelled to discover its parentage in its own development; that the phases through which an animal passes in its progress from the egg to the adult are no accidental freaks, no mere matters of developmental convenience, but represent more or less closely, in more or less modified manner, the successive ancestral stages through which the present condition has been acquired. Evolution tells us that each animal has had a pedigree in the past. Embryology reveals to us this ancestry, because every animal in its own development repeats this history, climbs up its own genealogical tree. Such is the Recapitulation Theory, hinted at by Agassiz,[335] and suggested more directly in the writings of von Baer,[336] but first clearly enunciated by Fritz Müller, and since elaborated by many, notably by Balfour, and by Ernst Haeckel. It is concerning this theory, which forms the basis of the science of Embryology, and which alone justifies the extraordinary attention this science has received, that I venture to address you this morning. A few illustrations from different groups of animals will best explain the practical bearings of the theory, and the aid which it affords to the zoologist of to-day, while these will also serve to illustrate certain of the difficulties which have arisen in the attempt to interpret individual

---

335   Louis, the elder Agassiz.

336   Note here as elsewhere (Lectures 1 and 10, and the Interlude) that von Baer rejected recapitulation as an explanation for the similarity of embryos of different species.

development by the light of past history—difficulties which I propose to consider at greater length.

A very simple example of recapitulation is afforded by the eyes of the sole, plaice, turbot, and their allies. These "flat fish" have their bodies greatly compressed laterally; and the two surfaces, really the right and left sides of the animal {are,} unlike, one being white, or nearly so, and the other coloured. The flat fish has two eyes, but these in place of being situated, as in other fish, one on each side of the head, are both on the coloured side. The advantage to the fish is clear, for the natural position of rest of a flat fish is lying on the sea bottom; with the white surface downwards and the coloured one upwards. In such a position an eye situated on the white surface could be of no use to the fish, and might even become a source of danger, owing to its liability to injury from stones or other hard bodies on the sea bottom. No one would maintain that flat fish were specially created as such. The totality of their organisation shows clearly enough that they are true fish, akin to others in which the eyes are symmetrically placed one on each side of the head, in the position they normally hold among vertebrates. We must therefore suppose that flat fish are descended from other fish in which the eyes are normally situated.

The Recapitulation Theory supplies a ready test. On employing it—*i.e.*, on studying the development of the flat fish—we obtain a conclusive answer. The young sole on leaving the egg is shaped just as any ordinary fish, and has the two eyes placed symmetrically on the two sides of the head. It is only after the young fish has reached some size, and has begun to approach the adult in shape, and to adopt its habit of resting on one side on the sea bottom, that the eye of the side on which it rests becomes shifted forwards, then rotated on to the top of the head, and finally twisted completely over to the opposite side.[337]

The brain of a bird differs from that of other vertebrates in the position of the optic lobes, these being situated at the sides instead of on the dorsal surface. Development shows that this lateral position is a secondarily acquired one, for throughout all the earlier stages the optic lobes are, as in other vertebrates, on the dorsal surface, and only shift down to the sides shortly before the time of hatching.

Crabs differ markedly from their allies, the lobsters, in the small size and rudimentary condition of their abdomen or "tail." Development however affords abundant evidence of the descent of crabs from macrurous ancestors, for a young crab at what is termed the Megalopa stage has the abdomen as large as a lobster or prawn at the same stage.

Molluscs afford excellent illustrations of recapitulation. The typical gasteropod has a large spirally coiled shell; the limpet however has a large conical shell, which in the adult gives no sign of spiral twisting, although the structure of the animal shows clearly its affinity to forms with spiral shells. Development solves the riddle at once,

---

337   The glaring omission from this example is any attempt at a mechanistic explanation for the positioning of flatfish eyes. In fairness, AMM's purpose is merely to report that the strange arrangement is a feature which appears at a late stage of development, a state of affairs which he considers a demonstration of recapitulation. Naïve enquirers might ask whether it happens *because* the fish is lying on the sea bed or whether the fish's benthic habit is *facilitated by* the deformation (but see note 232, Lecture 10).

telling us that in its early stages the limpet embryo has a spiral shell, which is lost on the formation subsequently of the conical shell of the adult.

Recapitulation is not confined to the higher groups of animals, and the Protozoa themselves yield most instructive examples. A very striking case is that of Orbitolites, one of the most complex of the procellanous Foraminifera, in which each individual during its own growth and development passes through the series of stages by which the cyclical or discoidal type of shell was derived from the simpler spiral form.

In *Orbitoites tenuissima*, as Dr. Carpenter has shown,[338] *"the whole transition is actually presented during the successive stages of its growth. For it begins life as a Cornuspira, ... its shell forming a continuous spiral tube, with slight interruptions at the points at which its successive extensions commence; while its sarcodic body consists of a continuous coil with slight constrictions at intervals. The second stage consists in the opening out of its spire, and the division of its cavity at regular intervals by transverse septa, traversed by separate pores, exactly as in Peneroplis. The third stage is marked by the subdivision of the "peneropline" chambers into chamberlets, as in the early forms of Orbiculina. And the fourth consists in the exchange of the spiral for the cyclical plan of growth, which is characteristic of Orbitolites; a circular disc of progressively increasing diameter being formed by the addition of successive annular zones around the entire periphery."*

The shells both of Foraminifera and of Mollusca afford peculiarly instructive examples for the study of recapitulation. As growth of the shell is effected by the addition of new shelly matter to the part already existing, the older parts of the shell are retained, often unaltered, in the adult; and in favourable cases, as in *Orbitolites tenuissima*, all the stages of development can be determined by simple inspection of the adult shell.

It is important to remember that the Recapitulation Theory, if valid, must apply not merely in a general way to the development of the animal body, but must hold good with regard to the formation of each organ or system, and with regard to the later equally with the earlier phases of development.[339] Of individual organs, the brain of birds has been already cited. The formation of the vertebrate liver as a diverticulum from the alimentary canal, which is at first simple, but by the folding of its walls becomes greatly complicated, is another good example; as is also the development of the vomer[340] in Amphibians as a series of toothed plates, equivalent morphologically to the placoid scales[341] of fishes, which are at first separate, but later on fuse together and lose the greater number of their teeth.

Concerning recapitulation in the later phases of development and in the adult animal, the mode of renewal of the nails or of the epidermis generally is a good example, each cell commencing its existence in an indifferent form in the deeper

---

338   {M} W. B. Carpenter, "On an Abyssal Type of the Genus Orbitolites," *Phil. Trans.* 1883, part ii. p. 553.

339   This sentence appears to presage a subtle change of direction: from using the Recapitulation Theory to explain observations, to using observations as a test of the theory and its universality. However, the examples that follow are presented as further strange anatomical features which require the theory to explain them.

340   The bone which separates the right and left halves of the nasal cavity.

341   Enamel-covered scales of cartilaginous fish, considered by some as analogous to teeth.

layers of the epidermis, and gradually acquiring the adult peculiarities as it approaches the surface, through removal of the cells lying above it.

The above examples, selected almost haphazard, will suffice to illustrate the Theory of Recapitulation.

The proof of the theory depends chiefly on its universal applicability to all animals, whether high or low in the zoological scale, and to all their parts and organs. It derives also strong support from the ready explanation which it gives of many otherwise unintelligible points.

Of these latter a familiar and most instructive instance is afforded by rudimentary organs—i.e., structures which, like the outer digits of the horse's leg, or the intrinsic muscles of the ear of a man, are present in the adult in an incompletely developed form, and in a condition in which they can be of no use to their possessors—or else structures which are present in the embryo, but disappear completely before the adult condition is attained, for example the teeth of whalebone whales, or the branchial clefts of all higher vertebrates.

Natural Selection explains the preservation of useful variations, but will not account for the formation and perpetuation of useless organs; and rudiments such as those mentioned above would be unintelligible but for Recapitulation, which solves the problem at once, showing that these organs, though now useless, must have been of functional value to the ancestors of their present possessors, and that their appearance in the ontogeny of existing forms is due to repetition of ancestral characters. Such rudimentary organs are, as Darwin pointed out, of larger relative or even absolute size in the embryo than in the adult, because the embryo represents the stage in the pedigree in which they were functionally active.

Rudimentary organs are extremely common, especially among the higher groups of animals, and their presence and significance are now well understood. Man himself affords numerous and excellent examples, not merely in his bodily structure, but by his speech, dress and customs. For the silent letter *b* in the word doubt, or the *w* of answer, or the buttons on his elastic side boots are as true examples of rudiments, unintelligible but for their past history, as are the ear muscles he possesses but cannot use, or the gill-clefts, which are functional in fishes and tadpoles, and are present though useless in the embryos of all higher vertebrates, which in their early stages the hare and the tortoise alike possess, and which are shared with them by cats and by kings.

Another consideration of the greatest importance arises from the study of the fossil remains of the animals that formerly inhabited the earth. It was the elder Agassiz who first directed attention to the remarkable agreement between the embryonic growth of animals and their palaeontological history. He pointed out the resemblance between certain stages in the growth of young fish and their fossil representatives, and attempted to establish, with regard to fish, a correspondence between their palaeontological sequence and the successive stages of embryonic development. He then extended his observations to other groups, and stated his conclusions in these words:[342] *"It may therefore be considered as a general fact, very likely to be more fully illustrated as investigations cover a wider ground, that the phases of development of all*

---

342   {M} L. Agassiz, "Essay on Classification," 1859, p. 115.

living animals correspond to the order of succession of their extinct representatives in past geological times."

This point of view is of the utmost importance. If the development of an animal is really a repetition of its ancestral history, then it is clear that the agreement or parallelism which Agassiz insists on between the embryological and palaeontological records must hold good.[343] Owing to the attitude which Agassiz subsequently adopted with regard to the theory of Natural Selection,[344] there is some fear of his services in this respect failing to receive full recognition, and it must not be forgotten that the sentence I have quoted was written prior to the clear enunciation of the Recapitulation Theory by Fritz Müller.

The imperfection of the geological record has been often referred to and lamented. It is very true that our museums afford us but fragmentary pictures of life in past ages; that the earliest volumes of the history are lost, and that of others but a few torn pages remain to us; but the later records are in far more satisfactory condition. The actual number of specimens accumulated from the more recent formations is prodigious; facilities for consulting them are far greater than they were; the international brotherhood of science is now fully established, and the fault will be ours if the material and opportunities now forthcoming are not rightly and fully utilised.

By judicious selection of groups in which long series of specimens can be obtained, and in which the hard skeletal parts, which alone can be suitably preserved as fossils, afford reliable indications of zoological affinity, it is possible to test directly this correspondence between palaeontological and embryological histories, while in some instances a single lucky specimen will afford us, on a particular point, all the evidence we require.

Great progress has already been made in this direction, and the results obtained are of the most encouraging description. By Alexander Agassiz a detailed comparison was made between the fossil series and the developmental stages of recent forms in the case of the Echinoids, a group peculiarly well adapted for such an investigation. The two records agree remarkably in many respects, more especially in the independent evidence they give as to the origin of the asymmetrical forms from more regular ancestors. The gradually increasing compilation {complication} in some of the historic series is found to be repeated very closely in the development of their existing representatives; and with regard to the whole group, Agassiz concludes

---

343   In *Ontogeny and Phylogeny* (Cambridge, MA: Belknap Press, 1977) Stephen Jay Gould sums up Louis Agassiz's extension of the Recapitulation Theory to include palaeontological evidence thus: *"Before Agassiz, recapitulation had been defined as a correspondence between two series: embryonic stages and adults of living species. Agassiz introduced a third series: the geological record of fossils. An embryo repeats both a graded series of living, lower forms and the history of its type as recorded by fossils. There is a 'threefold parallelism' of embryonic growth, structural gradation, and geologic succession."*

344   This is presumably a reference to Agassiz's advocacy of *orthogenesis*, a popular theory first suggested in the 1870s by Theodor Eimer that evolution is linear, driven by internal forces rather than selective pressures from the environment. This clearly ran counter to Darwin's theory (which AMM espoused) but it enabled Agassiz to rationalize the process of evolution with his idealism and religious beliefs.

that,[345] *"comparing the embryonic development with the paleontological one, we find a remarkable similarity in both, and in a general way there seems to be a parallelism in the appearance of the fossil genera and the successive stages of the development of the Echini."* Neumayr has followed similar lines, and by him, as by {and between him and} other authorities on the group, there seems to be general agreement as to the parallelism between the embryological and palaeontological records, not merely for Echini, but for other groups of Echinodermata as well.

The Tetrabranchiate Cephalopoda[346] are an excellent group in which to study the problem, for though no opportunity has yet occurred for studying the embryology of the only surviving member of the group, the pearly nautilus, yet owing to the fact that growth of the shell is effected by addition of shelly matter to the part already present, and to the additions being made in such manner that the older part of the shell persists unaltered, it is possible, from examination of a single shell—and in the case of fossils the shells are the only part of which we have exact knowledge—to determine all the phases of its growth; just as in the shell of Orbitolites all the stages of development are manifest on inspection of an adult specimen. In such a shell as Nautilus or Ammonites the central chamber is the oldest or first formed one, to which the remaining chambers are added in succession. If therefore the development of the shell is a repetition of ancestral history, the central chamber should represent the palaeontologically oldest form, and the remaining chambers in succession, forms of more and more recent origin. Ammonite shells present, more especially in their sutures, and in the markings and sculpturing of their surface, characters that are easily recognised, and readily preserved in fossils; and the group consequently is a very suitable one for investigation from this standpoint. Würtenberger's admirable and well-known researches[347] have shown that in the Ammonites such a correspondence between historic and embryonic development does really exist; that, for example, in Aspidoceras the shape and markings of the shells in young specimens differ greatly from those of adults, and that the characters of the young shells are those of palaeontologically older forms. Another striking illustration of the correspondence between palaeontological and developmental records is afforded by the antlers of deer, in which the gradually increasing complication of the antler in successive years agrees singularly closely with the progressive increase in size and complexity shown by the fossil series from the Miocene age to recent times.

Of cases where a single specimen has sufficed to prove the palaeontological significance of a developmental character, Archaeopteryx affords a typical example. In recent birds the metacarpals are firmly fused with one another, and with the distal series of carpals; but in development the metacarpals are at first, and for some time, distinct. In Archaeopteryx this distinctness is retained in the adult, showing that what is now an embryonic character in recent birds, was formerly an adult one. Other examples might easily be quoted, but these will suffice to show that the relation between Palaeontology and Embryology, first enunciated by Agassiz, and required by

---

345 {M} A. Agassiz, "Palaeontological and Embryological Development," an Address before the American Association for the Advancement of Science, 1880.

346 Nautilus and related molluscs with four gills.

347 {M} L. Würtenberger, „*Studien über die Stammesgeschichte der Ammoniten. Ein geologischer Beweis far die Darwin'sche Theorie.*" Leipzig, 1880.

the Recapitulation Theory, does in reality exist. There is much yet to be done in this direction. A commencement, a most promising commencement, has been made, but as yet only a few groups have been seriously studied from this standpoint.

It is a great misfortune that palaeontology is not more generally and more seriously studied by men versed in embryology, and that those who have so greatly advanced our knowledge of the early development of animals should so seldom have tested their conclusions as to the affinities of the groups they are concerned with by direct reference to the ancestors themselves, as known to us through their fossil remains. I cannot but feel that for instance the determination of the affinities of fossil Mammalia, of which such an extraordinary number and variety of forms are now known to us, would be greatly facilitated by a thorough and exact knowledge of the development, and especially the later development, of the skeleton in their existing descendants, and I regard it as a reproach that such exact descriptions of the later stages of development should not exist even in the case of our commonest domestic animals.

The pedigree of the horse has attracted great attention and has been worked at most assiduously, and we are now, largely owing to the labours of American palaeontologists, able to refer to a series of fossil forms commencing in the lowest Eocene beds, and extending upwards to the most recent deposits, which show a complete gradation from a more generalised mammalian type to the highly specialised condition characteristic of the horse and its allies, and which may reasonably be regarded as indicating the actual line of descent of the horse. In this particular case, more frequently cited than any other, the evidence is entirely palaeontological. The actual development of the horse has yet to be studied, and it is greatly to be desired that it should be undertaken speedily. Klever's[348] recent work on the development of the teeth in the horse may be referred to as showing that important and unexpected evidence is to be obtained in this way.

A brilliant exception to the statement just made as to the want of exact knowledge of the later development of the more highly organised animals is afforded by the splendid labours of Professor W. K. Parker {Professor Kitchen Parker} whose recent death has deprived zoology of one of her most earnest and single-minded students, and zoologists, young and old alike, of a true and sincere friend. Professor Parker's extraordinarily minute and painstaking investigations into the development of the vertebrate skull rank among the most remarkable of zootomical achievements and afford a rich mine of carefully recorded facts, the full value and bearing of which we are hardly yet able to appreciate.

If further evidence as to the value and importance of the Recapitulation Theory were needed, it would suffice to refer to the influence which it has had on the classification of the animal kingdom. Ascidians and Cirripedes may be quoted as important groups, the true affinities of which were first revealed by embryology; and in the case of parasitic animals the structural modifications of the adult are often so great that, but for the evidence yielded by development, their zoological position could not be determined. It is now indeed generally recognised that in doubtful cases embryology affords the safest of all clues, and that the zoological position of such forms can hardly

---

348   {M} Klever, „Zur Kenntniss der Morphogenese des Equidengebisses," *Morphologisches Jahrbuch*, xv. 1889, p. 308.

be regarded as definitely established unless their development, as well as their adult anatomy, is ascertained.

It is owing to this Recapitulation Theory that Embryology has exercised so marked an influence on zoological speculation. Thus the formation in most, if not in all, animals of the nervous system and of the sense organs from the epidermal layer of the skin, acquired a new significance when it was recognised that this mode of development was to be regarded as a repetition of the primitive mode of formation of such organs; while the vertebral theory of the skull affords a good example of a view, once stoutly maintained, which received its death-blow through the failure of embryology to supply the evidence requisite in its behalf. The necessary limits of time and space forbid that I should attempt to refer to even the more important of the numerous recent discoveries in embryology, but mention may be very properly made here of Sedgwick's determination of the mode of development of the body cavity in Peripatus, a discovery which has thrown most welcome light on what was previously a great morphological puzzle.

We must now turn to another side of the question. Although it is undoubtedly true that development is to be regarded as a recapitulation of ancestral phases, and that the embryonic history of an animal presents to us a record of the race-history, yet it is also an undoubted fact, recognised by all writers on embryology, that the record so obtained is neither a complete nor a straightforward one. It is indeed a history, but a history of which entire chapters are lost, while in those that remain many pages are misplaced and others are so blurred as to be illegible; words, sentences, or entire paragraphs are omitted, and, worse still, alterations or spurious additions have been freely introduced by later hands, and at times so cunningly as to defy detection.

Very slight consideration will show that development cannot in all cases be strictly a recapitulation of ancestral stages. It is well known that closely allied animals may differ markedly in their mode of development. The common frog is at first a tadpole, breathing by gills, a stage which is entirely omitted by the West Indian Hylodes. A crayfish, a lobster, and a prawn are allied animals, yet they leave the egg in totally different forms. Some developmental stages, as the pupa condition of insects, or the stage in the development of a dogfish in which the oesophagus is imperforate, cannot possibly be ancestral stages. Or again, a chick embryo of say the fourth day is clearly not an animal capable of independent existence, and therefore cannot correctly represent any ancestral condition, an objection which applies to the developmental history of many, perhaps of most animals.

Haeckel long ago urged the necessity of distinguishing in actual development between those characters which are really historical and inherited, and those which are acquired or spurious additions to the record. The former he termed palingenetic or ancestral characters, the latter cenogenetic or acquired.[349] The distinction is

---

349   Stephen Jay Gould (*Ontogeny and Phylogeny*, see footnote 343) states that Palingenesis had two interpretations: "*1. For Bonnet, the ontogenetic unfolding of individuals already preformed in the egg—referring especially to the 'end' stage of 'ontogeny,' the resurrection of each body at the end of time from a 'germ of restitution' preformed in the original homunculus within the egg. 2. For Haeckel, the true repetition of past phylogenetic stages in ontogenetic stages of descendants.*" The difference is helpful in understanding AMM's usage. Preformationism had been largely abandoned by the middle

undoubtedly a true one, but an exceedingly difficult one to draw in practice. The causes which prevent development from being a strict recapitulation of ancestral characters, the mode in which these came about, and the influence which they respectively exert, are matters which are greatly exercising embryologists, and the attempt to determine which has as yet met with only partial success.

The most potent and the most widely spread of these disturbing causes arise from the necessity of supplying the embryo with nutriment. This acts in two ways. If the amount of nutritive matter within the egg is small, then the young animal must hatch early, and in a condition in which it is able to obtain food for itself. In such cases there is of necessity a long period of larval life, during which natural selection may act so as to introduce modifications of the ancestral history, spurious additions to the text.

If on the other hand, the egg contain within itself a considerable quantity of nutrient matter, then the period of hatching can be postponed until this nutrient matter has been used up. The consequence is that the embryo hatches at a much later stage of its development, and if the amount of food material is sufficient may even leave the egg in the form of the parent. In such cases the earlier developmental phases are often greatly condensed and abbreviated; and as the embryo does not lead a free existence, and has no need to exert itself to obtain food, it commonly happens that these stages are passed through in a very modified form, the embryo being, as in a four-day chick, in a condition in which it is clearly incapable of independent existence.

The nutrition of the embryo prior to hatching is most usually effected by granules of nutrient matter, known as food yolk, and embedded in the protoplasm of the egg itself; and it is on the relative abundance of these granules that the size of the egg chiefly depends. Large size of eggs implies diminution of number of the eggs, and hence of the offspring; and it can be well understood that while some species derive advantage in the struggle for existence by producing the maximum number of young, to others it is of greater importance that the young on hatching should be of considerable size and strength, and able to begin the world on their own account. In other words, some animals may gain by producing a large number of small eggs, others by producing a smaller number of eggs of larger size—i.e., provided with more food yolk.

---

of the 19th century, if only because evidence from the microscope would not sustain it. Haeckel's more limited meaning refers to the embryological repetition of what AMM calls "ancestral characters" without the implication that they were cryptically present before appearing. Gould also provides two meanings for Cenogenesis: "1. *According to Haeckel, exceptions to the repetition of phylogeny in ontogeny, produced by heterochrony (temporal displacement), heterotopy (spatial displacement), or larval adaptation. 2. According to de Beer, adaptations introduced into juvenile stages that do not affect the subsequent course of ontogeny.*" The British embryologist and evolutionary biologist Sir Gavin de Beer FRS (1899–1972), is quoted by Gould as writing (1930): "*If only the recapitulationists would abandon the assertion that that which is repeated is the* adult *condition of the ancestor, there would be no reason to disagree with them.*" This is precisely the issue which AMM is trying to grapple with here and which, in a variety of ways, occupies him for much of the rest of the lecture.

The immediate effect of a large amount of food yolk is to mechanically retard the processes of development; the ultimate result is to greatly shorten the time occupied by development. This apparent paradox is readily explained. A small egg, such as that of Amphioxus, starts its development rapidly, and in about eighteen hours gives rise to a free swimming larva, capable of independent existence, with a digestive cavity and nervous system already formed; while a large egg like that of the hen, hampered by the great mass of food yolk by which it is distended, has in the same time made but very slight progress. From this time however other considerations begin to tell. Amphioxus has been able to make this rapid start owing to its relative freedom from food yolk. This freedom now becomes a retarding influence, for the larva, containing within itself but a very scanty supply of nutriment, must devote much of its energies to hunting for and to digesting its food, and hence its further development will proceed more slowly.

The chick embryo, on the other hand, has an abundant supply of food in the egg itself; it has no occasion to spend time searching for food, but can devote its whole energies to the further stages of its development. Hence, except in the earliest stages, the chick develops more rapidly than Amphioxus, and attains its adult form in a much shorter time.

The tendency of abundant food yolk to lead to shortening or abbreviation of the ancestral history, and even to the entire omission of important stages, is well known.[350] The embryo of forms well provided with yolk takes short cuts in its development, jumps from branch to branch of its genealogical tree, instead of climbing steadily upwards. Thus the little West Indian frog, Hylodes, produces eggs which contain a larger amount of food yolk than those of the common English frog. The young Hylodes is consequently enabled to pass through the tadpole stage before hatching, to attain the form of a frog before leaving the egg; and the tadpole stage is only imperfectly recapitulated, the formation of gills for instance being entirely omitted.

The influence of food yolk on the development of animals is closely analogous to that of capital in human undertakings. A new industry, for example that of pen-making, has often been started by a man working by hand and alone, making and selling his own wares; if he succeed in the struggle for existence, it soon becomes necessary for him to call in others to assist him, and to subdivide the work; hand labour is soon superseded by machines, involving further differentiation of labour; the earlier machines are replaced by more perfect and more costly ones; factories are built, agents engaged, and in the end a whole army of work-people employed. In later times a man commencing business with very limited means will start at the same level as the original founder, and will have to work his way upwards through much the same stages—i.e., will repeat the pedigree of the industry.

The capitalist on the other hand, is enabled like Hylodes, to omit these earlier stages, and after a brief period of incubation, to start business with large factories equipped with the most recent appliances, and with a complete staff of work people —i.e., to spring into existence fully fledged.

---

350   See footnote 249, Lecture 10, for a discussion of this strange way of looking at the effect of yolk.

There is no doubt that abundance of food yolk is a direct and very frequent cause of the omission of ancestral stages from individual development; but it must not be viewed as {the} sole cause. It is quite impossible that any animal, except perhaps in the lowest zoological groups, should repeat all the ancestral stages in the history of the race; the limits of time available for individual development will not permit this. There is a tendency in all animals towards condensation of the ancestral history, towards striking a direct path from the egg to the adult.[351] This tendency is best marked in the higher, the more complicated members of a group—*i.e.*, in those which have a longer and more tortuous pedigree—and, though greatly strengthened by the presence of food yolk in the egg, is apparently not due to this in the first instance.

Thus the simpler forms of Orbitolites, as O. *tenuissima,* repeat in their development all the stages leading from a spiral to a cyclical shell; but in the more complicated species, as Dr. Carpenter has pointed out, there is a tendency towards precocious development of the adult characters, the earlier stages being hurried over in a modified form; while in the most complex examples, as in O. *complanata* the earlier spiral stages may be entirely omitted, the shell acquiring almost from its earliest commencement the cyclical mode of growth. There is no question here of relative abundance of food yolk, but merely of early or precocious appearance of adult characters.

The question of the relations and influence of food yolk, involving as it does the larger or smaller size of the egg, is however merely a special side of the much wider question of the nutrition of the embryo, one of the most potent of the disturbing elements affecting development. Speaking generally, we may say that large eggs are more often met with in the higher than the lower groups of animals. Birds and Reptiles are cases in point, and if Mammals do not now produce large eggs, it is because a more direct and more efficient mode of nourishing the young by the placenta has been acquired by the higher forms, and has replaced the food yolk that was formerly present, and is now retained in quantity by Monotremes alone. Molluscs afford another good example, the eggs of Cephalopoda being of larger size than those of the less highly organised groups. The large size of the eggs of Elasmobranchs, and perhaps that of Cephalopods also, may possibly be associated with the carnivorous habits of the animals; for it is of importance that forms which prey on other animals should hatch of considerable size and strength.

The influence of habitat must also be considered. It has long been noticed as a general rule that marine animals lay small eggs, while their freshwater allies have eggs of much larger size. The eggs of the salmon or trout are much larger than those of the cod or herring; and the crayfish, though only a quarter the length of a lobster, lays eggs of actually larger size.[352] This larger size of the eggs of fresh-water forms appears to be dependent on the nature of the environment to which they are exposed. Considering the geological instability of the land as compared with the ocean, there

---

351   Condensation—the continual shortening or compression of recapitulated ontological stages—was accepted by all recapitulationists from Haeckel's time onwards as the only way of explaining why the development of more complex organisms did not come with the penalty of ever-lengthening periods of embryogenesis.

352   One wonders whether these examples are just selectively fortuitous illustrations of the principle and whether it would really be borne out by a more extensive survey.

can be no doubt that fresh-water fauna {are}, speaking generally, derived from the marine fauna; and the great problem with regard to fresh-water life is to explain why it is that so many groups of animals which flourish abundantly in the sea should have failed to establish themselves in fresh water. Sponges and Coelenterates abound in the sea, but their fresh-water representatives are extremely few in number; Echinoderms are exclusively marine; there are no fresh-water Cephalopods, and no Ascidians; and of the smaller groups of Worms, Molluscs, and Crustaceans, there are many that do not occur in fresh water.

Direct experiment has shown that in many cases this distribution is not due to inability of the adult animals to live in fresh water; and the real explanation appears to be that the early larval stages are unable to establish themselves under such conditions. This interesting suggestion, which has been worked out in detail by Professor Sollas,[353] undoubtedly affords an important clue. To establish itself permanently in fresh water an animal must either be fixed, or else be strong enough to withstand and make headway against the currents of the streams or rivers it inhabits, for otherwise it will in the long run be swept out to sea, and this consideration applies to larval forms equally with adults.

The majority of marine Invertebrates leave the egg as minute ciliated larvae, and such larvae are quite incapable of holding their own in currents of any strength. Hence it is only forms which have got rid of the free swimming ciliated larval stage, and which leave the egg of considerable size and strength, that can establish themselves as freshwater animals. This is effected most readily by the acquisition of food yolk—hence the large size of the eggs of fresh-water animals—and is often supplemented, as Sollas has shown, by special protective devices of a most interesting nature. For this reason fresh-water forms are not so well adapted as their marine allies for the study of ancestral history as revealed in larval or embryonic development.

Before leaving the question of food yolk, reference must be made to the proposal of the brothers Sarasin,[354] to regard the yolk cells as forming a distinct embryonic layer, the lecithoblast,[355] distinct from the blastoderm. I do not desire to speak dogmatically on a point the full bearings of which are not yet apparent, but I venture to think that this suggestion will not commend itself to embryologists. The distinction between the yolk granules and the cells in which they are embedded is a real and fundamental one; but I see no reason for regarding the yolk cells as other than originally functional endoderm cells in which yolk granules have accumulated to such an extent that they have in extreme cases become devoted solely to the storing of food for the embryo.[356]

Of all the causes tending to modify development, tending to obscure or falsify the ancestral record, food yolk is the most frequent and the most important; its position in the egg determines the mode of segmentation, and its relative abundance

---

353 {M} W. J. Sollas, " On the Origin of Freshwater Fauna," *Scientific Transactions of the Royal Dublin Society, vol. iii. Ser. II, 1886.*

354 Although frequently referred to thus, they were in fact second cousins.

355 {M} P. and F. Sarasin, "Ergebnisse naturwissenschaftlicher Forschunge auf Ceylon," Bd. ii. Heft iii. 1889.

356 {M} *Cf.* E. B. Wilson, "The Development of Renilla," *Phil. Trans,* 1883, p. 755.

affects profoundly the entire embryonic history, and decides at what particular stage, and of what size and form, the embryo shall hatch.

The loss of food yolk is another disturbing element, the full influence of which is as yet imperfectly understood, but the possibility of which must be always kept in mind. It is best known in the case of mammals, where it has led to apparent, though very deceptive, simplification of development; and it will probably not be until the embryology of the large-yolked monotremes is at length described, that we shall fully understand the formation of the germinal layers in the higher placental mammals. Amongst invertebrates we know but little as yet concerning the effects of loss of food yolk. It has been suggested that the extraordinary nature of the segmentation of the egg of *Peripatus capensis* made known to us through Mr. Sedgwick's admirable researches, may be due to loss of food yolk: a suggestion which receives support from the long duration of uterine development in this case. Our knowledge is very imperfect as to the ease with which food yolk may be acquired or lost; but until our information is more precise on this point it seems unwise to lay much stress on suggested pedigrees which involve great and frequent alternations in the amount of food yolk present.

Of causes other than food yolk, or only indirectly connected with it, which tend to falsify the ancestral history, many are now known, but time will only permit me to notice the more important. These are distortion, whether in time or space; sudden or violent metamorphosis; a series of modifications, due chiefly to mechanical causes, and which may be spoken of as developmental conveniences; the important question of variability in development; and finally the great problem of degeneration.[357]

Concerning distortions in time, all embryologists have noticed the tendency to anticipation or precocious development of characters which really belong to a later stage in the pedigree. The early attainment of the cyclical form in the shell of *Orbitolites complanata* is a case in point; and Würtenberger has specially noticed this tendency in Ammonites. Many early larvae show it markedly, the explanation in this case being that it is essential for them to hatch in a condition capable of independent existence- *i.e.*, capable at any rate of obtaining and digesting their own food.

Anachronisms, or actual reversal of the historical order of development of organs, or parts, occur frequently. Thus the joint surfaces of bones acquire their characteristic curvatures before movement of one part or another is effected, and before even the joint cavities are formed. Another good example is afforded by the development of the mesenterial filaments in Alcyonarians. Wilson has shown in the case of Renilla that in the development of an embryo from the egg the six endodermal filaments appear first, and the two long ectodermal filaments at a later period; but that in the formation of a bud this order of development is reversed, the ectodermal filaments being the first formed. He suggests in explanation, that as the endodermal filaments are the digestive organs, it is of primary importance to the free embryo that they should be formed quickly. The long ectodermal filaments are chiefly concerned with maintaining currents of water through the colony; in bud-development they

---

357   AMM lists here a series possible reasons why the observed ontogeny of an organism may appear not to recapitulate its phylogeny (its ancestry). The overall strategy is clear: finding reasons why recapitulation may not be evident, rather than evidence against recapitulation itself.

appear before the endodermal filaments, because they enable the bud during its early stages to draw nutrient matter from the body fluid of the parent; while the endodermal filaments cannot come into use until the bud has acquired both mouth and tentacles. The completion of the ventricular septum in the heart of higher vertebrates before the auricular septum is a well-known anachronism, and every embryologist could readily furnish many other cases. A curious instance is afforded by the development of the teeth in mammals, if recent suggestions as to the origin of the milk dentition are confirmed, and the milk dentition prove to be a more recent acquisition than the permanent one.[358,359]

But the most important case in reference to distortion in time concerns the reproductive organs. If development were a strict and correct recapitulation of ancestral history, then each stage would possess reproductive organs in a mature condition. This is not the case, and it is clearly of the greatest importance that it should not be. It is true that the first commencement of the reproductive organs may occur at a very early {larval} stage, or even that the very first step in development may be a division of the egg into somatic and reproductive cells; and it is possible that, as maintained by Weismann, this latter condition is a primitive one. Still even in these cases the reproductive organs merely commence their development at these early stages, and do not become functional until the animal is adult.

Exceptionally in certain animals, and as a normal occurrence in others, precocious maturation of the reproductive organs takes place, and a larval form becomes capable of sexual reproduction. This may lead to arrest of development, either at a late larval period as in the Axolotl,[360] or at successively earlier and earlier stages, as in the gonophores of the Hydromedusae, until finally the extreme condition seen in Hydra is produced. We do not know the causes that determine the period, whether late or early, at which the reproductive organs ripen, but the question is one of great interest and importance, and deserves careful attention. The suggestion has been made that entire groups of animals, such as the Mesozoa,[361] are merely larvae, arrested through such precocious acquiring of reproductive power, and it is conceivable that this may be the case. Mesozoa are a puzzling group in which the life history, though

---

358 {M} *Cf.* Thomas Oldfield, "On the Homologies and Succession of the Teeth in the Dasyuridae, with an attempt to trace the history of the evolution of the Mammal and Teeth in general," *Phil. Trans.* 1887.

359 See footnote 257, Lecture 10.

360 Some 30 years earlier (around 1862) a Parisian zoologist, August Duméril, had caused the metamorphosis of water-living, externally gilled axolotls into internally lunged *salamanders* (called ambystoma) experimentally by maintaining them in restricted quantities of water. Although they proved to be non-reproductive, there was much debate at the time and subsequently as to whether this represented the direct transformation of one species into another. Weismann, who partially corroborated Duméril's result, resisted this view, concluding that there was no change of species, merely an unexplained sterility in a more developed form of the same animal. AMM's description of the axolotl as being in a state of *arrested development*, albeit fertile, indicates how the uncertainty had by his time become resolved.

361 A polyphyletic collection of worm-like parasites of invertebrates, of uncertain taxonomy but possibly related to platyhelminthes.

known with tolerable completeness, has as yet given us no reliable clue concerning their affinities to other animals, a tantalising distinction that is shared with them by Rotifers and Polyzoa.

Distortion of a curious kind is seen in cases of abrupt metamorphosis, where, as in the case of many Echinoderms, of Phoronis, and of the metabolic[362] insects, the larva and the adult differ greatly in form, habits, mode of life, and very usually in the nature of their food and the mode of obtaining it; and the transition from one stage to the other is not a gradual but an abrupt one, at any rate so far as external characters are concerned. Sudden changes of this kind, as from the free swimming Pluteus to the creeping Echinus, or from the sluggish leaf-eating caterpillar to the dainty butterfly, cannot possibly be recapitulatory, for even if small jumps are permissible in nature, there is no room for bounds forward of this magnitude. Cases of abrupt metamorphosis may always be viewed as due to secondary modifications, and rarely, if ever, have any significance beyond the particular group of animals concerned. For example, a Pluteus larva may be recognised as belonging to the group of Echinoidea before the adult urchin has commenced to be formed within it, and the Lepidopteran caterpillar is already an unmistakable insect. Hence, for the explanation of the meta-morphoses in these cases it is useless to look outside the groups of Echinoidea and Insecta respectively.

Abrupt metamorphosis is always associated with great change in external form and appearance, and in mode of life, and very usually in mode of nutrition. A gradual transition in such cases is inadmissible, because in the intermediate stages the animal would be adapted to neither the larval nor the adult condition; a gradual conversion of the biting mouth parts of the caterpillar to the sucking proboscis of a moth would inevitably lead to starvation. The difficulty is evaded by retaining the external form and habits of one particular stage for an unduly long period, so that the relations of the animal to the surrounding environment remain unchanged, while internally preparations for the later stages are in progress. Cinderella and the princess are equally possible entities, each being well adapted to her environment. The exigences of the situation do not permit however of a gradual change from one to the other: the transformation, at least as regards external appearance, must be abrupt.

KIeinenberg has recently directed attention to cases in which the larval and adult organs develop independently—the larval nervous system for instance, aborting completely and forming no part of that of the adult. I am not sure that I fully under-stand Kleinenberg's argument, but it seems very possible that such cases, which are probably far more numerous than is yet admitted, may be due to what may be termed the telescoping of ancestral stages one within another,[363] which takes place in actual development, and may accordingly be grouped under the head of develop-mental convenience. Undue prolongation of an early ancestral stage, as in cases of abrupt metamorphosis, must involve modification, especially in the muscular and nervous systems; in such cases a telescoping of ancestral stages takes place as we

---

362   See footnote 259, Lecture 10.

363   This is difficult to imagine but seems to suggest a form of condensation which goes beyond the mere compression assumed by all recapitulationists (see Lecture 10 and Interlude; discussed in detail by Gould and note 349, Lecture 13)—a kind of Russian doll arrangement.

have seen, the adult being developed within the larva. Such telescoping must distort the recapitulatory history, and as the shape of the larva and adult may differ widely, an independent origin of organs, especially the muscular and nervous systems, may be acquired secondarily.

The stage in the development of Squilla,[364] in which the three posterior maxillipedes disappear completely, to reappear at a later stage in a totally different form, is not to be interpreted as meaning that the adult maxillipedes are entirely new structures unconnected historically with those of the larva. Neither is the annual shedding of the antlers of deer to be regarded as the repetition of an ancestral hornless condition intercalated historically between successive stages provided with antlers. In both cases the explanation is afforded by convenience, whether of the embryo or adult.

Many embryological modifications or distortions may be attributed to mechanical causes, and may fairly be considered under the head of developmental conveniences. The amnion of higher vertebrates is a case in point, and is probably rightly explained as due in the first instance to sinking or depression of the embryo into the yolk, in order to avoid distortion through pressure against a hard unyielding eggshell. A similar device is employed, presumably for the same reason, in the early development of many insect embryos; and the depression of the Taenia head within the cyst is a phenomenon of very similar nature.

Restriction of the space within which development occurs often causes displacement or distortion of organs whose growth, restricted in its normal direction, takes place along the lines of least resistance. The telescoping of the limbs and other organs within the body of an insect larva is a simple case of such distortion; and a more complicated example, closely comparable in many ways to the invagination of the Taenia head, is afforded by the remarkable inversion of the germinal layers in Rodents, first described by Bischoff in the guinea pig, and long believed to be peculiar to that animal, but subsequently and simultaneously discovered by three independent observers, Kupffer, Selenka, and Fraser, to occur in varying degrees in rats, mice, and in other rodents.[365]

One of the most recent attempts to explain developmental peculiarities as due to mechanical causes is Mr. Dendy's suggestion with regard to the pseudogastrula stage in the development of calcareous sponges. It is well known that while the larva is in the amphiblastula stage, and still imbedded in the tissues of the parent, the granular cells become invaginated within the ciliated cells, giving rise to the pseudogastrula stage. At a slightly later stage, when the larva becomes free, the invaginated granular cells become again everted, and the larva spherical in shape; while still later invagination occurs once more, the ciliated cells being this time invaginated within the granular cells. The significance of the pseudogastrula stage has hitherto been undetermined, but Mr. Dendy points out that the larva always occupies a definite position with reference to the parental tissues; that the ciliated half of the larva is covered by a soft and yielding wall, while the opposite half, composed of the granular

---

364 Mantis Shrimp.

365 Fraser's identity is unknown but he reported incomplete studies to the British Association in 1882 (*Nature* **26**: 493). The matter of germinal layer inversion is discussed in Wilson JT, On the question of the interpretation of the structural features of the early blastocyst of the guinea-pig. *Journal of Anatomy* 1928; **62**: 346–358.

cells, is covered by a layer stiffened with rigid spicules; and his observations on the growth of the larva lead him to think that the pseudogastrula stage is brought about mechanically by flattening of the granular cells through pressure against this rigid wall of spicules.

Embryology supplies us with many unsolved problems, and it is not to be wondered at that this should be the case. Some of these may fairly be spoken of as mere curiosities of development, while others are clearly of greater moment. I do not propose to catalogue these, but will merely mention two or three which I happen to have recently run my head against and remember vividly. The solid condition of the oesophagus, in Elasmobranch embryos, first noticed by Balfour, is a very curious point. The oesophagus has at first a well-developed lumen, like the rest of the alimentary canal; but at an early period, stage K of Balfour's nomenclature, the part of the oesophagus overlying the heart, and immediately behind the branchial region, becomes solid, and remains solid for a long time, the exact date of reappearance of the lumen not being yet ascertained. Mr. Bles and myself have recently noticed that a similar solidification of the oesophagus occurs in tadpoles of the common frog.[366] In young free swimming tadpoles the oesophagus is perforate, but in tadpoles of about 7 mm. length it becomes solid and remains so until a length of about 10½ mm. has been attained. The solidification occurs at a stage closely corresponding with that in which it first appears in the dogfish, and a curious point about it is that in the frog the oesophagus becomes solid just before the mouth-opening is formed, and remains solid for some little time after this important event. This closing of the oesophagus clearly cannot be recapitulation, but the fact that it occurs at corresponding periods in the frog and dogfish suggests that it may possibly, as Balfour hinted, *"turn out to have some unsuspected morphological bearing."*

Another developmental curiosity is the duplication of the gill slits by growth downwards of tongues from their dorsal margins; a duplication which is described as occurring in Amphioxus and in Balanoglossus, but in no other animal; and the occurrence of which, in apparently closely similar fashion, is one of the strongest arguments in favour of a real affinity between these two forms. It is hardly possible that such a modification should have been acquired independently twice over.

A much more litigious question is the significance of the neurenteric canal of vertebrates, that curious tubular communication between the central canal of the nervous system and the hinder end of the alimentary canal that is conspicuously present in the embryos of lower vertebrates, and retained in a more or less disguised condition in the higher groups as well. The neurenteric canal was discovered by that famous embryologist Kowalevsky in Ascidians and in Amphioxus. He drew special attention to the occurrence of a stage in both Ascidians and in Amphioxus in which the larva is free swimming and in which the sole communication between the alimentary cavity and the exterior is through the neurenteric canal and the central

---

366   Bles and AMM presented a joint paper entitled "On Variability in Development" to the same meeting at which this President's Address was given. Edward Bles's status at the time is unclear; he became a junior demonstrator in AMM's department some two years after the address was presented, and his BSc degree also came later (from Cambridge University). It is likely that he was a medical student.

canal of the nervous system; and suggested[367] that animals may have existed or may still exist in which the nerve tube fulfilled a non-nervous function, and possibly acted as part of the alimentary canal; a suggestion that has recently been revived in a somewhat extravagant form. A passage of food particles into the alimentary cavity through the neural tube has not yet been seen and probably does not occur, as the larva still possesses sufficient food yolk to carry on in its development. It is therefore permissible to hold that the neurenteric canal may be a mere embryological device, and devoid of any deep morphological significance.[368]

The question of variation in development is one of very great importance, and has perhaps not yet received the attention it deserves. We are in some danger of assuming tacitly that the mode of development of allied animals will necessarily agree in all important respects or even in details, and that if the development of one member of a group be known, that of the others may be assumed to be similar. The more recent progress of embryology is showing us that such inferences are not safe, and that in allied genera or species, or even in different individuals of the same species, variations of development may occur affecting important organs and at almost any stage in their formation.[369]

Great individual variations in the earliest processes of development—i.e., the segmentation of the egg—have been described by different writers. In Renilla, Wilson found an extraordinary range of variation in the segmentation of eggs from which apparently identical embryos were produced. In some cases the egg divided into two in the norm manner; in other cases it divided at once into eight, sixteen, or thirty-two segments, which in different specimens were approximately equal or markedly unequal in size. Sometimes a preliminary change of form occurred without any further result, the egg returning to its spherical shape, and pausing for a time before recommencing the attempt to segment. Segmentation sometimes commenced at one pole, as in telolecithal eggs, with the formation of four or five small segments, the rest of the egg breaking up later, either simultaneously or progressively, into segments about equal in size to those first formed; while lastly, in some instances segmentation was very irregular, following no apparent law. It is noteworthy that the variability in the case of Renilla is apparently confined to the earliest stages, for whatever the mode of segmentation, the embryos in their later stages were indistinguishable from one another. Similar modifications in the segmentation of the egg have been described in the oyster by Brooks, in Anodon and other Mollusca, in Hydra, and in Lumbricus, in which last Wilson has recently shown that marked differences occur in the eggs even of the same individual animal. In the different species of Peripatus there appear also to be considerable variations in the details of segmentation.

In the early embryonic stages after the completion of segmentation very considerable variation may occur in allied species or genera. Among Coelenterates for

---

367   {M} A. Kowalevsky, "Weitere Studien über die Entwickelungsgeschichte des Amphioxus lanceolatus," Archiv. fur mikroskopisch Anatomie, Bd. xiii. 1877, p. 201.

368   A current view is that the neurenteric canal, as a transient connection between the amniotic sac and the yolk sac, serves to equilibrate the pressures in the two cavities as they change size.

369   This paragraph and the examples which follow clearly imply that simple recapitulation is wanting as a basis for understanding developmental differences.

instance the mode of formation of the hypoblast presents most perplexing modifications: it may arise as a true gastrula invagination; as cells budded off from one pole
of the blastula into its cavity; as cells budded off from various parts of the wall of the
blastula; by delamination or actual division of each cell of the blastula wall; or it may
be present from the first {start} as a solid mass of cells enclosed by the epiblast cells.
It is in connection with these variations that controversy has arisen as to the primitive
mode of development of the gastrula, a point to which I shall return later on.

Among the higher Metazoa or Coelomata the extraordinary modifications in
the position and in every conceivable detail of formation of the mesoblast in different
and often in closely allied forms have given rise to ardent discussion, and have led
to the proposal of theory after theory, each rejected in turn as only affording a partial
explanation, and now culminating in Kleinenberg's protest against the use of the
term mesoblast at all, at any rate in a sense implying any possibility of comparison
with the primary layers, epiblast and hypoblast, of Coelenterata.

This is not the place to attempt to decide so difficult and technical a point, even
were I capable of so doing; but we may well take warning from this extraordinary
diversity of development, the full extent of which I believe we as yet realise most
imperfectly, that in our attempts to reconstruct ancestral history from ontogenetic
development we have taken in hand no light task. To reconstruct Latin from modern
European languages would in comparison be but child's play.

Of the readiness with which special developmental characters are acquired by
allied animals the brothers Sarasin[370] have given us evidence in the extraordinary modifications presented by the embryonic and larval respiratory organs of Amphibians.
Confining ourselves to those forms which do not lay their eggs in water, and in which
consequently development takes place within the egg, we find that Ichthyophis and
Salamandra have three pairs of specially modified external gills. Nototrema has two
pairs; Alytes and Typhlonectes have only a single pair, which in the latter genus take
the form of enormous leaf-like outgrowths from the sides of the neck. In Hylodes
and Pipa there are no gills, the tail acting as the larval respiratory organ; and in *Rana
opisthodon,* according to Boulenger, larval respiration is effected by nine pairs of folds
of the skin of the ventral surface of the body.

Most of these extraordinarily diversified organs are clearly secondarily acquired
structures; it is possible that they all are, and that external gills, as was suggested
by Balfour for Elasmobranchs, are to be regarded as embryonal respiratory organs
acquired by the larva and of no ancestral value. The point however cannot be considered settled, for on this view the external gills of Elasmobranchs and Amphibians
would be independently acquired and not homologous structures, a view contradicted
by the close agreement in their relations in the two groups, as well as by the absence
of any real break between external and internal gills in Amphibians.

It is well known that the frog and the newt differ greatly in important points of
their development. The two-layered condition of the epiblast in the frog is a marked
point of difference, which involves further changes in the mode of formation of the
nervous system and sense organs. The kidneys and their ducts differ considerably in
their development in the two forms, as do also the blood-vessels. Concerning the early

---

370 {M} P. and F. Sarasin "Ergebnisse naturvissenschaftlicher Forschungen auf Ceylon,"
vol. ii. chap. i. pp. 24–38.

development of the blood-vessels, there are considerable differences even between allied species of frogs. In *Rana esculenta* Maurer finds that there is at first in each branchial arch a single vessel or aortic arch, running directly from the heart to the aorta: from the cardiac end of this aortic arch a vessel grows out into the gill as the afferent bronchial vessel, the original aortic arch losing its connection with the heart, and becoming the efferent branchial vessel. Afferent and efferent branchial vessels become connected by capillaries in the gill, and the course of the circulation, so long as gill-breathing is maintained, is from the heart through the truncus arteriosus to the afferent branchial vessel, then through the gill capillaries to the efferent branchial vessel, and then on to the aorta. When the pulmonary circulation is thoroughly established the bronchial circulation is cut off by the efferent vessel reacquiring its connection with the heart, when the blood naturally takes the direct passage along it to the aorta, and so escapes the gill capillaries.

In *Rana temporaria* the mode of development is very different: the afferent and efferent vessels arise in each arch independently and almost simultaneously; the afferent vessel soon acquires connection with the heart, but unlike *R. esculenta*, the efferent vessel has no connection with the heart until the gills are about to atrophy. In other words the continuous aortic arch, from heart to aorta, is present in *R. esculenta* prior to the development of the gills; it becomes interrupted while the gills are in functional use, but is re-established when these begin to atrophy. In *R. temporaria*, on the other hand, there is no continuous aortic arch until the gills begin to atrophy.

The difference is an important one, for it is a matter of considerable morphological interest to determine whether the continuous aortic arch is primitive for vertebrates—*i.e.*, whether it existed prior to the development of gills. This point could be practically settled if we could decide which of the two frogs, *R. esculenta* and *R. temporaria*, has most correctly preserved its ancestral history in this respect.[371] About this there can be little doubt. The development of the vessels in the newts, a less modified group than the frogs, agrees with that of *R. esculenta*, and interesting confirmation is afforded by a single aberrant specimen of *R. temporaria*, in which Mr. Bles and myself found the vessels developing after the type of *R. esculenta*—*i.e.*, in which a complete aortic arch was present before the gills were formed. We are therefore justified in concluding that as regards the development of the branchial blood-vessels, *R. esculenta* has retained a primitive ancestral character which is lost in *R. temporaria;* and it is interesting to note that were our knowledge of the development of amphibians confined to the common frog, the most likely form to be studied, we should in all probability have been led to wrong conclusions concerning the ancestral condition of the blood-vessels in a point of considerable importance.

A matter which at present is attracting much attention is the question of degeneration. Natural selection, though consistent with and capable of leading to steady upward progress and improvement, by no means involves such progress as a necessary consequence. All it says is that those animals will, in each generation, have the best chance of survival which are most in harmony with their environment, and such animals will not necessarily be those which are ideally the best or most perfect.

If you go into a shop to purchase an umbrella, the one you select is by no means necessarily that which most nearly approaches ideal perfection, but the one which

---

371   Surely, this a circular argument?

best hits off the mean between your idea of what an umbrella should be and the amount of money you are prepared to give for it: the one in fact that is on the whole best suited to the circumstances of the case or the environment for the time being. It might well happen that you had a violent antipathy to a crooked handle, or else were determined to have a catch of a particular kind to secure the ribs, and this might lead to the selection—i.e., the survival—of an article that in other and even in more important respects was manifestly inferior to the average.

So is it also with animals: the survival of a form that is ideally inferior is very possible. To animals living in profound darkness the possession of eyes is of no advantage, and forms devoid of eyes would not merely lose nothing thereby, but would actually gain, inasmuch as they would escape the dangers that might arise from injury to a delicate and complicated organ. In extreme cases, as in animals leading a parasitic existence, the conditions of life may be such as to render locomotor, digestive, sensory, and other organs entirely useless; and in such cases those forms will be best in harmony with their surroundings which avoid the waste of energy resulting from the formation and maintenance of these organs.

Animals which have in this way fallen from the high estate of their forefathers, which have lost organs or systems which their progenitors possessed, are commonly called degenerate. The principle of degeneration, recognised by Darwin as a possible, and under certain conditions a necessary consequence of his theory of natural selection, has been since advocated strongly by Dohrn, and later by Lankester in an Evening Discourse delivered before the Association at the Sheffield Meeting in 1879. Both Dohrn and Lankester suggested that degeneration occurred much more widely than was generally recognised.

In animals which are parasitic when adult, but free swimming in their early stages, as in the case of the Rhizocephala whose life-history was so admirably worked out by Fritz Müller, degeneration is clear enough. So also is it in the case of the solitary Ascidians, in which the larva is a free swimming animal with a notochord, an elongated tubular nervous system, and sense organs, while the adult is fixed, devoid of the swimming tail, with no notochord, and with a greatly reduced nervous system and aborted sense organs. In such cases the animal when adult is, as regards the totality of its organisation, at a distinctly lower morphological level, is less highly differentiated than it is when young, and during individual development there is actual retrograde development of important systems and organs.

About such cases there is no doubt; but we are asked to extend the idea of degeneration much more widely. It is urged that we ought not to demand direct embryological evidence before accepting a group as degenerate. We are reminded of the tendency to abbreviation or to complete omission of ancestral stages of which we have quoted examples above; and it is suggested that if such larval stages were omitted in all the members of a group we should have no direct evidence of degeneration in a group that might really be in an extremely degenerate condition. Supposing for instance the free larval stages of the solitary Ascidians were suppressed, say through the acquisition of food yolk, then it is urged that the degenerate condition of the group might easily escape detection. The supposition is by no means extravagant; food yolk varies greatly in amount in allied animals, and cases like Hylodes, or amongst Ascidians Pyrosoma, show how readily a mere increase in the amount of food yolk in the egg may lead to the omission of important ancestral stages.

The question then arises whether it is not possible, or even probable, that animals which now show no indication of degeneration in their development are in reality highly degenerate, and whether it is not legitimate to suppose such degeneration to have occurred in the case of animals whose affinities are obscure or difficult to determine. It is more especially with regard to the lower vertebrates that this argument has been employed; and at the present day zoologists of authority, relying on it, do not hesitate to speak of such forms as Amphioxus and the Cyclostomes as degenerate animals, as wolves in sheep's clothing, animals whose simplicity is acquired and deceptive rather than real and ancestral. I cannot but think that cases such as these should be regarded with some jealousy. There is at present a tendency to invoke degeneration rather freely as a talisman to extricate us from morphological difficulties; and an inclination to accept such suggestions, at any rate provisionally, without requiring satisfactory evidence in their support.

Degeneration of which there is direct embryological evidence stands on a very different footing from suspected degeneration, for which no direct evidence is forthcoming; and in the latter case the burden of proof undoubtedly rests with those who assume its existence. The alleged instances among the lower vertebrates must be regarded particularly closely, because in their case the suggestion of degeneration is admittedly put forward as a means of escape from difficulties arising through theoretical views concerning the relation between vertebrates and invertebrates. Amphioxus itself, so far as I can see, shows in its development no sign of degeneration, except possibly with regard to the anterior gut diverticula, whose ultimate fate is not altogether clear. With regard to the earlier stages of development, concerning which, thanks to the patient investigations of Kowalevsky and Hatschek, our knowledge is precise, there is no animal known to us in which the sequence of events is simpler or more straightforward. Its various organs and systems are formed in what is recognised as a primitive manner; and the development of each is a steady upward progress towards the adult condition. Food yolk, the great cause of distortion in development, is almost absent, and there is not the slightest indication of the former possession of a larger quantity. Concerning the later stages our knowledge is incomplete, but so much as has been ascertained gives no support to the suggestion of general degeneration.

Our knowledge of the conditions leading to degeneration is undoubtedly incomplete, but it must be noticed that the conditions usually associated with degeneration do not occur. Amphioxus is not parasitic, is not attached when adult, and shows no evidence of having formerly possessed food yolk in quantity sufficient to have led to the omission of important ancestral stages. Its small size, as compared with other vertebrates, is one of the very few points that can be referred to as possibly indicating degeneration, and will be considered more fully at a later point in my address.

A consideration of much less importance, but deserving of mention, is that in its mode of life Amphioxus not merely differs as already noticed from those groups of animals which we know to be degenerate, but agrees with some at any rate of those which there is reason to regard as primitive or persistent types. Amphioxus, like Balanoglossus, Lingula, Dentalium, and Limulus, is marine, and occurs in shallow water, usually with a sandy bottom, and, like the three smaller of these genera, it lives habitually buried almost completely in the sand, into which it burrows with great rapidity.

I do not wish to speak dogmatically. I merely wish to protest against a too ready assumption of degeneration; and to repeat that so far as I can see, Amphioxus has not yet, either in its development, in its structure, or in its habits, been shown to present characters that suggest, still less that prove, the occurrence in it of general or extensive degeneration. In a sense, all the higher animals are degenerate; that is they can be shown to possess certain organs in a less highly developed condition than their ancestors, or even in a rudimentary state. Thus a crab as compared with a lobster is degenerate in the matter of its tail, a horse as compared with Hipparion in regard to its outer toes; but it is neither customary nor advisable to speak of a crab as a degenerate animal compared to a lobster; to do so would be misleading. An animal should only he spoken of a degenerate when the retrograde development is well marked, and has affected not one or two organs only, but the totality of its organization.[372]

It is impossible to draw a sharp line in such cases, and to limit precisely the use of the term degeneration. It must be borne in mind that no animal is at the top of the tree in all respects. Man himself is primitive as regards the number of his toes, and degenerate in respect to his ear muscles; and between two animals even of the same group it may be impossible to decide which of the two is to be called the higher and which the lower form. Thus to compare an oyster with a mussel: the oyster is more primitive than the mussel as regards the position of the ventricle of the heart and its relations to the alimentary canal; but is more modified in having but a single adductor muscle, and almost certainly degenerate in being devoid of a foot.

Care must also be taken to avoid speaking of an animal as degenerate in regard to a particular organ merely because that organ is less fully developed than in allied animals. An organ is not degenerate unless its present possessor has it in a less perfect condition than its ancestors had. A man is not degenerate in the matter of the length of his neck as compared with a giraffe, nor as compared with an elephant in respect of the size of his front teeth, for neither elephant nor giraffe enters into the pedigree of man. A man is however degenerate, whoever his ancestors may have been, in regard to his ear muscles; for he possesses these in a rudimentary and functionless condition, which can only be explained by descent from some better equipped progenitor.

Closely connected with the question of degeneration is that of the size of animals,[373] and its bearing on their structure and development—a problem noticed by many writers, but which has perhaps not yet received the attention it merits.

If we are right in interpreting the eggs of Metazoa as representing the unicellular or protozoon stage in their ancestry, then the small size of the egg may be viewed as recapitulatory. But the gradual increase in size of the embryo, and its growth up to the adult condition, can only be regarded as representing in a most general way, if at all, the actual or even the relative sizes of the intermediate ancestral stages of the pedigree.

---

372   This paragraph is notable for the way it moves the application of the word *degenerate* from the loss of an organ or feature to a general description of the whole organism. The transfer is noted explicitly two paragraphs further on, after a curious elision between degeneracy and hierarchical position.

373   See Lecture 8.

It is quite true that animals belonging to the lower groups are, as a general rule, of smaller size than those of higher grade; and also that the giants are met with among the highest members of each division. Cephalopoda are the highest molluscs, and the largest cephalopods greatly exceed in size any other members of the group; decapods are at once the highest and the largest crustaceans; and whales, the hugest animals that exist, or so far as we know, that ever have existed, belong to the highest group of all, the mammalia. It would be easy to quote exceptions, but the general rule obtains admittedly. However, although there may be, and probably is, a general parallelism between the increase in size from the egg to the adult, and the historical increase in size during the passage from lower to higher forms; yet no one could maintain that the sizes of embryos represent at all correctly those of the ancestors; that for instance the earliest birds were animals the size of a chick embryo at a time when avian characters first declare themselves, or that the ancestral series in all cases presented a steady progression in respect of actual magnitude.

In the lower animals—*e.g.*, in Orbitolites—the actual size of the several ancestral stages is probably correctly recapitulated during the growth of the adult; and it is very possible that it is so also in such forms as the solitary sponges. In higher animals, except in the early stages of those forms which are practically devoid of food yolk, and which hatch as pelagic larvae, this certainly does not obtain. This is clear enough, but is worth pointing out, for if as most certainly is the case the embryos of animals are actually smaller than the ancestral forms they represent, it is possible that the smallness of the embryo may have had some influence on its organisation, and be responsible for some of the modifications in the ancestral history; and more especially for the disappearance of ancestral organs in free swimming larvae.

In adult animals the relation between size and structure has been very clearly pointed out by Herbert Spencer. Increased size involves by itself increased complexity of structure; the determining consideration being that while the surface area of the body increases as the squares of the linear dimensions, the mass of the body increases as their cubes. If for example we imagine two animals of similar shape and proportions, but of different size; for the sake of simplicity, we may suppose them to be spherical, and that the diameter of one is twice that of the other; then the larger one will have four times the extent of surface of the smaller, but eight times its mass or bulk and it is quite possible that while the extent of surface, or skin, in the smaller animal might suffice for the necessary respiratory and excretory interchanges, it would be altogether insufficient in the larger animal, in which increased extent of surface must be provided by foldings of the skin, as in the form of gills. To take an actual instance; Limapontia is a minute nudibranchiate, or sea-slug, about the sixth of an inch in length; it has a smooth body, totally devoid of respiratory processes, while forms allied to it, but of larger size, have their extent of surface increased by branching processes, which often take the form of specialised gills. This is a peculiarly instructive case, because Limapontia in its early developmental stages possesses a large spirally coiled shell, and shows other evidence of descent from forms with specialised breathing organs. We are certainly right in associating the absence of respiratory organs in the adult with the small size of the animal; and comparison with allied forms suggests very strongly that there has been in its pedigree an actual reduction of size, which has led to the degeneration of the respiratory organs.

This is an important conclusion: it is a well-known fact that the smaller members of a group are as a rule more simply organised than the larger members, especially with regard to their respiratory and circulatory systems; but if we are right in concluding that reduction in size may be an actual cause of simplification or degeneration in structure, then we must be on our guard against assuming hastily that these smaller and simpler animals are necessarily primitive in regard to the groups to which they belong. It is possible for instance that the simplification or even absence of respiratory organs seen in Pauropus,[374] in the Thysanura,[375] and in other small Tracheata,[376] may be a secondary character, acquired through reduction of size.

An interesting illustration of the law discussed above is afforded by the brains of mammals; it has been noticed by many anatomists that the extent of convolution, or folding of the surface of the cerebral hemispheres in mammals, is related not to the degree of intelligence of the animal, but to its actual size, a beaver having an almost smooth brain and a cow a highly complicated one. Jelgersma, and independently of him Professor Fitzgerald,[377] have explained this as due to the necessity of preserving the due proportion between the outer layer of grey matter or cortex, which is approximately uniform in thickness, and the central mass of white matter. But for the foldings of the surface the proportion of white matter to grey matter would be far higher in a large than in a small brain.[378]

It must not be forgotten on the other hand, that many zoologists hold the view, in favour of which the evidence is steadily increasing, that the primitive or ancestral members of each group were of small size. Thus Fürbringer remarks with regard to birds that on the whole small birds show more primitive and simpler conditions of structure than the larger members of the same group. He expresses the opinion that the first birds were probably smaller than Archaeopteryx, and notes that reptiles and mammals also show in their earlier and smaller types more primitive features than do their larger descendants. Finally, Fürbringer concludes that "*it is therefore the study of the smaller members within given groups of animals which promises the best results as to their phylogeny.*"

Again, one of the most striking points with regard to the pedigree of the horse, as agreed on by palaeontologists, is the progressive reduction in size which we meet

---

374   A group of multi-legged arthropods, possibly related to centipedes.

375   Silverfish and some related wingless insects.

376   A grouping, suggested Haeckel but now largely obsolete, encompassing spiders, myriapods and insects.

377   {M} Cf. *Nature*, June 5, 1890, p, 125.

378   Professor Fitzgerald was either George F Fitzgerald (1846–1899), Lord Kelvin Professor of Natural Philosophy, University of Glasgow, or more probably George Francis Fitzgerald FRS FRSE (1851–1901), Professor of Natural and Experimental Philosophy, Trinity College, Dublin. The footnote 377 reference is to "The influences at work in producing the cerebral convolutions", a letter to *Nature* (1890; 42, no. 1075) by DJ Cunningham of the Anatomy School, Trinity College Dublin, discussing the causes of folds in the surface of the brains of large (= complex) animals. Cunningham cites two papers by Jelgersma in which the folding is explained (as AMM indicates) by the need to accommodate a higher surface:volume ratio in support of greater brain activity. Cunningham claims to have reached similar conclusions about the folds independently, referring to "*my colleague, George F Fitzgerald*" to confirm the geometrical relationship.

with as we pass backwards in time from stage to stage. The Pliocene Hipparion was smaller than the existing horse, in fact about the size of a donkey; the Miocene Mesohippus about equalled a sheep; while Eohippus, from the Lower Eocene deposits, was no larger than a fox. Not only is there good reason for holding that as a rule larger animals are descended from ancestors of smaller size, but there is also much evidence to show that increase in size beyond certain limits is disadvantageous, and may lead to destruction rather than to survival. It has happened more than once in the history of the world, and in more than one group of animals, that gigantic stature has been attained immediately before extinction of the group, a final and tremendous effort to secure survival, but a despairing and unsuccessful one. The Ichthyosauri, Plesiosauri, and other extinct reptilian groups, the Moas, and the huge extinct Edentates, are well-known examples, to which before long will be added the elephants and the whales, and it may be iron-clads[379] as well. The whole question of the influence of size is of the greatest possible interest and importance, and it is greatly to be hoped that it will not be permitted to remain in its present uncertain and unsatisfactory condition.

It may be suggested that Amphioxus is an animal which has undergone reduction in size, and that its structural simplicity may, like that of Limapontia, be due in part at least to this reduction. Such evidence as we have tells against this suggestion; the first system to undergo degeneration in consequence of a reduction in size is the respiratory, and the respiratory organs of Amphioxus, though very simple, are also for a vertebrate unusually extensive.

We have now considered the more important of the influences which are recognised as affecting developmental history in such a way as to render the recapitulation of ancestral stages less complete than it might otherwise be, which tend to prevent ontogeny from correctly repeating the phylogenetic history. It may at this point reasonably be asked whether there is any way of distinguishing the palingenetic history from the later cenogenetic modifications grafted on to it; any test by which we can determine whether a given larval character is or is not ancestral. Most assuredly there is no one rule, no single test that will apply in all cases; but there are certain considerations which will help us, and which should be kept in view.

A character that is of general occurrence among the members of a group, both high and low,[380] may reasonably be regarded as having strong claims to ancestral rank; claims that are greatly strengthened if it occurs at corresponding developmental periods in all cases; and still more if it occurs equally in forms that hatch early as free larvae, and in forms with large eggs, which develop directly into the adult. As examples of such characters may be cited the mode of formation and relations of the notochord, and of the gill clefts of vertebrates, which satisfy all the

---

379 Presumably pachyderms such as rhinoceros.

380 Once again, hierarchical terminology slightly obscures the meaning. Assuming *high and low* mean *more highly evolved* and *less highly evolved* or *later and earlier in the pedigree*, we have to take *group* to mean that which we would now call a clade—a collection of all the organisms descended from a single ancestor (and including the ancestor). However, the argument now becomes circular, for if a particular characteristic (a synapomorphy, in modern parlance) is possessed by all organisms in a clade it is by definition ancestral.

conditions mentioned. Characters that are transitory[381] in certain groups, but retained throughout life in allied groups, may with tolerable certainty be regarded as ancestral for the former; for instance the symmetrical position of the eyes in young flat fish, the spiral shell of the young limpet, the superficial positions of the madreporite in Elasipodous Holothurians, or the suckerless condition of the ambulacral feet in many Echinoderms.

A more important consideration is that if the developmental changes are to be interpreted as a correct record of ancestral history, then the several stages must be possible ones, the history must be one that could actually have occurred—i.e., the several steps of the history as reconstructed must form a series, all the stages of which are practicable ones. Natural selection explains the actual structure of a complex organ as having been acquired by the preservation of a series of stages, each a distinct if slight advance on the stage immediately preceding it, an advance so distinct as to confer on its possessor an appreciable advantage in the struggle for existence. It is not enough that the ultimate stage should be more advantageous than the initial or earlier condition, but each intermediate stage must also be a distinct advance. If then the development of an organ is strictly recapitulatory, it should present to us a series of stages, each of which is not merely functional, but a distinct advance on the stage immediately preceding it.[382] Intermediate stages—e.g., the solid oesophagus of the tadpole—which are not and could not be functional, can form no part of an ancestral series; a consideration well expressed by Sedgwick[383] thus: *"Any phylogenetic hypothesis which presents difficulties from a physiological standpoint must be regarded as very provisional indeed."*

A good example of an embryological series fulfilling these conditions is afforded by the development of the eye in the higher Cephalopoda. The earliest stage consists in the depression of a slightly modified patch of skin; round the edge of the patch the epidermis becomes raised up as a rim; this gradually grows inwards from all sides, so that the depressed patch now forms a pit, communicating with the exterior through a small hole or mouth. By further growth the mouth of the pit becomes still more narrowed, and ultimately completely closed, so that the pit becomes converted into a closed sac or vesicle; at the point at which final closure occurs formation of cuticle takes place, which projects as a small transparent drop into the cavity of the sac; by formation of concentric layers of cuticle this drop becomes enlarged into the spherical transparent lens of the eye, and the development is completed by histological changes in the inner wall of the vesicle, which convert it into the retina, and by the formation of folds of skin around the eye, which become the iris and the eyelids respectively.

---

381 i.e., they appear during development and then disappear.

382 As elsewhere, we are forced to take *advanced* as meaning *more evolved* or *later in the pedigree*, rather than *closer to a developmental target*, otherwise the evolutionary process becomes teleological. Yet if we follow Darwinian reasoning, characters are retained because they provide survival advantage, not because their presence indicates some abstract superiority. The examples which follow need to be read in this light. A snail's eye may appear *clumsy* but snails are successful animals.

383 {M} Sedgwick, "On the Early Devolopment of the Anterior Part of the Wolffian Duct and Body in the Chick," *Quarterly Journal of Microscopical Science*, Vol. xxi. 1881, p. 456.

Each stage in this developmental history is a distinct advance, physiologically, on the preceding stage, and furthermore each stage is retained at the present day as the permanent condition of the eye in some member of the group Mollusca. The earliest stage, in which the eye is merely a slightly depressed and slightly modified patch of skin, represents the simplest condition of the Molluscan eye, and is retained throughout life in Solen. The stage in which the eye is a pit, with widely open mouth, is retained in the limpet; it is a distinct advance on the former, as through the greater depression the sensory cells are less exposed to accidental injury. The narrowing of the mouth of the pit in the next stage is a simple change, but a very important step forwards. Up to this point the eye has served to distinguish light from darkness, but the formation of an image has been impossible. Now, owing to the smallness of the aperture, and the pigmentation of the walls of the pit, which accompanies the change, light from any one part of an object can only fall on one particular part of the inner wall of the pit or retina and so an image, though a dim one, is formed. This type of eye is permanently retained in the Nautilus. The closing of the mouth of the pit by a transparent membrane will not affect the optical properties of the eye, and will be a gain, as it will prevent the entrance of foreign bodies into the cavity of the eye. The formation of the lens by deposit of cuticle is the next step. The gain here is increased distinctness and increased brightness of the image, for the lens will focus the rays of light more sharply on the retina, and will allow a greater quantity of light, a larger pencil of rays from each part of the object, to reach the corresponding part of the retina. The eye is now in the condition in which it remains throughout life in the snail and other gasteropods. Finally the formation of the folds of skin known as iris and eyelids provide for the better protection of the eye, and is a clear advance on the somewhat clumsy method of withdrawal seen in the snail.

The development of the vertebrate liver is another good but simpler example. The most primitive form of the liver is that of Amphioxus, in which it is present as a simple saccular diverticulum of the intestinal canal, with its wall consisting of a single layer of cells, and with bloodvessels on its outer surface. The earliest stage in the formation of the liver in higher vertebrates—the frog, for instance—is practically identical with this. In the frog the next stage consists in folding of the wall of the sac, which increases the efficiency of the organ by increasing the extent of surface in contact with bloodvessels. The adult condition is attained simply by a continuance of this process; the foldings of the wall becoming more and more complicated, but the essential structure remaining the same—a single layer of epithelial cells in contact on one side with bloodvessels, and bounding on the other directly or indirectly the cavity of the alimentary canal.

It is not always possible to point out the particular advantage gained at each step even when a complete developmental series is known to us, but in such cases, as for instance in Orbitolites, our difficulties arise chiefly from ignorance of the particular conditions that confer advantage in the struggle for existence in the case of the forms we are dealing with.[384]

The early larval stages in the development of animals, and more especially those that are marine and pelagic in habit, have naturally attracted much attention, since in the absence, probably inevitable, of satisfactory palaeontological evidence, they afford

---

384   In this paragraph, the language becomes comfortably Darwinian once again.

us the sole available clue to the determination of the mutual relations of the large groups of animals, or of the points at which these diverged from one another.

In attempting to interpret these early ontogenetic stages as actual ancestral forms, beyond which development at one time did not proceed, we must keep clearly in view the various disturbing causes which tend to falsify the ancestral record; such as the influence of food yolk, or of habitat, and the tendency of diminution in size to give rise to simplification of structure, a point of importance if it be granted that these free larvae are of smaller size than the ancestral forms to which they correspond. If on the other hand, in spite of these powerful modifying causes, we do find a particular larval form occurring widely and in groups not very closely akin, then we certainly are justified in attaching great importance to it, and in regarding it as having strong claims to be accepted as ancestral for these groups.

Concerning these larval forms, and their possible ancestral significance, our knowledge has made no great advance since the publication of Balfour's memorable chapter on this subject;[385] and I propose merely to allude briefly to a few of the more striking instances. The earliest, the most widely spread, and the most famous of larval forms is the gastrula, which occurs in a simple or in a modified form in some members of each of the large animal groups. It is generally admitted that its significance is the same in all cases, and the evidence is very strong in favour of regarding it as a stage ancestral for all Metazoa. The difficulty arising from its varying mode of development in different forms is however still unsolved, and embryologists are not yet agreed whether the invaginate or delaminate form is the more primitive. In favour of the former is its much wider occurrence; in favour of the latter the fact that it is easy to picture a series of stages leading gradually from a unicellular protozoon to a blastula, a diblastula, and ultimately a gastrula, each stage being a distinct advance, both morphological and physiological, on the preceding stage; while in the case of the invaginate gastrula it is not easy to imagine any advantage resulting from a flattening or slight pitting in of one part of the surface, sufficient to lead to its preservation and further development.

Of larval forms later than the gastrula, the most important by far is the Pilidium larva,[386] from which it is possible, as Balfour has shown, that the slightly later Echinoderm larva, as well as the widely spread Trochosphere larva, may both be derived.[387] Balfour concludes that the larval forms of all Coelomata, excluding the crustacea and vertebrates, may be derived from one common type, which is most nearly represented now by the Pilidium larva and which *was an organism something like a Medusa, with a radial symmetry.* The tendency of recent phylogenetic speculations is to accept this in full, and to regard as the ancestor of Turbellarians and of

---

385   *A Treatise on Comparative Embryology*, Vol. 2 (London: Macmillan, 1881), Chapter XIII.

386   The larvae of ribbon worms (Nemertea). They are *hat* or *cap* shaped and the worm develops from within.

387   Modern understanding, in which ribbon worms are protostomes and echinoderms are deuterostomes, makes such a derivation impossible. Ditto with crustacea and vertebrates in the next sentence. Trochosphere larvae are those of annelids and molluscs, and may thus share ancestry with ribbon worms, although the relationship between larvae envisaged by Balfour is not clear.

all higher forms, a jelly-fish or Ctenophoran, which in place of swimming freely has taken to crawling on the sea-bottom.

Of the two groups excluded above, the Crustacea and the Vertebrata, the interest of the former centres in the much discussed problem of the significance of the Nauplius larva.[388] There is now a fairly general agreement that the primitive crustacea were types akin to the phyllopods—i.e., forms with elongated and many-segmented bodies, and a large number of pairs of similar appendages. If this is correct, then the explanation of the Nauplius stage must be afforded by the phyllopods them-selves, and it is no use looking beyond this group for it. A Nauplius larva occurs in other crustacea merely because they have inherited from their phyllopod ancestors the tendency to develop such a stage, and it is quite legitimate to hold that higher crustaceans are descended from phyllopods, and that the Nauplius represents in more or less modified form an earlier ancestor of the phyllopods themselves.

As to the Nauplius itself the first thing to note is that though an early larval form, it cannot be a very primitive form, for it is already an unmistakable crustacean; the absence of cilia, the formation of a cuticular investment, the presence of jointed schizopodous limbs, together with other anatomical characters, proving this point conclusively. It follows therefore either that the earlier and more primitive stages are entirely omitted in the development of crustacea, or else that the Nauplius represents such an early ancestral stage with crustacean characters which properly belong to a later stage, thrown back upon it and precociously developed. The latter explanation is the one usually adopted; but before the question can be finally decided more accurate observations than we at present possess are needed concerning the stages inter-mediate between the egg and the Nauplius. The absence of a heart in the Nauplius may reasonably be associated with the small size of the larva.

Concerning the larval forms of vertebrates, it is only in Amphioxus and the Ascidians that the earliest larval stages are free-living, independent animals. In both groups the most characteristic larval stage is that in which a notochord is present, and a neural tube, open in front, and communicating behind through a neurenteric canal with the digestive cavity, which has no other opening to the exterior. This is a very early stage, both in Amphioxus and Ascidians; but so far as we know, it cannot be compared with any invertebrate larva. It is customary, in discussions on the affinities of vertebrates, to absolutely ignore the vertebrate larval forms, and to assume that their peculiarities are due to precocious development of vertebrate characteristics.[389]

---

388   Described in Lecture 6.

389   In the HL volume, a marginal note by Duckworth beside this sentence says: "*Gaskell did not make this mistake.*" The identity of Gaskell is unclear but the note provokes a historical consideration of what this difficult paragraph is actually saying: AMM clearly understands Amphioxus and Ascidians (sea squirts; tunicates, or urochordates as Balfour labelled them) to be vertebrates, although we would now describe them as non-vertebrate chordates for they lack any hard tissue directly surrounding the notochord. Larvae of both groups have a notochord, but whereas Amphioxus retains this in the adult form, the sessile ascidians lose it except for a cerebral ganglion. AMM's uncertainty about their ancestry seems to derive from the assumption, commonly held at the time, that all animals with pelagic larvae are related. Of course, most of such larvae are invertebrate and, in the 19th century, there was furthermore

It may turn out that this view of the matter is correct; but it has certainly not yet been proved to be so, and the development of both Amphioxus and Ascidians is so direct and straightforward that evidence of some kind may reasonably be required before accepting the doctrine that this development is entirely deceptive with regard to the ancestry of vertebrates.

Zoologists have not quite made up their minds what to do with Amphioxus: apparently the most guileless of creatures, many view it with the utmost suspicion, and not merely refuse to accept its mute protestations of innocence, but regard and speak of it as the most artful of deceivers. Few questions at the present day are in greater need of authoritative settlement.

That ontogeny really is a repetition of phylogeny must I think be admitted, in spite of the numerous and various ways in which the ancestral history may be distorted during actual development. Before leaving the subject, it is worth while inquiring whether any explanation can be found of recapitulation. A complete answer can certainly not be given at present, but a partial one may perhaps be obtained. Darwin himself suggested that the clue might be found in the consideration that at whatever age a variation first appears in the parent, it tends to reappear at a corresponding age in the offspring; but this must be regarded rather as a statement of the fundamental fact of embryology than as an explanation of it. It is probably safe to assume that animals would not recapitulate unless they were compelled to do so: that there must be some constraining influence at work, forcing them to repeat more or less closely the ancestral stages. It is impossible, for instance, to conceive what advantage it can be to a reptilian or mammalian embryo to develop gill clefts which are never used, and which disappear at a slightly later stage; or how it can benefit a whale, that in its embryonic condition it should possess teeth which never cut the gum, and which are lost before birth. Moreover, the history of development in different animals or groups of animals offers to us, as we have seen, a series of ingenious, determined, varied, but more or less unsuccessful efforts to escape from the necessity of recapitulating, and to substitute for the ancestral process a more direct method. A further consideration of importance is that recapitulation is not seen in all forms of development, but only in sexual development; or at least only in development from the egg. In the several forms of asexual development, of which budding is the most frequent and most familiar, there is no repetition of ancestral phases; neither is there in cases of regeneration of lost parts, such as the tentacle of a snail, the arm of a starfish, or the

---

no phylogenetic distinction between protostomes and deuterostomes (see also Lecture 8). Thus the question in mind is whether amphioxus and ascidian larvae mislead us by anticipating (and in the case of ascidians then losing) the definitive vertebrate characteristic. The deception falls away if the error of common (recent) ancestry of vertebrates and invertebrates is removed. We would certainly now reject it and it could be that Gaskell, whoever he/she was, similarly avoided being misled in this way. AMM's position is, nevertheless, a perfectly reasonable one, for it lies squarely in the two major issues which pervade this Lecture, Lecture 10 and others: the embryological recapitulation of phylogeny and the difficult matter of degeneration. One can sense his discomfort at being required to *ignore* evidence, and his hesitation is characteristic of his unwillingness to accept theories which observations do not support.

tail of a lizard; in such regeneration it is not a larval tentacle, or arm, or tail, that is produced, but an adult one.

The most striking point about the development of the higher animals is that they all alike commence as eggs. Looking more closely at the egg and the conditions of its development, two facts impress us as of special importance: first, the egg is a single cell, and therefore represents morphologically the Protozoon, or earliest ancestral phase; secondly, the egg, before it can develop, must be fertilised by a spermatozoon, just as the stimulus of fertilisation by the pollen grain is necessary before the ovum of a plant will commence to develop into the plant-embryo.

The advantage of cross-fertilisation in increasing the vigour of the offspring is well known, and in plants devices of the most varied and even extraordinary kind are adopted to ensure that such cross-fertilisation occurs. The essence of the act of cross-fertilisation, which is already established among Protozoa, consists in combination of the nuclei of two cells, male and female, derived from different individuals. The nature of the process is of such a kind that two individual cells are alone concerned in it; and it may I think be reasonably argued that the reason why animals commence their existence as eggs—i.e., as single cells—is because it is in this way only that the advantage of cross-fertilisation can be secured, an advantage admittedly of the greatest importance, and to secure which natural selection would operate powerfully.

The occurrence of parthenogenesis, either occasionally or normally, in certain groups is not I think a serious objection to this view. There are very strong reasons for holding that parthenogenetic development is a modified form, derived from the sexual method. Moreover, the view advanced above does not require that cross-fertilisation should be essential to individual development, but merely that it should be in the highest degree advantageous to the species, and hence leaves room for the occurrence, exceptionally, of parthenogenetic development.

If it be objected that this is laying too much stress on sexual reproduction, and on the advantage of cross-fertilisation, then it may be pointed out in reply that sexual reproduction is the characteristic and essential mode of multiplication among Metazoa; that it occurs in all Metazoa, and that when asexual reproduction, as by budding, &c., occurs, this merely alternates with the sexual process which sooner or later becomes essential. If the fundamental importance of sexual reproduction to the welfare of the species be granted, and if it be further admitted that Metazoa are descended from Protozoa, then we see that there is really a constraining force of a most powerful nature compelling every animal to commence its life-history in the unicellular condition, the only condition in which the advantage of cross-fertilisation can be obtained—i.e., constraining every animal to begin its development at its earliest ancestral stage, at the very bottom of its genealogical tree.

On this view the actual development of any animal is strictly limited at both ends; it must commence as an egg, and it must end in the likeness of the parent. The problem of recapitulation becomes thereby greatly narrowed; all that remains being to explain why the intermediate stages in the actual development should repeat the intermediate stages of the ancestral history. Although narrowed in this way, the problem still remains one of extreme difficulty.

It is a consequence of the Theory of Natural Selection that identity of structure involves community of descent: a given result can only be arrived at through a given sequence of events: the same morphological goal cannot be reached by two independent paths. A negro and a white man have had common ancestors in the past;

and it is through the long-continued action of selection and environment that the two types have been gradually evolved. You cannot turn a white man into a negro merely by sending him to live in Africa: to create a negro the whole ancestral history would have to be repeated; and it may be that it is for the same reason that the embryo must repeat or recapitulate its ancestral history in order to reach the adult goal. I am not sure that we can at present get much further; but the above considerations give opportunity for brief notice of what is perhaps the most noteworthy of recent embryological papers, Kleinenberg's remarkable monograph on Lopadorhynchus.[390]

Kleinenberg directs special attention to what is known to evolutionists as the difficulty with regard to the origin of new organs, which is to the effect that although natural selection is competent to account for any amount of modification in an organ after it has attained a certain size, and become of functional importance, yet that it cannot account for the earliest stages in the formation of an organ before it has become large enough or sufficiently developed to be of real use. The difficulty is a serious one; it is carefully considered by Mr. Darwin, and met completely in certain cases; but as Kleinenberg correctly states, no general explanation has been offered with regard to such instances.

As such general explanation Kleinenberg proposes his theory of the development of organs by substitution. He points out that any modification of an organ or tissue must involve modification, at least in functional activity, of other organs. He then continues by urging that one organ may replace or be substituted for another, the replacing organ being in no way derived morphologically from the replaced or preceding organ but having a genetic relation to it of this kind: that it can only arise in an organism so constituted, and is dependent on the prior existence of the replaced organ, which supplies the necessary stimulus for its formation. As an example he takes the axial skeleton of vertebrates. The notochord, formed by change of function from the wall of the digestive canal, is the sole skeleton of the lowest vertebrates, and the earliest developmental phase in all the higher forms. The notochord gives rise directly to no other organ, but is gradually replaced by other and unlike structures by substitution. The notochord is an intermediate organ, and the cartilaginous skeleton which replaces it is only intelligible through the previous existence of the notochord; while, in its turn, the cartilaginous skeleton gives way, being replaced, through substitution, by the bony skeleton.

The successive phases in the evolution of weapons might be quoted as an illustration of Kleinenberg's theory. The bow and arrow are {is} a better weapon than a stick or stone; they are used for the same purpose, and the importance or need for a better weapon led to the replacement of the sling by the bow. The bow does not arise by further development or increasing perfection of the sling; it is an entirely new weapon, towards the formation of which the older and more primitive weapons have acted as a stimulus, and which has replaced these latter by substitution, while the substitution at a later date of firearms for the bow and arrow is merely a further instance of the same principle.[391]

---

390   *Die Entstehung des Annelids aus der Larve von Lopadorhynchus Nebst bemerkungen über die Entwicklung anderer Polychaeten* (Leipzig: Wilhelm Engelmann, 1886).

391   This armorial analogy only works to a limited extent. It might have been better, surely, to find a change of function based on a pre-existing device. Perhaps the bow was

It is too early yet to realise the full significance of Kleinenberg's most suggestive theory; but if it be really true that each historic stage in the evolution of an organ is necessary as a stimulus to the development of the next succeeding stage, then it becomes clear why animals are constrained to recapitulate. Kleinenberg suggests further that the extraordinary persistence in embryonic life of organs which are rudimentary and functionless in the adult may also be explained by his theory, the presence of such organs in the embryo being indispensable as a stimulus to the development of the permanent structures of the adult. It would be easy to point out difficulties in the way of the theory. The omission of historic stages in the actual ontogenetic development, of which almost all groups of animals supply striking examples, is one of the most serious; for if these stages are necessary as stimuli for the succeeding stages, then their omission requires explanation; while if such stimuli are not necessary, the theory would appear to need revision. Such objections may however prove to be less serious than they appear at first sight; and in any case Kleinenberg's theory may be welcomed as an important and original contribution, which deserves—indeed demands—the fullest and most careful consideration from all morphologists, and which acquires special interest from the explanation it offers of recapitulation as a mechanical process, through which alone it is possible for an embryo to attain the adult structure.

That recapitulation does actually occur, that the several stages in the development of an animal are inseparably linked with and determined by its ancestral history, must be accepted. *"To take any other view is to admit that the structure of animals and the history of their development form a mere snare to entrap our judgment."* Embryology however is not to be regarded as a master-key that is to open the gates of knowledge and remove all obstacles from our path without further trouble on our part; it is rather to be viewed and treated as a delicate and complicated instrument, the proper handling of which requires the utmost nicety of balance and adjustment, and which, unless employed with the greatest skill and judgment, may yield false instead of true results.

Embryology is indeed a most powerful and efficient aid, but it will not, and cannot, provide us with an immediate or complete answer to the great riddle of life. Complications, distortions, innumerable and bewildering, confront us at every step, and the progress of knowledge has so far served rather to increase the number and magnitude of these pitfalls than to teach us how to avoid them. Still there is no cause for despair—far from it.

---

derived from something already to hand—for example, a device for rotating a stick to create fire. It has been a problem throughout the history of anatomy and physiology (and one which students need to be forewarned of even today) that we become misled by the names and functional descriptors we attach to organs or other body systems, thinking that their principal or only function is that which we initially ascribe to them. This is the basis of the problem raised by Kleinenberg: an organ which is not sufficiently formed to carry out its mature function may yet be capable of doing something else of value to the organism. *Change of use* is only a meaningful concept if we allow ourselves the conceit of believing that function equates to purpose. In reality, nature is opportunistic (as Darwin surely appreciated). But that conclusion begins to destroy the logic of Kleinenberg's position which AMM discusses in the next paragraph.

If our difficulties are increasing, so also are our means of grappling with them; if the goal appears harder to reach than we thought, on the other hand its position is far better defined, and the means of approach, the lines of attack, are more clearly recognised. One thing above all is apparent—that embryologists must not work single-handed, and must not be satisfied with an acquaintance, however exact, with animals from the side of development only; for embryos have this in common with maps, that too close and too exclusive a study of them is apt to disturb a man's reasoning power.

Embryology is a means, not an end. Our ambition is to explain in what manner and by what stages the present structure of animals has been attained. Towards this embryology affords most potent aid; but the eloquent protest of the great anatomist of Heidelberg[392] must be laid to heart, and it must not be forgotten that it is through comparative anatomy that its power to help is derived. What would it profit us, as Gegenbaur justly asks, to know that the higher vertebrates when embryos, have slits in their throats, unless through comparative anatomy we were acquainted with forms now existing in which these slits are structures essential to existence? Anatomy defines the goal, tells us of the things that have to be explained; embryology offers a means, otherwise denied to us, of attaining it. Comparative anatomy and palaeontology must be studied most earnestly by those who would turn the lessons of embryology to best account, and it must never be forgotten that it is to men like Johannes Müller, Stannius, Cuvier, and John Hunter —the men to whom our *exact* knowledge of comparative anatomy is due—that we owe also the possibility of a science of embryology.

---

392   This undoubtedly refers to Carl Gegenbauer and appears more cryptic than it should merely because it anticipates the mention of his name in the next sentence. Gegenbauer became Professor of Anatomy in Heidelberg in 1873. He believed that the pedigree of organisms and the homology of organs was more reliably revealed in the similarity of mature forms than by embryology and that functional similarities should not be overlooked. He expressed his views (the *eloquent protest*) in a forthright criticism of the ideas of Alexander von Götte in 1876 (*Morphologisches Jarbuch* 1: 299–345); the previous year, von Götte had described in exhaustive detail the embryology of a species of frog and used his findings to make sweeping generalizations about the phylogenetic relationships of vertebrates. To further contextualize AMM's statement, we can note that the first edition of his *The Frog: An Introduction to Anatomy, Histology and Embryology* was published in 1882.

INTERLUDE

A REVEALING BOOK REVIEW

As CFM explains in his Preface, Lecture 13 was chronologically out of order and covers material dealt with in earlier lectures, especially Lecture 10, but was included in the collection because of its importance. Some 18 months after delivering it as the 1890 BAAS Biology Section President's Lecture, AMM published in *Nature* (24 March 1892; **45**(1169): 482–483) a review of a new (4th) edition of Ernst Haeckel's study of human evolution *Anthropogenie, oder Entwickelungsgeschichte des Menschen* (Leipzig, W. Engelmann). The review appears above initials rather than a full name but the content and forthright style leave no doubt about authorship.

The review is so instructive of AMM's stage of biological thought that it is appropriate to add it to the collection at this point. It is reproduced here in its entirety, without annotations. It reveals his views on several controversial matters and allows us to take stock of his position.

By way of context, Haeckel, one of the loudest voices in post-Darwinian 19th-century theoretical biology and champion of the recapitulation theory, was 58 years old at the time of the review and had 27 more years to live. (His exact contemporary August Weismann predeceased him by five years.) AMM was 40 and had but 21 months remaining. Some 14 years earlier, AMM's friend and colleague Balfour (1851–1882) had crossed swords with Haeckel over the misrepresentation of one of the illustrations in his (Balfour's) study of the development of elasmobranch fishes. Balfour, had he also lived, would have led the charge towards a new, thoroughly Darwinian reading of embryology and the interpretation of developmental relationships between animals. One gets a strong sense from the lectures that AMM felt the need to pick up the baton of Balfour's foreshortened legacy, and this may partly explain the polite animosity which emerges as the review moves to its conclusion.

Regarding the recapitulation theory, Stephen Jay Gould (*Ontogeny and Phylogeny*, Cambridge, MA, Belknap Press, 1977) insists that a correct interpretation of Haeckel's position requires that animals pass through the *adult* stages of earlier forms, rather than just their embryonic forms. It is clear from this review (fifth paragraph), as also from Lectures 10, 13, 17 and 20, that AMM's acceptance of the theory does not rest on this stricture. Like Balfour (*A Treatise on Comparative Embryology*, Vol. I: Introduction; London, Macmillan, 1880), he saw ontogenic development as involving passage through earlier forms including earlier embryonic forms, even though there are many examples (Lecture 13) in which this does not happen. Von Baer, a persistent critic of Haeckel, used the existence of exceptions as one of the reasons why he found the recapitulation theory unacceptable. Like AMM, von Baer viewed the biological world in terms of common ancestry rather than linear ontogeny. On the other hand, von Baer accepted Darwinian ideas only with a teleological spin (animal development directed towards a predetermined state of perfection), which AMM clearly refuses do (despite using some unfortunate linguistic conventions; see Lectures 10 and 13).

AMM encapsulates recapitulation in a typically pithy phrase. The review illuminates a number of other matters covered in the Lectures, including degeneration

(Lectures 17, 19), inheritance through the germ-plasma (Lectures 7 and 9), the origin of vertebrates (Lecture 20), and the inheritance of acquired characteristics (Lecture 14).

Further evidence of his position comes in paragraphs 6 and 8: Weismann's germ-plasma (criticized by both Haeckel and AMM) can be seen as the direct forerunner of modern gene-based inheritance; the (Lamarckian) inheritance of acquired characteristics (accepted by Haeckel, not by AMM) is now completely discredited (except in the epigenetic sense: see Lecture 1); degeneration (accepted by AMM, not by Haeckel) is a useful way of describing how characters no longer advantageous to survival become lost (by gradual accumulation of gene mutations, as we would now say); detailed accounts of human gastrulation (accepted by Haeckel but evidently not by AMM, or perhaps not in the form Haeckel describes) can be found in any modern textbook of embryology. Further on, AMM criticizes as anachronistic Haeckel's overuse of animal embryos in understanding human development.

Overall therefore, one can argue that AMM straddles at least two conflicting late 19th-century positions: he was unable to abandon recapitulation despite its flaws (also recognized by Weismann), and felt secure in promoting it as the current embryological dogma, yet he understood and promoted a pedigree-based, purpose-free, pure "survivalist" interpretation of Darwin which later biologists have come to accept. (The Lectures from the second volume provide abundant evidence in support of this second point.) One can also conclude that whilst he chose what eventually turned out to be the losing side on some embryological arguments, he took, perhaps intuitively, the winning side on others.

It is also tempting to read the review not only as representing a generational leap in biological thought but also as a reflection of a cultural-geographical transition—from the well-established European/German academic tradition towards the rising influence of UK and American science, with its emphasis on experiment and deduction in place of observation, theoretical assumptions and interpretation.

THE EVOLUTION OF MAN

THE importance of the subject-matter of the book, and the length of time that has elapsed since the appearance of the former editions, and the prominent position held by the author, seem to call for more detailed notice of this work than is usually accorded to a new edition.

The "Anthropogenie," which was first published in 1874, is the third of the series of books in which Prof. Haeckel has attempted to determine the laws governing the form, structure, and mutual relations of living things, and to establish the general principles of biological science. The first of these, under the title "Generale Morphologie," appeared in 1866, almost simultaneously with the completion of Herbert Spencer's "Principles of Biology." It is a comprehensive and ambitious work, which, in the author's words, *"constituted the first attempt to apply the general doctrine of development to the whole range of organic morphology,.... and to introduce the Darwinian theory of descent into the systematic classification of animals and plants, and to found a "natural system" on the basis of genealogy; that is, to construct hypothetical pedigrees for the various species of organisms."*

It contains also the first systematic attempts to deal in detail with the ancestry of man, as regards the groups of animals lower than mammals. This is, perhaps, the most solid piece of work Prof. Haeckel has done; it contains much matter of great value, and discussions and speculations of extreme ingenuity and interest. Later discoveries have rendered much of it obsolete, but it still remains the most important work of its kind; and but for its somewhat cumbersome terminology, would be widely read even now. The "Natural History of Creation," first published in 1886, goes over a good deal of the same ground as the earlier work, but is written in a much more popular style, and aims at presenting in a form suited to the general reader the main arguments on which the Darwinian theory is based, together with a detailed application of the theory to the principal groups of animals, and an attempt to determine their mutual relations and lines of descent. The "Anthropogenie," the book now before us, is a more elaborate application of the same principles to the special problem of the evolution of man.

In the new edition the general arrangement remains much the same as before; but, in order to include the results of more recent investigations, a great part of the book has been re-written, and new chapters have been added on subjects that have attracted especial attention of late years, such as the Gastrea theory, the Coelom theory, and the nature and origin of segmentation. A large number of new figures have been inserted, and the genealogical tables, for which Prof. Haeckel has a special fondness, have been greatly increased in number, and in elaboration of detail.

The book, which is adapted rather for the general reader than the scientific student, is written in an attractive and popular style, and presents the main facts of vertebrate embryology in an intelligible and well illustrated form. As might be expected from his former writings, the main feature of Prof. Haeckel's work is a detailed exposition and vigorous defence of what he has named "the fundamental law of biogenesis," better known in this country as the recapitulation theory, according to which, the actual or ontogenetic development of an animal is a repetition of the ancestral or phylogenetic development of the species; or, to put it more simply, animals in their development climb up their own genealogical trees. This is now generally accepted by embryologists, but it has not always been so, and Haeckel reminds us, with justice, that he was one of the first to realize and teach the doctrine.

Prof. Haeckel has much to say on other points of theoretical interest. He protests strongly against Weismann's views with regard to heredity, pointing out that the very existence of germ-plasma is as yet a mere assumption, and maintaining that acquired characters may be and are actually transmitted. He objects equally strongly to the views as to the widespread occurrence of degeneration, which were first put forward by Dr. Dohrn; and on the much debated question of the origin of vertebrates he sides with those who fully accept the evidence afforded by the anatomy and development of Amphioxus; he admits the Vertebrate affinities of Balanoglossus and looks for the ancestors of Vertebrates among the unsegmented Turbellarian worms. As a controversialist, Prof. Haeckel is impressive rather than convincing. He hits hard and with effect, but prefers to counter rather than to parry the blows of his opponent. It is impossible to pass over without protest the terms in which he writes—it must be admitted under provocation—of the opinions and work of other investigators.

Prof. Haeckel's fondness for genealogical trees, and his facility in constructing them, are well known and have been much criticised, perhaps a little unfairly. Acceptance of the doctrine of evolution involves the recognition of a blood-relationship, near or remote, between any one animal and any other; and the only true classification is one which places this fact in the forefront and adopts it as the basis on which the scheme is to rest. Genealogical tables undoubtedly stimulate enquiry, and so long as it is realized that they are necessarily in great part tentative or provisional, they will probably do more good than harm. It would be easy to take exception to many points in Prof. Haeckel's numerous and elaborate pedigrees, but it will be generally admitted that they are instructive, and often extremely suggestive, even though the conclusions may not meet with general acceptance.

The least satisfactory part of the book is that which deals with human embryology. No attempt whatever is made to explain the earlier stages of development; the special difficulties of the problem are absolutely ignored; the human gastrula is spoken of in a confident way, as though such a stage really existed; and the accounts of the development of the several organs and systems are too often taken from other animals. A student who relied on Prof. Haeckel's descriptions would obtain an entirely erroneous idea of the actual course of development of the human embryo.

Owing to the difficulty of obtaining material in proper condition for microscopical examination, our acquaintance with human embryology long remained imperfect; and the descriptions in text-books were largely based on our knowledge of other Vertebrates, and illustrated by figures from embryos of dogs, pigs, rabbits, or even chicken and dogfish. The time for all this is now past. During the last ten years our knowledge has advanced wonderfully; and although the earliest stages are still unknown, it is not too much to say that our knowledge of the development of the human embryo, from a stage corresponding to a chick embryo at the commencement of the second day onwards, is as satisfactory, as complete, and as well illustrated as that of any other mammal.

For this great advance we are indebted almost entirely to the labours of German embryologists, and notably to the splendid work of Prof. His. Prof. Haeckel has in his volume many hard things to say of Prof. His but is indebted to him for the only really good figures of human embryos which he gives, and would have materially improved his book had he studied more carefully the admirable descriptions of the Leipzig Professor. It is a matter for great regret that a book of 900 pages, having for its title, "Anthropogenie, oder Entwickelungsgeschichte des Menschen," should be allowed to appear in which the account of the actual development of the human embryo is so inadequate or even erroneous.

A.M.M.

LECTURES ON THE DARWINIAN THEORY

Although CFM describes the lectures in the second volume as dating to 1893, records held in the University of Manchester Archives show that AMM gave a series of three public lectures on *The Darwinian Theory* ten years earlier as part of the 1882–83 Owens College *Museum Lectures* series ("open to both Ladies and Gentlemen"; 1s for a single lecture or 2s 6d for the course):

Lecture I, 7 February: *Development of the Theory*
Lecture II, 14 February: *Applications of the Theory*
Lecture III, 21 February: *Objections to the Theory*

No separate written versions have been traced but it could be these to which CFM refers in the first two paragraphs of his Preface. (The possibility of a date misprint—1893 for 1883—might also be considered.) The title of Lecture 19 matches that of the February 21st lecture and the style throughout certainly suggests they were intended for a general rather than specialist audience.

In the winter of 1886, AMM presented "a series of eight University Extension Lectures on Natural History" as part of programme of Sunday afternoon lectures run by the Recreation in Ancoats society (a Manchester-based charitable organization devoted to the education of working people and public wellbeing). According to the advertisement:

"The object of the course is to explain the principles on which the Theory of Evolution (more familiarly known as the Darwinian Theory) is based, and to illustrate these principles by reference to the leading features of the structure and life history of the various groups of animals. No previous study of the subject will be assumed, but, inasmuch as the mode of treatment is in some respects novel, it is hoped that the course will prove of interest to those who have already devoted some attention to some branch of Natural History, as well as to those less versed in the subject."

He gave a second course on Natural History to the same organization in January and February 1887 called *The Evolution of Man and Animals*:

"The course will form a direct continuation of the one delivered to Ancoats during the Michaelmas Term of 1886 and will include a further and more detailed consideration of the laws determining the structure and distribution of animals. The first lecture will be devoted to an examination of the arguments on which the Theory of Natural Selection—better known as the Darwinian Theory—is based: and to a consideration of some of the more important applications of this theory, and of the principal objections that have been urged against it. In the fifth Lecture a detailed application will be made of the principles, established during the preceding lectures, to a special problem—the origin of vertebrate animals, including Man. In the three remaining Lectures the Geographical Distribution of Animals will be considered,

and the general and special characters and mutual relations of the terrestrial, fresh water, and marine faunas of various parts of the earth will be described and explained. An account will also be given of recent marine dredging expeditions."

It is not difficult to imagine that these also formed the basis of the some of the published lectures. (Connections with other public lectures which AMM gave from time to time are noted in footnote comments.)

The programme for the 1883 *Museum Lectures* series provided content summaries for each lecture, much as CFM does for the collected lectures (reproduced below). Note that in the Contents for Lecture 14, CFM gives incorrect dates for St Hilaire: 1771–1840, rather than 1772–1844. Alfred Russel Wallace died in 1913.

LECTURES

ON THE

DARWINIAN THEORY

DELIVERED BY THE LATE

ARTHUR MILNES MARSHALL,

M.A. , M. D., D.Sc., F. R. S.
PROFESSOR OF ZOOLOGY IN OWENS COLLEGE ;
FORMERLY FELLOW OF ST. JOHN'S COLLEGE, CAMBRIDGE.

EDITED BY

C. F. MARSHALL

M.D., B.Sc., F.R.C.S,

WITH 37 ILLUSTRATIONS
MOSTLY FROM ORIGINAL DRAWINGS AND PHOTOGRAPHS

NEW YORK
MACMILLAN AND CO.
LONDON: D. NUTT
1894

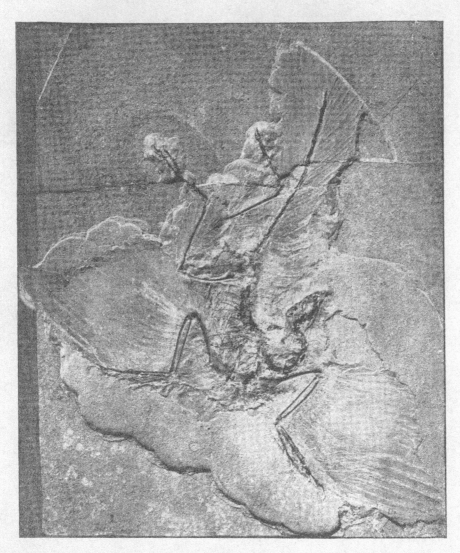

ARCHÆOPTERYX

PREFACE

THIS volume consists of a series of lectures delivered by the late Professor Marshall in connection with the Extension Lectures of the Victoria University during the year 1893. These have been amplified in some places where occasion required from other unpublished lectures by the same author.

Owing to the variable nature of the original MS., some parts being fully written, while others were in the form of notes, there is a consequent variability in the fulness of the text, and any errors or discrepancies should be attributed to the editor and not the author.

It is greatly to be regretted that these Lectures were not elaborated and prepared for publication by the author himself; but, in spite of the shortcomings of the book, I trust that it may form a useful contribution to the literature of Darwinism, since the Lectures were delivered by one of Darwin's most earnest disciples.

The large majority of the illustrations are taken from original drawings by the author, or from photographs from nature. Some of the drawings are modified from other sources, and in these cases the source has been acknowledged. The blocks for the illustrations have been prepared with great care by Messrs. Walker & Boutall.

In preparing this book I have received valuable assistance from Mr. W. E. Hoyle, Dr. C. H. Hurst, Professors Hickson and Herdman, and Mr. W. Garstang. I must also thank Mr. Francis Darwin for permission to reproduce the extracts in Lecture VIII.

C. F. M.

LONDON, *October 1894*

CONTENTS

LECTURE 14
HISTORY OF THE THEORY OF EVOLUTION

Evidence in support of the descent of the various breeds of domestic pigeons from the wild rock-pigeon, *Columba livia.*

The principles and practice of breeding animals. Variation with heredity. The power and efficiency of Artificial Selection. The breeding of horses, cattle, sheep, dogs, fowl, &c.

NATURAL SELECTION.

There are in nature causes which act in much the same way as man acts, when selecting artificially the best animals for breeding purposes; causes which must lead to structural modifications.

The tendency to rapid increase of numbers. Rapid multiplication of rabbits in Australia.

The causes tending to keep the numbers stationary on the average. Effect of removal of any check to increase.

The struggle for existence applies to all animals and at all times. Competition is keenest between the most closely similar forms. Competition amongst men.

Variation: its universal occurrence. Variation occurs in all directions, and affects all organs and parts. The importance of small variations in the breeding of domestic animals.

Natural Selection: the keynote of the Darwinian Theory. The fittest to survive are not necessarily those which are ideally most perfect.

Changes in environment. Geological evidence of changes of climate in England and in Greenland.

Comparison between Natural and Artificial Selection. Natural Selection acts for the good of the Species. Artificial Selection is directed solely towards the benefit or the pleasure of Man.

LECTURE 16
THE ARGUMENT FROM PALAEONTOLOGY.

The Structure of the Earth's Crust. A series of superposed layers or strata, composed of various materials, originally deposited for the most part under water, and subsequently raised above the surface.

Fossils are the actual remains, as bones, teeth, shells, etc., or the indications, as, *e.g.,* footprints, of the animals that formerly inhabited the earth. Amongst them, were our collections complete, would be the entire series of the ancestors of all living animals. The actual age of Fossils cannot be determined, but their relative ages are known from the position of the strata in which they are found.

The Imperfection of the Geological Record. Only certain parts of certain animals can be preserved as fossils: special conditions are necessary for their preservation, and for their subsequent exposure at places accessible to man. It is irrational to expect fossils to yield continuous series of forms, save under exceptional circumstances. "A Chapter of Accidents."

The Geological Evidences of Evolution. Life on the earth has been continuous. Cuvier's Cataclysms have no place in nature. In passing from the older to the more recent strata, there is a general advance in organisation, and a gradual approach towards the existing condition. Fossils are "generalized forms" rather than directly intermediate links. The geological history of the Horse, of Birds and Reptiles, and of Paludina.

The Extinction of Species. Persistent types. Comparison of the Geological Record with the History of Man.

THE GEOGRAPHICAL DISTRIBUTION OF ANIMALS.

The Problems of Geographical Distribution; and the methods of attacking them. The clue to the present condition is afforded by the study of past history.

The Geographical Distribution of Camels, Marsupials, and Tapirs.

LECTURE 17
THE ARGUMENT FROM EMBRYOLOGY.

Embryology as a clue to Zoological affinities. The development of an animal often gives us evidence, otherwise unattainable, as to its relations with other animals or groups of animals. Good illustrations of this are afforded by Ascidians, by Barnacles, and by many groups of Parasitic animals.

The Recapitulation Theory. The doctrine of Evolution tells us that animals, like men, have pedigrees. The study of Embryology reveals to us this ancestry, because every animal in its development tends to repeat the history of the race.

Illustrations of Recapitulation. The development of Flat fish, Crabs, Prawns and Barnacles.

Embryology and Palaeontology. Examples of Recapitulation as seen in Fossils. The Shells of Foraminifera and of Mollusca. The Embryology of Ammonites. The Antlers of Deer.

Rudimentary or Vestigial Organs: structures which are present in a condition in which they can be of no use to their possessors. Natural Selection will not account for the formation or perpetuation of such structures; but the Recapitulation Theory explains these at once, as organs which were of functional value to the ancestors of their present possessors, and which appear in the development of existing forms owing to the tendency to repeat ancestral characters. Examples of Vestigial Organs: their Zoological importance.

Causes tending to falsify the Ancestral History as preserved in actual development.
1. *The tendency to condensation of the Ancestral History.*
2. *The tendency to the omission of Ancestral Stages.* The structure of an egg. Germ yolk and food yolk. The causes regulating the number and size of the eggs produced by different animals.
3. *The tendency to distortion of Ancestral Stages.*
4. *The tendency to the accentuation or undue prolongation of Particular Stages:* best seen in cases of abrupt transformation or metamorphosis.
5. *The tendency to the acquisition of new characters,* through the action of Natural Selection.

Tests of Recapitulation. All stages must be possible ones, and each stage must be an improvement on the preceding one. The eyes of Cuttle-fish.

Embryonic Stages regarded as Ancestral Forms. The egg is the equivalent of the Protozoon stage.

LECTURE 18
THE COLOURS OF ANIMALS AND OF PLANTS.

Colour may be non-significant, as in the case of the redness of the blood of many animals, or the colours of animals living in darkness. More usually, however, colour can be shown to have a direct relation to the welfare of the individual or species, and to be attributable to the action of Natural Selection.

Of late years our knowledge on this subject has advanced greatly, mainly through the observations of Mr. Wallace, supplemented by those of Mr. Bates, Mr. Trimen, Mr. Poulton, and others.

COLOURS OF ANIMALS.

1. *Apatetic Coloration*

Apatetic Coloration serves to hinder recognition: it may be considered under three heads:—

(a) *Protective resemblances:* aiding escape from enemies. The resemblances are usually either to plants, as in the case of the leaf insects, and stick insects, and of the green coloured caterpillars and other frequenters of plants or trees; or else to inanimate objects, as in the case of the whiteness of the Arctic Hare, and of other defenceless Arctic animals.

A peculiar and interesting form of protective resemblance is afforded by the cases of mimicry, in which a defenceless butterfly or other animal escapes attack through its superficial resemblance to a noxious or venomous animal.

(b) *Aggressive resemblances* are cases in which the object gained, like that of the wolf in sheep's clothing, is to facilitate approach to the prey through a superficial resemblance to other objects. Examples are afforded by the whiteness of the Polar Bear and other predacious Arctic animals, or the colouring of the Lion or Tiger.

(c) *Alluring resemblances* are cases in which the coloration is such as to cause the animal to resemble a flower or other attractive object, and so to entice the approach of prey.

2. *Sematic Coloration.*

Sematic Coloration is the direct opposite of Apatetic Coloration and aims at securing recognition; it is of two chief kinds:-

(a) *Warning Colours.* Insects or other animals which are inedible owing to an unpleasant flavour or other cause, are usually very conspicuously coloured: the object being to advertise their inedibility, and to secure instant recognition, lest they should be killed by mistake.

(b) *Recognition Marking,* such as the white tail of the Rabbit and the markings on certain Deer, are believed to aid recognition by members of the same species.

3. *Epigamic Coloration.*

Under this head all the cases of Sexual Coloration are included, in which, as in the Peacock, the bright colouring is confined to one sex, or is at any rate more marked in it, and is displayed for the purpose of attracting the opposite sex.

COLOURS OF PLANTS.

The bright colours of Flowers and of Fruits serve to attract the insects which fertilise the flowers, and birds and mammals which secure the dispersal of the seeds.
Cross-Fertilisation. The methods of ensuring cross-fertilisation in *Orchids.*
Attractive, and Protective Fruits.

LECTURE 19
OBJECTIONS TO THE DARWINIAN THEORY.

Natural Selection explains the present structure of animals as due to the slow accumulation of small variations, which are transmitted by inheritance from generation to generation. Natural Selection acts primarily for the good of the species, not of the individual.

"Natural Selection tends only to make each organic being as perfect as, or slightly more perfect than, the other inhabitants of the same country with which it comes into competition." "It will not produce absolute perfection."

"Natural Selection cannot possibly produce any modification in a species exclusively for the good of another species."

Missing Links. Erroneous ideas as to the true nature of "links," and as to their supposed absence. "Links "are nearly always indirect, rarely direct; they may combine the special characters of both the forms they connect, but more usually have the characters of neither. The true link between any two forms is afforded by the common ancestor from whom both alike are descended. Examples of links of various kinds.

Persistent Types. Many instances arc known of genera of animals which have persisted, without appreciable modifications in structure, for enormously long periods, and in some cases from Silurian times to the present day. The occurrence of such persistent types is in no way opposed to the Theory of Natural Selection.

Degeneration or Retrograde Development. An animal may be less highly organised when adult than it is in its earlier stages of existence. During development, organs such as eyes, legs, etc., that are present in the young animals may disappear or become vestigial. This again is not opposed to the theory of Natural Selection. Natural Selection tends to preserve those forms which are best adapted to their environment, and not necessarily those which are ideally most perfect. Examples of degeneration of individual organs, and of entire animals.

Difficulty as to the persistence of lowly organised animals alongside the higher forms.

Alleged uselessness of small variations. The whole theory and practice of breeding domestic animals depend on selecting the right animals by scrupulous attention to minute differences. The objection that the right variation may not be present is met by the fact that variation affects all organs and occurs in all directions.

Difficulty as to the earliest commencement of organs. Natural Selection can only act on an organ after it has already attained sufficient size to be of practical importance and utility. *The Theory of Change of Function*: an organ may lose its original purpose and yet persist because it is of use for some other purpose: one of these purposes may predominate at one time, another at another, and the organ may undergo structural modification in consequence. Examples of change of function: the wing of the bat; lungs and swimming bladders; electric organs of fish.

Other Objections. Insufficiency of Time. Evidence of Design and Forethought. Instinct.

LECTURE 20
THE ORIGIN OF VERTEBRATED ANIMALS.

The evidence available for the determination of genetic affinities is of three chief kinds:—

1. That afforded by comparison of structure.
2. That afforded by development.
3. That afforded by fossils.

THE CRANIATE VERTEBRATES.

The Zoological characters of the higher Vertebrates. The brain and spinal cord; the vertebral column and skull.

The characters of Fish and Amphibians. The mutual relations of the two groups. The transition from the water-breathing to the air-breathing condition. The evolution of lungs.

The Zoological characters of Reptiles, Birds, and Mammals. The relations of these groups to Fish and Amphibians. Evidence in favour of the descent of Reptiles, Birds, and Mammals from aquatic ancestors.

The mutual relations of Reptiles and Birds. The special characters of Birds. Fossil forms intermediate between the two groups. Embryological evidence. The development of the pelvic girdle.

The special characters of Mammals. The development of Mammals. Indications of the former presence of large eggs. The eggs of *Echidna*. Paleontological evidence in favour of the descent of Mammals from Reptiles.

THE ACRANIATE VERTEBRATES.

The Anatomical and Embryological characters of Amphioxus. Relation of Amphioxus to other Vertebrates.

The characters of *Ascidians*. Supposed ancestral relations to other Vertebrates. The development of Ascidians. Evidence of degeneration. Evidence in favour of a pelagic marine ancestry for all Vertebrates. Supposed relations with Invertebrates.

THE DESCENT OF MAN.

Man distinctly an animal; moreover a Vertebrate and a Mammal. Comparison of the structure and development of man with other animals. Points of resemblance of man to the higher apes. Rudimentary organs present in man. The tendency to Reversion.

The few structures found in man which are peculiar to him.

Other characteristics of man compared with lower animals.

Application of the Darwinian Theory to the Language of Man.

Conclusion. Causes known which account for the structure, language, and habits of man. The possibility of other attributes not explained by these laws.

LECTURE 21
THE LIFE AND WORK OF DARWIN.

SUMMARY OF LEADING EVENTS.

Birth . . . . 12th February, 1809.
At School at Shrewsbury . . . . 1818–1825.
At University of Edinburgh . . . . 1825–1827.
At University of Cambridge . . . . 1828–1831.
Voyage of H.M.S. Beagle . . . . 1831–1836.
First Note-Book on "Origin of Species" commenced 1st July, 1837.
Marriage . . . . 29th January, 1839.
"Journal of Researches" published . . . . 1839.
"Zoology and Geology of Voyage of *Beagle*" . . . . 1840–1846.
"First outline of "Origin of Species "written . . . . 1842.
"Monograph on the Cirripedia" . . . . 1846–1854.
Announcement of the Theory of Natural Selection by Darwin and Wallace 1st July, 1858.
"Origin of Species" published . . . . 1859.
"Fertilisation of Orchids" . . . . 1862.
"Variation of Animals and Plants under Domestication" . . . . 1868.
"Descent of Man" . . . . 1871.
"Expression of the Emotions" . . . . 1872.
"Movements and Habits of Climbing Plants" . . . . 1875.
"Insectivorous Plants" . . . . 1875.
"Cross and Self-Fertilisation in Plants" . . . . 1876.
"The different Forms of Flowers in Plants of the same species" . . . . . 1877.
"The Power of Movement in Plants" . . . . 1880.
"Vegetable Mould and Earthworms" . . . . 1881.
Death . . . . 19th April, 1882.

The Darwin Family: Erasmus Darwin, 1731–1802: author of "Zoonomia." and grand-father of Charles Darwin.

Life of Charles Darwin: School and College career; found lectures "intolerably dull." The Voyage of the *Beagle*, 1831–1836; Outline of the Voyage; its purpose and its results; its influence on Darwin—" by far the most important event in my life." Return to England; lived for nearly six years in London. In 1842 settled at Down, near Beckenham, where the greater part of his work was accomplished, and the rest of his life spent.

Darwin's Work: The *Beagle* Reports, 1840—1846; the Theory of the Structure and Distribution of Coral Reefs, 1842. The Monograph on the Cirripedia, 1846—1854.

The "Origin of Species." Preliminary Work; unwillingness to publish until completed. Independent discovery of Natural Selection by Darwin and Wallace. Influence of Malthus' Essay on Population. "The chief work of my life."

Darwin's later works, "The Descent of Man," and "Expression of the Emotions."" The great series of Botanical works.

Personal Characteristics: Daily life and habits. Chronic ill-health. System of work: Direct observation and experiment: Simple methods and few instruments:

Dogged industry. His last book. "Vegetable Mould and Earthworms," considered as a sample of his mode of work. Darwin's estimate of his own powers: "My mind seems to have become a kind of machine for grinding general laws out of large collections of facts." Avoidance of controversy. Darwin's influence on Scientific Thought. The great importance of little things.

LECTURE 14

HISTORY OF THE THEORY OF EVOLUTION

This first lecture in the second series establishes the historical context for Darwin's work with a rich assemblage of key figures, many brought to life as personalities and by their scientific pedigrees. There is a strong sense of discovery and advancement but with due regard to several of the controversies that preceded Darwin's magnificent conceptual achievement. As a non-specialist lecture, this is a more accessible presentation of the history of evolutionary biology than that given to Owens College students in 1879 (Lecture 1).

Of particular interest is AMM's analysis of Lamarck's contribution. It is easy to dismiss Lamarck for being on the wrong side of history, based on his belief in what has come to be known, perhaps superficially, as the *inheritance of acquired characteristics*. The phrase oversimplifies matters for it fails to distinguish the result from the mechanism. (Lamarck's reputation also suffered in the first half of the 20th century by association with the ideologically driven failures of the Lysenko regime in the USSR.) Lamarck understood perfectly well that characters and their variants are inherited and that such inheritance was the basis for evolution. He also acknowledged an important role for the environment in determining which characters come to be established. His error, and that of several contemporaries and followers, was to believe in the transmission between generations of alterations acquired during life as a direct result of use, experience or environmental influences. AMM makes the distinction very clear in this account, giving direct quotes from Lamarck's work to emphasize the point.

Having said that, AMM then rather oversimplifies subsequent events by implying that Darwin came along and set the story straight to everyone's satisfaction. In fact, theories of the heritable transmission of (what we would now call) somatically acquired characters continued to be believed right to the end of the 19th century and even beyond, and it is often argued that Darwin himself, in his theory of pangenesis and gemmules (Lecture 7), was attracted to the idea.

It was August Weismann who drew a clear and purposeful distinction between the functional body (*soma* or *somatoplasm*) and the continuity of the germ-plasm and showed that transmission of characteristics from the one to the other was impossible (as later confirmed mechanistically in Francis Crick's exposition of the Central Dogma). Weismann published his ideas as they developed (the germ-plasm emerged in 1885) but his magnum opus *Das Keimplasma: Eine Theorie der Vererbung* first appeared (in German) in 1892. This was the year of AMM's review of Haeckel's *Vertebrate Embryology*, a decade after the *Museum* extension lectures on *The Darwinian Theory* were prepared, and a year before his death.

AMM does not mention Weismann in this lecture, nor in any of those which follow, despite discussing his early work on inheritance in Lectures 7 (from 1888) and 9 (1890) and mentioning the germ-plasm in his critique of Haeckel's book. There is nothing in the Lectures or archival material to suggest that AMM was aware of Weismann's complete theory.

HISTORY OF THE THEORY OF EVOLUTION

VERY early in human history the necessity arose for collective names to indicate the various groups and kinds of animals; to distinguish those that were useful for food, clothing, or weapons, &c., from those which were useless or dangerous. An early mode of classification was according to habitat; thus Solomon divided animals into beasts, fowls, creeping things, and fishes. This classification was hardly improved upon till the time of ARISTOTLE, 384–322 (?) B.C., a man for whose intellectual power the word stupendous seems barely adequate. Aristotle made many shrewd and acute observations which were not understood at the time, and were rediscovered 2000 years later: he was a man far ahead of his age.

After this follows a great gap in the history. Facts were steadily accumulating, but there was no system or governing principle. Nothing was known of the history of life on earth, and there was indeed no idea that there had been such a history. Scattered facts imperfectly ascertained were mixed with much superstition, and associated rather with witchcraft than with science.

LINNAEUS, 1707–1778, was the founder of modern scientific natural history. He was a self-made man, and had a long-continued struggle with poverty during the early part of his career. A botanist from his birth, he was never tired of learning facts about plants. Linnaeus was originally intended for the Church, but got on so badly at the schools that in 1727 his father proposed to apprentice him to a shoemaker; however, a friend interested in the young botanist persuaded the father, a poor pastor, to let him learn medicine. With this object in view Linnaeus went to Lund, and in 1728 to Upsala. In 1732 he was sent by the Literary and Scientific Society of Upsala to Lapland. In 1735 he went to Holland, and was introduced to Boerhaave, the celebrated physician of Leyden, and by him was introduced to a rich banker, who became his patron and in 1736 sent him to England. He afterwards went to Paris and eventually to Stockholm, where he gained his living by practising as a physician. In 1741 he was appointed Professor of Medicine at Upsala, and at the end of the year exchanged the chair of Medicine for that of Botany and Natural History. His poverty was now over and his fame well established.

Linnaeus was a man of extraordinary industry, and sent out his pupils in all directions, thereby collecting information and specimens from every part of the world, with the minute description, arrangement, and classification of which he charged himself. The first edition of his "Systema Naturae" appeared in 1735, and the twelfth edition in 1766. The great merit of Linnaeus lay in the fact that he was a man of method, and his strength lay in the orderly arrangement of his knowledge. He introduced a clear, precise, and definite terminology, and that this might be universal he employed Latin, the universal scientific language even at the present day. Still more important than this, he limited and defined the use, not only of words but of names, and established the binomial system. For instance, all roses were called Rosa: the common dog-rose was described as *Rosa sylvestris vulgaris, flore adorata incarnato,* which means, "The common rose of the woods, with a flesh-coloured sweet-scented flower." This system was sufficient perhaps, but clumsy. As we with our friends find it convenient to use a double name, so with plants and animals Linnaeus gave a double name—a *generic* and a *specific,* or, as he called it, a "trivial" name. Larger groups of families and orders were defined by a few easily recognised points, such as the number of stamens and pistils in the case of flowers, and the mode of their attachment.

This precise and accurate terminology and orderly arrangement for the first time made it possible to determine the number of different kinds of animals and plants, and to define their characters. This was a service of the utmost importance, and through Linnaeus botany first became a science; so also with zoology, though for a time less perfectly than botany. For the first time it became possible to think or speak of the animal or vegetable kingdom comprehensively, and when a name was used, to know precisely what it signified. This greater exactness at once brought men face to face with the problem—What is a species? What do we mean by species of animals or plants?

I will illustrate this by examples of what Linnaeus meant by species. For instance, the jackdaw, the raven, the rook, and the crow, are all species of the genus *Corvus.* These birds are clearly more like one another than either of them is like a starling or an eagle; in other words, there is a certain resemblance or affinity between them. Yet, they always differ in certain slight peculiarities of structure, form, and habits; also, they always produce their own kind, and do not interbreed—at any rate, as a rule. The jackdaw *(Corvus monedula)* is grey at the sides of its neck, and builds its nest in holes or cavities in rocks, churches, chimneys, and uninhabited houses. It feeds chiefly on insects, and is much the smallest of the four birds. The raven *(Corvus corax)* is the largest of the four, and is black all over. It makes bulky nests on crags or in trees. The rook *(Corvus frugilegus)* is a trifle smaller than the crow, and lives in noisy flocks. It is black, with a grey forehead and throat. Its nests are found in small trees, often near human habitations. The crow *(Corvus corone)* is smaller than the raven, and is black, tinged with green on the neck and throat, and purple on the back.[393] So also the lion, tiger, leopard, and domestic cat are all species of one genus, differing in points of structure, size, and habits. Again, the pine, fir, and larch are all species of the genus *Pinus.*

The idea of a species is therefore one separated by distinctive characters, which are constant and reproduced in the offspring. Jackdaws do not lay eggs from which crows are hatched, nor do cats give birth to lions. Linnaeus' idea of species was that they always had been distinct; that at the original stocking of the earth one pair of each kind or species was created, and that the existing species of animals and plants are the direct descendants of these original inhabitants. The initial objection to this view is that, if there were only one pair of each species to start with, they would immediately have eaten one another up, or have themselves died; herbivorous animals devouring plants, and being devoured in their turn by carnivorous animals. Linnaeus does not

---

393   Wallace also makes use of this example to illustrate the meaning of *species* in his *Darwinism; An Exposition of the Theory of Natural Selection with Some of its Applications* (London: Macmillan, 1889).

seem to have troubled himself much with the problem, and appears to have adopted the current views of the day without inquiry. The case was, however, different with some of his contemporaries.

BUFFON, 1707–1788, was a contemporary of Linnaeus, but a man of very different stamp. He came of a wealthy family, and enjoyed the best education that France could give him. At the age of twenty-one he succeeded to a handsome property, at the very time when Linnaeus, with an allowance of £8 a year from his father, was a struggling student at the University of Upsala, putting folded paper into the soles of his old shoes to keep out the damp and cold. Buffon was a man with a very keen interest in Natural History, and of remarkable industry and perseverance. In 1739 he was appointed Superintendent of the Jardin des Plantes at Paris, a post which he held till his death. In his great book on Natural History he gave a comprehensive account of all that was known concerning the distribution, habits, instincts, and structure of animals. Buffon strongly disapproved of the sharp lines and rigid systems of Linnaeus, and was more fond of general theories than of minute details. His materials were derived rather from extensive reading than from direct observation, and, having weak eyesight himself, most of his anatomical work was done by assistants. His great reputation was due very largely to the exceedingly attractive style in which he wrote, and the charm with which he invested the whole subject. He led many to think about and take an interest in Natural History, and to add to it by their own observations, who would not otherwise have done so. Buffon must not, however, be thought of as a mere popular writer; he really opened up new fields and led the way in many matters of the utmost importance. For instance, concerning the homologies of the mammalian skeleton, he was the first to compare the arm of man with the fore-leg of the horse. He also paid much attention to geographical distribution, and laid stress on the resemblance between the fauna of Northern Europe and that of America, and explained this by the existence of a former land connection wide enough to permit migrations. Moreover, the natural history of the various races of man was for the first time treated scientifically. Buffon, while at first a believer in the absolute fixity of species, later on was led to suggest that plants and animals may not be bound by fixed and immovable limits of species, but may vary freely, so that one kind may gradually and slowly be evolved by natural causes from the type of another. He points out the fundamental likenesses of type in many animals, underlying the external diversities of character and shape, which strongly suggest the notion of descent from some common ancestor. He goes further than this, and, granting the possibility of modification, sees no reason to fix its limits. He even suggests that from a single primordial being Nature has been able in the course of time to develop the whole continuous series of existing animal and vegetable life. This view is often expressed in a studiously guarded manner, and denied in half-ironical terms a few pages further on. This was the starting-point of the great idea of Evolution,[394] and Buffon's name

---

394   The first application of this Greek-based word to biology, signifying the transition of one organismal form into another, is usually attributed not to Buffon but to his near-contemporary Charles Bonnet (in *Considérations sur les corps organisés*, 1762). Thus its currency long precedes Darwin's *Origin of Species* (1859). Its etymological meaning is "to unroll", which allowed its adoption by pre-formationists seeking to describe

well deserves a place in its history. Before pursuing the progress of the theory of Evolution we must consider the work and personality of one of the greatest of all zoologists, Cuvier, and of his contemporaries.

CUVIER, 1769–1832, was a man of extraordinary industry and ability, and of commanding power—one of the giants of biological science. He was born in France, of Swiss parentage, and originally intended for the Church. From 1788 to 1794 he was engaged as private tutor to a French family near Caen. At this time he commenced to study the fossil brachiopods, which led him to compare them with the living species. Having by this time attracted the notice of influential people, he was in 1794 invited to Paris, and was appointed Assistant Professor, and in 1802 Professor, of Comparative Anatomy at the Jardin des Plantes. After this he rapidly rose to posts of great importance and distinction. Cuvier was specially distinguished by the zeal with which he applied himself to the actual dissection of large numbers of animals of many groups, by the clearness with which he kept himself free from the vague theories of his day; also by the way in which he kept to his facts, drew his own conclusions from them, and absolutely rejected any theories that were opposed to them.

Cuvier may justly be regarded as the father of Comparative Anatomy. His most important service was the demonstration of the true nature of fossils. About the time of his arrival in Paris attention was being directed to the skeletons and parts of skeletons of animals which were being disinterred round about Paris, especially at Montmartre. These constituted a great puzzle at the time, the bones of many of them being immensely large and unlike those of any known animals. Cuvier eagerly set to work at this subject, and the minute knowledge he had obtained of living animals rendered the work comparatively easy, while the law of correlation—a single bone giving the clue to the structure, position, habits, food, &c., of the animal—helped him greatly. Cuvier was soon able to point out that the animals of which these were the remains were in many cases not like those now living in Europe; in other cases they were unlike animals living anywhere on the earth at the present time. Some, such as the elephant, rhinoceros, and opossum were no longer European; others, such as the mammoth, were animals which at the present time are extinct. Attention was thus directed to this subject, and inquiry in other directions stimulated. In the older rocks other fossil remains still more unlike existing forms were discovered.

Cuvier has exercised the greatest influence on the study of zoology down to the time of Darwin. He was the founder of Comparative Anatomy, and penetrated into

---

the unfolding of pre-existing forms. Amongst pre-Darwinian users of the word, with various biological interpretations, were Charles Lyell (in the 1830s, relating to fossils emerging from sequential geological strata), Herbert Spencer (whose own theory of evolution, based on the work of Lamarck, was published in 1858) and even Darwin's own grandfather Erasmus Darwin (in *Zoonomia*, his all-encompassing, cosmological view of life's journey from a *one living filament* towards complex perfection, published 1794–96). As has often been noted, Darwin preferred to refer to *transmutation*, perhaps to avoid confusion or association with the views of others (including his grandfather) with which he disagreed. *Evolution* is famously absent from original versions of *The Origin of Species* although it does appear in the amended 6th edition (1872–76).

the subject much more deeply than Linnaeus, his object being not merely to system-
atise but to study the animals themselves.

The real nature of fossils was known to Aristotle, but the knowledge was
forgotten and lost. The current doctrine of the Middle Ages, and even so late as the
eighteenth century, was that they were freaks of Nature, and they were regarded as
unsuccessful creative attempts or models into which life had never been breathed.
Cuvier overthrew all this finally, and proved that they were the remains of animals
formerly dwelling on earth, and of different kinds or species to those now living.
He, moreover, showed that the farther back we go in time, the more do they differ
from recent animals. This was a tremendous step forwards, yet he stopped short, and
held back on the brink of a great and comprehensive theory. He was struck with the
differences rather than the resemblances among animals, and strenuously denied the
possibility of any relation between recent and fossil forms, stating that "the immutability of species is a necessary consequence of the existence of scientific natural
history." He formulated the doctrine of *Catastrophism*, or the periodical annihilation
of existing animals followed by re-creation in a modified form. The question of the
Origin of Species was now becoming a burning one.

GOETHE, 1790, in an "Essay on the Metamorphosis of Plants" showed the principle
of fundamental unity, and demonstrated that all parts of a flower are really modified
leaves or stem. In the cultivated rose the stamens and pistils are turned into petals,
and gardeners find it possible to cultivate a plant so that it shall be all leaves and no
flower, or shall have a gorgeous flower while the leaves remain small and insigni-
cant. It is a pleasant reflection to a naturalist that the keenest and brightest intellects
of all ages have not been unmindful of the charms of Natural History, and that they
have taken delight in, and have themselves made contributions of great value to the
subject. It is grateful to acknowledge this indebtedness to men more widely known for
their labours in other directions. Aristotle's first classification of animals on scientific
principles—viz., into Vertebrates and Invertebrates—holds good to the present day;
he also recognised the true nature of fossils as the remains of formerly living animals.
To poets of all ages and every nation we owe many shrewd and accurate observations,
especially on the habits of birds and flowers. Goethe was much struck with the power
of modification or adaptation, which led him to the conception of blood relationship
between animals, and of the descent from a common original type.

The most famous of the opponents of Cuvier, and upholders of the doctrine of
mutability of species, were two of his own countrymen, who were colleagues in Paris
for many years, Lamarck and St. Hilaire.[395]

LAMARCK, 1744–1829, was originally intended for the Church, like Linnaeus,
Cuvier, St. Hilaire, and Darwin. He, however, had a passion for the army, and on
the death of his father in 1760 set off for Germany, where the French were then
fighting. There he distinguished himself as a volunteer, but owing to being accidentally disabled by a comrade had to abandon his career. He then went to Paris, and
commenced the study of medicine, while holding a post as a banker's clerk. He was
for a long time interested in botany, and in 1779 published three small volumes on

---

395   See Lecture 1 for a more detailed account of these arguments.

the Flora of France. This attracted the notice of Buffon, and in 1793 Lamarck obtained an appointment at the Jardin des Plantes, the year before Cuvier, and applied himself with great vigour to the study of zoology. Lamarck and Cuvier worked practically side by side for many years. Cuvier, working mainly at Vertebrates and at fossils, was impressed by the differences between species, especially those between recent and fossil species. Lamarck worked mainly at the lower Invertebrates—jelly-fish, worms, and snails. He, on the other hand, was struck by their resemblances rather than the differences between them, and was impressed by the difficulty of settling which were distinct species and which might have come from the same parents. On this point he observes: *"The more we know of animals and plants, the more difficult we find it to settle which are related to each other and which are not."*

Lamarck was also impressed by the variability of animals and plants, according to their surroundings, and by the influence of drier soil or mountain habitat in causing stunting of growth and other alterations. In his *"Philosophie Zoologique,"* published in 1809, and in the *"Histoire Naturelle des Animaux sans Vertébres,"* published in 1815, he upholds the doctrine that all species of animals, including man, are descended from other species. With respect to the manner of modification, he attributes something to the direct action of environment, and much to use and disuse—i.e., to the effects of habit; for example, the use of the long neck of the giraffe for browsing on trees. He writes *"The systematic divisions of classes, orders, families, genera, and species are the arbitrary and artificial productions of man. Species arise out of varieties. In the first beginning only the very simplest and lowest animals and plants came into existence; those of a more complex organisation only at a later period. The course of the earth's development and that of its organic inhabitants was continuous, and not interrupted by violent revolutions. Life is purely a physical phenomenon."* If man can make such changes in a few hundred years, as for example, to produce the various domestic races of pigeons or rabbits, *"is it not possible that nature in all the long ages during which the world has existed may have produced the different kinds of plants and animals by gradually enlarging one part and diminishing another, to meet the wants of each?"*

This is a full statement of all the essential points; the unity of active causes in organic and inorganic nature; the ultimate explanation of these causes in the chemical and physical properties of matter; the derivation of all organisms from some few most simple forms; the coherent course of events in Nature, and the absence of cataclysmal revolutions. It is a full and complete statement of the doctrine of Evolution as held at the present day, man himself being included, both as regards his mental powers and his bodily structure. This is often confused with the Darwinian theory, but is really quite distinct. Lamarck tells us that animals are descended one from another, and have a common bond of union or blood relationship. This explains the affinities of animals and of man, but fails to explain how and why. The Darwinian theory explains how this came about, and why this progressive transformation of organic forms has taken place, and what causes effected the uninterrupted production of new forms. Lamarck's views, though perfectly correct, were mere speculations until Darwin supplied the reason and explained the mode of action. Lamarck considered the long neck of the giraffe as due to its constantly stretching its neck to pick leaves from high trees; the long tongue of the woodpecker, humming-bird, and ant-eater to the habit of fetching food out of deep or narrow crevices; the webbed toes of the frog to its constant endeavours to swim, and to the very movements of swimming. The true explanation, as we shall see afterwards, was furnished by Darwin's theory of natural selection.

ST. HILAIRE, 1771–1840, was educated as a priest, but owing to his passionate love for zoology was allowed to stay in Paris and work at the Jardin des Plantes. He was offered an appointment, and afterwards joined Lamarck at the Musée d'Histoire Naturelle in 1793. St. Hilaire was a great friend of Lamarck, and adopted his theory of descent. He believed that the transformations of animals were effected less by the action of the organism itself, than by change in the outer environment. A long and fierce controversy raged between the three friends, Cuvier, Lamarck, and St. Hilaire, chiefly between Cuvier and St. Hilaire, who, strangely enough, was thought an abler man than Lamarck. Shortly after Lamarck's death a formal and final discussion took place in the Academy of Sciences at Paris between Cuvier and St. Hilaire. This occurred on the 22nd of February 1830, and was renewed on the 19th of July, a bare week before the outbreak of the French Revolution. All Europe was excited by the controversy, and none more so than Goethe, then an old man, and a firm believer in the doctrine of Evolution.

Cuvier was far too strong for his opponent; a hard hitter, and a man of greater personal power and influence, and one who did not scruple to use his full strength. He urged the evidence of mummies and other buried remains, which, after a lapse of thousands of years, agree in the smallest details with existing species. If a changing environment causes alterations, why, he asked, are these not altered? He demanded evidence of connecting links between fossils and recent forms, and quoted his own unrivalled experiences as to its absence. Cuvier crushed his opponent by superior knowledge, by the better management of his case, and by personal authority. The verdict was a definite one, and the controversy was regarded as closed by final decision. The fixity of species was regarded as proved, and France has hardly yet recovered from the traditions of Cuvier.[396]

We now approach the final stage in the great controversy, and the scene of action shifts to our own country. Light was first afforded, not by zoology or botany, but by the sister science of geology, a peculiarly British study, and a very recent addition to the tree of knowledge.

HUTTON, 1726–1797, when sixty-two years old, published his "Theory of the Earth." The main motive of this book was to show that in order to understand how the earth's crust, with its component layers, was formed, and how fossils got into them, we must not guess, but must look for ourselves, and see what is now going on around us— how rivers and glaciers are carrying down earth and stones from the mountains to the sea, how the solid earth is being wasted every day, and new rocks formed by the disintegration of older ones.

---

396   This reads as a slight on the French biological tradition but it is not entirely clear that this is what AMM intended. Although he moves the spotlight to England in the next paragraph, he has to go back in time to do so. There is no doubt, as illustrated in the Lectures as a whole, that the vast majority of the contemporary anatomy, embryology and cell biology in which he was interested stemmed from Germany or German-speaking scientists. It could be that that he simply considered Cuvier's (erroneous) intervention as a prime cause of the shift of productive research away from France to other parts of Europe. (See also the final paragraph of Lecture 1 for another quasi-nationalist sentiment.)

CHARLES LYELL 1797–1875, like Hutton, was a Scotchman, and was born in the year of Hutton's death. Lyell worked out in detail Hutton's suggestions, and collected with great care all that was known of the changes now going on in the world, and of the causes that produce them; the rate of denudation and the amount carried down by streams; the mode in which plants and animals are buried in mud, peat, and sand. The first volume of his *"Principles of Geology"* was published in January 1830, seven months before the Paris controversy. In it he dealt with the changes continually taking place on the earth's surface; the rising up and the subsidence of the earth's crust; the action of rivers and volcanoes; and showed how the present configuration of the earth was due to these causes. He pointed out that causes now in action and influences now at work are not merely competent to produce the present state of things, but must inevitably have done so. He showed the history of the earth to be continuous and uninterrupted, and that to explain its present condition and past history we have simply to look around us.

Lyell's facts were numerous and his reasoning cogent; his conclusions therefore steadily gained supporters. It is a curious fact that, as regards fossils, Lyell himself declined to apply to them the principles he so justly insisted on for the crust of the earth. Yet it was precisely by applying a similar train of reasoning to the further problem that the final solution was attained, and by the study of what is going on around us at the present day, that principles were determined competent to account for the changes in all past time, and the death-blow given to Cuvier's views.

Evidence as to the reality of Evolution was now rapidly accumulating, not only through the work and writings of those already mentioned, but from many other sides. Professor HUXLEY in 1859 refers to the hypothesis that species living at any time are the result of the gradual modification of pre-existing species, as the *"only one to which physiology lends any countenance."* Sir JOSEPH HOOKER in 1859, in his *"Introduction to the Australian Flora,"* admits the truth of descent and modification of species, and supports the doctrine by many original observations.

HERBERT SPENCER'S essay in the *Leader*, 1852, constitutes "the high-water mark" of Evolution prior to Darwin: *"Even could the supporters of the development hypothesis merely show that the production of species by the process of modification is conceivable, they would be in a better position than their opponents. But they can do much more than this; they can show that the process of modification has effected and is effecting great changes in all organisms subject to modifying influences. . . . . They can show that any existing species—animal or vegetable—when placed under conditions different from its previous ones, immediately begins to undergo certain changes of structure fitting it for the new conditions. They can show that in successive generations these changes continue until ultimately the new conditions become the natural ones. They can show that in cultivated plants and domesticated animals, and in the several races of men, these changes have uniformly taken place. They can show that the degrees of difference so produced are often, as in dogs, greater than those on which distinctions of species are in other cases founded. They can show that it is a matter of dispute whether some of these modified forms are varieties or modified species. And thus they can show that throughout all organic Nature there is at work a modifying influence of the kind they assign as the cause of these specific differences; an influence which, though slow in its action, does in time, if the circumstances demand it, produce marked changes; an influence which, to all appearance, would produce in the millions of years, and under the great varieties of condition which geological records imply, any amount of change."*

It is impossible to depict better than this the condition prior to Darwin. In this essay there is full recognition of the fact of transition, and of its being due to natural influences or causes, acting now and at all times. Yet it remained comparatively unnoticed, because Spencer, like his contemporaries and predecessors, while advocating Evolution, was unable to state explicitly what these causes were.

We have now traced the main steps in the history of the doctrine of evolution, and have mentioned the names of the chief men with whom this history is most closely associated. This doctrine was rendered possible by Linnaeus by the introduction of definite and precise nomenclature in the language common to all civilised nations, which thereby enabled men to speak of animals and groups of animals with exactness and certainty. The difficulties of framing definitions based on facts, of which equally competent men took widely different views, led to the consideration of the question—Are species fixed or mutable? Buffon, Goethe, Lamarck, St. Hilaire, and Herbert Spencer are perhaps the most famous names among the supporters of Evolution. Lamarck and St. Hilaire are specially noteworthy, as they recognised the necessity of explaining the causes of the modification, and attempted, though with very partial success, to supply the explanation. Among the opponents of Evolution, Cuvier was by far the ablest.

The question first became a prominent one about the commencement of the present century. It was hinted at earlier by Buffon, but first obtained definite expression from Lamarck in 1801, and in more detail in the *"Philosophie Zoologique,"* published in 1809.

It is very commonly assumed that the doctrine that animals are not immutable, or the doctrine of Evolution, is of very recent origin, and that for it we are indebted to Darwin. Nothing can be more erroneous, for, as we have seen above, not only was it very clearly and emphatically maintained by several writers at the commencement of the present century, or the conclusion of the last, but the idea is found stated more or less explicitly by Aristotle over 2000 years ago.

The "Doctrine of Evolution" teaches that there is a relationship between the animals of successive periods or ages of exactly the same kind as that which exists between the men of successive generations or centuries—viz., a blood relationship. Just as men of each century are descendants of those of a preceding century, and progenitors of those of later ones, so it is with animals throughout all geologic history.

It is well to point out clearly the difficulty which has to be met. Animals of successive ages are unlike, and fossils do not give the intermediate series, nor satisfactory indications of them. The problem we have to consider is this: The men of successive generations are unlike in language, customs, dress, and appearance; now, are the differences between animals of successive ages of the same character as between men, though of wider nature; or are they of such a kind as to forbid the idea of descent one from another? In other words, are species immutable or variable? The doctrine of Evolution requires that they should be variable. In order to establish this doctrine it must be shown that there are causes, actually existent causes, competent to give rise to modifications of animals such as we find in passing from one geologic age to another. This is what is effected by the "Darwinian Theory," or the "Theory of Natural Selection," propounded independently and simultaneously on July 1, 1858, by Darwin and Wallace.

CHARLES DARWIN was born in 1809, and studied at Cambridge from 1827 to 1831. The voyage of the *Beagle* occupied from 1831 to 1836, the greater part of the time being spent on the east and west coasts of South America, and the voyage round the world being completed by way of New Zealand, Australia, Mauritius, St. Helena, and Brazil, and then by the Cape Verde Islands to England. The remainder of his life was devoted to work in England, work of extraordinary amount and most varied character. By this work he slowly accumulated facts, and especially the conditions under which the breeds of domesticated animals and cultivated plants come into existence, and are propagated or modified. I propose to consider more in detail the history of Darwin's life in the last lecture.

ALFRED RUSSEL WALLACE was born in Monmouthshire in 1823. As a boy he was an eager naturalist. From 1844 to 1845 he was English master at the Collegiate School at Leicester, and while there made the acquaintance of Mr. H. W. Bates, an ardent entomologist. A few years later the desire to visit tropical countries became too strong to resist, and a joint expedition took place to collect Natural History objects, and to *"gather facts towards solving the problem of the Origin of Species."*[397] In 1848 he started to the mouth of the Amazon, and worked with Bates till 1850, when Wallace moved to Rio Negro, finally returning in October 1852. His vessel was destroyed by fire, and he spent ten days in an open boat on the sea in the mid-Atlantic. From 1854 to 1862 he spent his time in the Malay Archipelago, where animal life was most luxuriant and least affected by man. In June 1858 Darwin received from Wallace the MS. of a paper *"On the Tendency of Varieties to depart indefinitely from the Original Type,"* this being the same conclusion at which Darwin himself had arrived. Darwin wished to publish

---

397   AMM is quoting here a line in Bates' reflective account of his and Wallace's travels in the Amazon Basin: *The Naturalist on the River Amazons: A Record of Adventures, Habits of Animals, Sketches of Brazillian and Indian Life and Aspects of Nature Under the Equator During Eleven Years of Travel* (London: John Murray) published in 1863. Bates reports that Wallace identified this as the purpose of their expedition in a letter to him of 1847, the year before they travelled. In fact, no letter containing the statement has been found and it is doubtful whether Wallace ever expressed such an aim. The avowed purpose of their joint venture was specimen collection with a view to commercial exploitation. The myth of the Wallace statement has been perpetuated ever since (including at https://en.wikipedia.org/wiki/Henry_Walter_Bates until the present author corrected it), probably because it allows a convenient but covert rewriting of history, exploiting a phrase (*origin of species*) which only later became ubiquitously associated with the Darwin–Wallace theory. For more on this gripping mystery, including a detailed analysis of why Wallace could not have made such a statement in 1847, see Van Wyhe J, A delicate adjustment: Wallace and Bates on the Amazon and 'The Problem of the Origin of Species'. *Journal of the History of Biology* 2014; **47**: 627–659. Van Wyhe speculates that Bates was using his memoir to stake a retrospective claim for partial ownership of the Theory of Evolution by Natural Selection, after it became so widely assigned to Wallace as well as Darwin. A particular curiosity in the present lecture version of the phrase are the initial capitals on *Origin* and *Species*. Was this what AMM really wrote or was it an unwitting editorial solecism by CFM born of familiarity with Darwin's famous book?

Wallace's paper at once, but on the urgent persuasion of Sir Joseph Hooker and Sir Charles Lyell he consented that extracts from one of his earlier MS. prepared in 1844, and from a letter to Professor Asa Gray in 1857, should be read at the same time. This took place at the Linnaean Society on July 1, 1858. Each man's discovery was perfectly independent, and neither knew on what lines the other was working. Full mutual recognition ensued, and most cordial intercourse and esteem.[398]

SUMMARY.—We have now traced the gradually increasing tendency towards a belief in Evolution. This was suggested in a tentative and almost cynical way by Buffon, and warmly supported by Goethe, and during the first half of the present century by Lamarck, St. Hilaire, Herbert Spencer, and others. Owing to one fatal flaw this belief failed to command anything like general acceptance. We know from fossils that the former dwellers on earth were unlike those now living, and that the doctrine of Evolution therefore involves modification. No one was able to point to the causes which could lead to such modification or to explain how it could have come about. This was the objection which was driven home with relentless force and persistency by Cuvier, supported by all the weight of his personal authority, and the influence rightly gained by his splendid contributions to the science of Comparative Anatomy. This is the objection which in 1830 proved fatal, and which led to the triumph of Catastrophism over Evolution. The supporters of Evolution were silenced, but not convinced. In France the defeat was complete, but not so in other countries. It was clear that the attack must be made along new lines if there were to be any prospect of success. This fatal objection must be met. Was it not possible to determine the causes? and if so, how? What could we hope to know of causes which could lead to modifications in animals of former geologic ages? Yet the answer was at hand, for seven months before Cuvier's final triumph at the Academy of Paris on July 19, 1830, appeared the first volume of Lyell's *"Principles of Geology,"* in which the true path was indicated, and the key to the past shown to be afforded by the study of the present. If we would know what happened in former times, we should look around us and see what is taking place before our eyes. In this way Lyell gauged the forces of Nature—the power of running water, the force of the tides, the effect of frost and heat, the slow movements of upheaval and subsidence, the slow change of climate due to astronomical causes. He was able to prove that causes now acting, and causes which must have been in action from immeasurably remote periods, were competent to produce the effects we wonder at—the upheaval of mountain ranges, the excavation of valleys, &c.—without any need for external or supernatural agencies, and indeed, leaving no room for such agencies, for then there was nothing further to accomplish. The final stroke was given by Darwin and Wallace, who, working perfectly independently, set themselves deliberately to attack the problem on the lines laid down by Lyell—viz., by prolonged and detailed study of the conditions under which animal life exists at the present day. After years of patient work they were led independently to identical conclusions, which were announced simultaneously from opposite sides of the globe.

---

398   A fuller account of Wallace and Darwin's gentlemanly conduct appears in Lecture 21.

LECTURE 15

ARTIFICIAL SELECTION AND NATURAL SELECTION

Having summarised the history that preceded the Darwin–Wallace theory, AMM now turns to the evidence. Many of the examples he presents appear in Darwin's work (especially *Origin of Species* and *The Variation of Animals and Plants under Domestication*) and also in writings by Wallace (including *Darwinism: An Exposition of the Theory of Natural Selection with Some Applications* and *Contributions to the Theory of Natural Selection*). The sources are sometimes evident from the lecture text and sometimes not, although it is not difficult to find them by searching Darwin-online.org.uk and Wallace-online.org.uk. Rather than litter the text with citations, in this lecture and the subsequent ones I have with a few exceptions only commented where there is something additional to say about AMM's exposition.

At the end of AMM's presentation of *The Struggle for Existence* element of Wallace's chart, and later at the end of *The Influence of Environment*, we find rare glimpses of his political position. They fit completely with his evident enthusiasm for social education amongst working people: besides the lecture series on which these Darwin lectures were based, material in the Manchester Archives reveals that between 1879 and 1892 he gave at least 13 lectures to workers' organisations in Bradford, Leeds and Manchester, in addition to the many university extension lectures associated with Owens College. Members of mechanics' or other institutions could attend the Owens College Museum lectures for a reduced fee. The annual Recreation in Ancoats events feature prominently in the list of public lectures; he kept copies of the Ancoats programmes, reports and financial accounts amongst his papers although whether he held any committee position with the Ancoats Brotherhood is not clear.

A report of the inquest into AMM's death, published in the *Manchester Guardian* (3 January 1894, p. 8), has the following addendum:

> *"Speaking at Sale yesterday, Mr Sadler, secretary to the Oxford University Extension Delegacy, lamented the loss that the University Extension movement had sustained in the death of Professor Milnes Marshall, of Owens College. The late Professor's personal distinction and high attainments had been an honour and a safeguard to University Extension, and he had assiduously served the movement both as a lecturer and as an administrator. It had, he believed, been not a [l]ittle owing to Dr Marshall that the Victoria University had thrown so large a part of its great energy into the Extension work. He had given cordial assent to a policy which secured good feeling and friendly emulation among the various authorities engaged in University Extension in the North of England. The formation and active life of the Lancashire and Cheshire Association of the University Extension centres was in considerable measure due to Dr Marshall's energy, influence and foresight. The Extension movement could ill afford to lose so firm a friend."*

LECTURE 15

ARTIFICIAL SELECTION AND NATURAL SELECTION

I NOW propose to consider the law[399] to which Darwin and Wallace were led, the evidence upon which it is founded, and the conclusions which follow from it. In the method of attack I propose to follow Darwin, and I would warn you against almost inevitable disappointment, for it is with commonplace things and facts of every-day occurrence that a great theory has to deal.

ARTIFICIAL SELECTION.

DOMESTIC PIGEONS.—Darwin early in his inquiry felt the importance of having individual animals under close observation, so that all conditions influencing them could be determined. For this purpose domestic animals were far more suitable than wild ones, and pigeons were selected for special study for these reasons:—(1) The evidence of their descent from a common ancestor is clear; (2) Their historical records extend back many centuries; (3) Their variations are very great, all kinds being easily kept in captivity and all breeding true.

There are probably at least 200 kinds of pigeons known which breed true, and these differ constantly from each other. The chief varieties are the following. (See Fig. 1.)

The *Pouter* is a large and upright bird with a long body and long legs, a moderate-sized beak, and a very large crop and oesophagus. It has the habit of inflating its crop, producing a "truly astonishing appearance," being then "puffed up with wind and pride."

The *Carrier* is a large bird with a very long beak. The skin round the eyes, over the nostrils, and on the lower jaw is much swollen, forming a prominent wattle.

The *Barb* has a short and broad beak, and a wattle of moderate size.

The *Fantail* has tail feathers to the number of 34 or even 42, twelve being the normal number. The tail is expanded and held erect. It has a peculiar gait, and a curious habit of trembling by convulsive movements of the neck. In a good specimen the tail should be long enough to touch the head.

The *Turbit* has a frill formed by divergent feathers along the front of the neck and breast. The beak is very short.

---

399 Although the Darwin–Wallace insight is now more commonly called the *theory* of evolution by natural selection, and this is indeed what AMM calls it elsewhere, *law* is not an inappropriate usage. Wallace's co-contribution was first expressed in a manuscript entitled "On the law which has regulated the introduction of new species" (*Annals and Magazine of Natural History*, **16**, 2nd series: 184–196). This was sent from Sarawak in 1855 and has become known as the Sarawak Law paper.

FIG. I.

The *Tumbler* has a small body and short beak. During flight it has the habit of turning involuntary back somersaults.

The *Jacobin* has long wings and tail and a moderately short beak. It has a hood formed by the feathers of the neck.

The *Trumpeter* has a tuft of feathers at the base of the beak, curling forwards. The feet are much feathered. The coo is very peculiar, and unlike that of any other pigeon, being rapidly repeated and continued for several minutes.

Among these forms there is thus great diversity in both form and colour. This diversity also affects the internal structure, for example the skull: the caudal and sacral vertebrae and also the ribs vary in number. The number of primary wing and tail feathers, the shape and size of the eggs, the manner of flight, and almost all other characters, also differ. If these birds were now found in a wild state, they would be considered to constitute distinct genera, yet they are known to be all descended from *Columba livia* the blue rock-pigeon of Europe, Africa, India, &c.

The arguments brought forward by Darwin to prove this are as follows:—

(i.) All domestic races are highly social, and none of them habitually build or roost in trees; hence it is in the highest degree probable that their ancestor was a social bird nesting on rocks.

(ii.) Only five or six wild species have these habits, and nearly all these but *Columba livia* can be ruled out at once.

(iii.) *Columba livia* has a vast range of distribution—from Norway to the Mediterranean, from Madeira to Abyssinia, and from India to Japan. It is very variable in plumage and very easily tamed. It is identical with the ordinary dove-cot pigeon, and except in colour practically identical with toy pigeons generally.

(iv.) There is no trace of domestic pigeons in the feral condition.

(v.) All races of domestic pigeons are perfectly fertile when crossed, and their mongrel offspring are also fertile. Hybrids between even closely allied *species* of pigeons are, on the other hand, sterile.

(vi.) All domestic pigeons have a remarkable tendency to *revert* in minute details of colouring to the blue rock-pigeon. This is of a slate-blue colour, with two bars on the wings, and a black bar near the end of the tail. The outer webs of the outer tail-feathers are edged with white: these markings are not seen together in any other species of the family. This tendency to *revert* was demonstrated by Darwin as follows: He first crossed a white fantail with a black barb; then a black barb with a red spot (a white bird with a red tail and a red spot on the forehead). He then succeeded in crossing the mongrel barb-fantail with the mongrel barb-spot, and the birds produced were blue, with markings on the tail and wings *exactly like those of the ancestral rock pigeon*. Thus two black barbs, a red spot, and a white fantail, produced as grand-children birds having every characteristic of *Columba livia,* including markings found in no other wild pigeon.

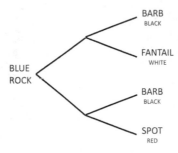

(vii.) All domestic pigeons resemble *Columba livia* in their habits. They all lay two eggs, and require the same time for hatching. They prefer the same food, and coo in the same peculiar manner, unlike other wild pigeons.

(viii.) *Columba livia* has been proved to be capable of domestication in Europe and in India.

(ix.) *Historical Evidence.*—Referring to Aldrovandi, who figured pigeons in the year 1600, we find the Jacobin with a less perfect hood; the Turbit apparently without its frill; the Pouter with shorter legs, and a less remarkable bird in all respects; the Fantail with fewer tail feathers, and a far less singular appearance; the Tumbler existed then, but none of the short-faced forms; the Carrier had a beak and wattle far less developed than the modern English Carrier. These were the same groups of pigeons, but with their distinctive characters less marked, thus showing convergence towards their common ancestor.

The mode of action of these changes is by *artificial selection,* or the power possessed by man of influencing the shape, size, and colour of animals by the accumulation of small differences in successive generations. This depends on two laws:

1. *The Law of Variation,* depending on the fact that no two animals are exactly alike.
2. *The Law of Inheritance,* or the tendency to hand down characters and peculiarities to descendants.

FIG. 2.

DOMESTIC POULTRY.—These afford another good example of artificial selection. (See Fig. 2.)

The *Gamecock* is characterised by its upright comb, strong beak, sharp spurs and by its great courage. Of all the different forms, this most resembles the wild *Gallus bankiva*.

The *Cochin* is of large size, and scarcely able to fly. The plumage is soft and downy; the legs thick and feathered; the comb and wattle well developed.

The *Dorking is* a large bird with a large comb and wattle, and possesses an extra toe.

The *Spanish* is tall and of stately carriage. The comb and wattle are very large.

The *Hamburgh* has a flat comb prolonged backwards, and a moderate-sized wattle.

The *Polish* is characterised by a large crest of feathers, the comb being either absent or very small. The wattle is sometimes replaced by a tuft of feathers.

The *Bantam* is of small size and bold erect carriage.

The *Silk-fowl* is a small bird with very silky feathers.

All these birds, differing so much among themselves, are descended from *Gallus bankiva,* the Jungle-fowl (Fig. 3), which is still found in a wild state in India and the Malay Islands. This bird was domesticated in India and China before 1400 B.C. and was introduced into Europe about 600 B.C. Several distinct breeds were known to the Romans about the commencement of the Christian era.

FIG. 3. *Gallus bankiva* (The Jungle-fowl).

THE ANCON SHEEP.—Another well-known example of artificial selection, and one of the few known instances in which new breeds have suddenly originated, is that of the *Ancon* Sheep, bred by Seth Wright, a farmer of Massachusetts.[400] In 1791 one of Seth Wright's sheep bore a male lamb which had very short and bandy legs. Now, as Wright was continually losing his sheep, owing to their jumping over his fences, it occurred to him that, if he could produce a breed of sheep with short bandy legs, he would lose none of them, as they would be unable to jump his fences. He therefore bred entirely from the short-legged ram when it had reached maturity, and after a few years succeeded in raising a considerable flock of this variety, which was known as the Ancon sheep.

The power of artificial selection is almost unlimited, and breeders of animals speak with the utmost confidence of being able to produce any desired result in the form of the body; and, in the case of poultry and pigeons, in the length of beak, the number of feathers, and even in the markings on particular feathers. Lord Spencer[401]

---

400   This example was mentioned by Darwin in both *Origin* and *Variation* and was used until quite recently as an example of rapid evolution taking place within a generation. In fact, it is a form of achondroplasia resulting from a gene mutation (see Gidney L, *International Journal of Osteoarchaeology* 2007; **17**: 318–321 and Bergman G, *Biology Forum* 2005; **98**: 435–448). One could argue that the mutation was beneficial insofar as it gave selective advantage within the environment created by Seth Wright. However, health problems associated with the disorder make it unlikely to persist in other than exceptional environmental circumstances.

401   John Charles Spencer, 1782–1845, British agriculturalist and politician who was a pioneer breeder of domestic cattle; first president of the Royal Agricultural Society.

says: *"It is therefore very desirable before any man commences to breed either cattle or sheep, that he should make up his mind as to the shape and qualities he wishes to obtain, and strictly pursue this object."* And speaking of Leicester sheep Lord Somerville remarks: *"It would seem as if they had just chalked on the wall a form perfect in itself, and then had given it existence."*[402] So also with plants; enormous changes are effected by cultivation—i.e., by selection, in fruits and flowers.

NATURAL SELECTION.

The theory of Natural Selection teaches that there are in Nature causes which act in much the same way as man acts when selecting artificially the best animals for breeding purposes; causes which must lead to structural modifications; and that this is the clue to the unlikeness between the fauna of successive geologic ages.

I propose first to give an outline of the argument and to consider the arrangement given in Mr. Wallace's chart of the Theory of Natural Selection.[403]

WALLACE'S CHART OF THE THEORY OF
NATURAL SELECTION.

PROVED FACTS		CONSEQUENCES
A. RAPID INCREASE OF NUMBERS	}	STRUGGLE FOR EXISTENCE
B. TOTAL NUMBERS STATIONARY		
C. STRUGGLE FOR EXISTENCE	}	SURVIVAL OF THE FITTEST
D. VARIATION WITH HEREDITY		
E. SURVIVAL OF THE FITTEST	}	STRUCTURAL MODIFICATIONS
F. CHANGE OF ENVIRONMENT		

---

402   This is taken directly from *Origin*. My colleague Julian Wiseman kindly points out that Perfect was a description of many new livestock breeds of the time, so Darwin, and then AMM, may have missed a hidden meaning. John Southey Somerville, Baron Somerville (1765–1819) was a member of the Board of Agriculture, the UK's foremost breeder of merinos and author of articles on animal husbandry and agricultural practice. For further details, see the *Dictionary of National Biography*: https://en.wiki-source.org/wiki/Somerville,_John_Southey_(DNB00).

403   Such a chart appears in the *Conclusion* to *Creation by Law*, Chapter VIII (pp. 264–301) of *Contributions to the Theory of Natural Selection*, published in 1870. The original essay with that title, which did not contain the chart, was published in *Quarterly Journal of Science*, October 1867. Wallace's original chart is headed *A Demonstration of the Origin of Species by Natural Selection*. AMM has modified the content a little and omitted Wallace's careful referencing of each *Proved Fact* and *Consequence* to chapters in Darwin's *Origin*, to Lyell's *Principles of Geology*, or to *"the pages in this volume where they are more or less fully treated"*.

A.   *All animals produce far more young than can survive.* Consider for instance the enormous number of eggs of fish or oysters.

B.   *The total numbers are on the average stationary.* As a necessary consequence of this there is a *struggle for existence,* because there is neither enough space nor food for all.

D.   *Variation with heredity.* No two animals are exactly alike, and their distinctive characters are transmitted from generation to generation. The consequence of this is the *survival of the fittest*—that is, that in the long run those best adapted to their circumstances and environment will have the best chance of surviving, and of leaving descendants who will hand down their peculiarities. Just as man selects artificially the forms best suited for his purpose, and by breeding from them produces great changes in structure and habit, so in Nature the best and fittest of each generation have an advantage and the best chance of survival.

F.   *Change of environment,* rendering old characters of less value and bringing new ones to the fore.

From this follow—*Structural modifications.*

Causes are always at work which must lead to change in structure, and this to an apparently unlimited extent.

Let us now examine the argument more closely.

A. RAPID INCREASE OF ORGANISMS.—*"There is no exception to the rule that every organic being, animal or plant, naturally increases at so high a rate that, if not destroyed, the earth would soon be covered by the progeny of a single pair."* Man himself has doubled his numbers in the United States in the course of twenty-five years, and at this rate in less than 1000 years there literally would not be standing-room on the earth for his progeny. Linnaeus showed that an annual plant producing two seeds only—and there is no plant so unproductive as this—and these each producing two in the following year, and so on, would in twenty-one years produce over a million plants, as shown in the following table:—

ANNUAL PLANT PRODUCING TWO SEEDS ONLY

YEAR.	NO. OF PLANTS.
1	1
2	2
3	4
5	16
7	64
9	256
11	1,024
13	4,096
15	16,384
17	65,536
19	262,144
21	1,048,576

The rate of increase of an animal, each pair producing ten pairs annually, and each animal living ten years, is shown in the following table:—

YEAR	PAIRS PRODUCED	PAIRS ALIVE AT END OF YEAR.
I	10	11
2	110	121
3	1,210	1,331
4	13,310	14,641
5	146,410	161,051
10		25,937,424,600
20		Over 700,000,000,000,000,000,000,000

Vast numbers of eggs are laid by some animals; the conger-eel, for instance, lays 15 millions; the herring 20,000; the oyster from half a million to 16 millions; and a very large oyster may produce even 60 millions of eggs. Supposing we start with one oyster and let it produce 16 million eggs, the average American yield, and let half, or eight millions, be female and go on increasing at the same rate; in the second generation we shall have 64 millions of millions of female oysters. In the fifth generation—i.e. the great-great-grandchildren of our first oyster—we should have 33 thousand millions of millions of millions of millions of millions of female oysters. If we add the same number of males we should have in all 66 + 33 noughts. If we estimate these as oyster-shells, we should have a mass more than eight times the size of the world.

A large number of eggs or young is, however, not essential. The *Fulmar petrel* lays only one egg, yet it is believed by Darwin to be one of the most numerous birds in existence. The *Passenger* pigeon again only lays two eggs, yet it is extraordinarily abundant in parts of North America, where its enormous migrating flocks darken the air for hours.[404] A remarkable account is quoted by Wallace of a wood in Kentucky, 40 miles in extent, where there was a perpetual tumult of crowding and fluttering pigeons, and where there were as many as a hundred nests on a single tree, the branches of which were often broken off by their weight, and the ground strewn with broken limbs of trees, eggs, and young birds, on which herds of hogs were fattening. Hawks, buzzards, and eagles were flying about in large numbers, seizing the young birds at pleasure; and numerous parties of men from all parts of the adjacent country were camping with their families for several days, felling trees to get the nests.

Another good example of rapid increase in numbers was seen in the rabbit pest of Australia in 1887. The common grey variety of wild rabbit introduced into Victoria in 1860, became so prolific as to overrun the greater portion of the colony and great sums of money were expended in endeavouring to exterminate it.

---

404   At the time when AMM was writing, the Passenger Pigeon (*Ectopistes migratorius*) was embarking on its precipitous and catastrophic decline, brought about by human hunting. It is said to have become extinct in 1914 when the last female in captivity died.

B. THE NUMBERS ARE STATIONARY AS A WHOLE.—Some forms increase while others diminish in numbers, and hence there is not actually this rapid increase of adult forms. Enormous numbers are devoured as food in their early stages, the seeds of plants being eaten by birds, and the young of various animals by other animals. Many animals again have their numbers kept down by parasites; for instance, the caterpillar of the large garden white butterfly is peculiarly liable to attacks from the ichneumon fly, which lays its eggs in the body of the caterpillar; and out of 533 larva collected by Mr. Poulton in 1888, 422 full-fed caterpillars died from the presence of ichneumon grubs—i.e., four out of five perished from this cause alone.

C. THE STRUGGLE FOR EXISTENCE—Every single organic being may be said to be striving to the utmost of its power to increase its numbers, while the vast majority of animals and plants that come into the world are doomed to die early. For instance, in the case of an annual plant producing 1000 seeds —which is no very large estimate— if the numbers remain stationary, only one of these 1000 can on the average come to maturity; and it may be said to struggle with plants of the same or other kinds which already clothe the ground. In fact, *"all the plants of a country are at war with each other."*

Again, the introduction of goats into St. Helena led to the entire destruction of the native forests, consisting of about a hundred distinct species of trees and shrubs, the young plants being devoured by the goats as fast as they grew up. A famous illustration of the nice balancing of forces between animals and plants is furnished by cats and clover. The red clover is fertilised almost exclusively by humble bees; and field mice destroy the nests of humble bees in large numbers. Newman[405] estimates that two-thirds of the total number of humble bees' nests in England are thus destroyed. Now, the number of mice depends largely on the number of cats, and hence the abundance of clover depends on the proper supply of cats. Darwin remarks that *"battle within battle must be continually recurring with varying success; and yet in the long run the forces are so nicely balanced that the face of Nature remains for a long time uniform, though assuredly the merest trifle would give the victory to one organic being over another."*

The real struggle is between the most closely allied, and therefore competing forms. The black rat, for example, was the common rat of Europe till the beginning of the eighteenth century, when it was driven out by the larger brown rat. Competition is keener in direct proportion to the closeness of interests, the two covering the same ground. That the struggle for existence is a very real one, and does actually lead to the extermination of less fit forms, is seen in the way in which the Maoris, for instance, are gradually becoming exterminated. The struggle is not necessarily one of actual warfare, the stronger killing the weaker, although this may occur; and it is rather to more rapid multiplication and greater power of endurance that survival is due. So in the case of commercial competition among men, they do not actually fight; yet all cannot succeed, and failure means bankruptcy and starvation, and leads to destruction as surely as actual hand-to-hand warfare. Industrialism is in fact war under the forms of peace.

---

405   Henry Wenman Newman (1788–1865) corresponded with Darwin in about 1861. Darwin's reference to Newman's work in the *Origin* was based on Newman HW, On the habits of the Bombinatrices, *Entomological Society of London Transactions* 1851; **24**, NS 1: 86–92, 109–112, 116–118.

D. VARIATION.—It is a well-known fact that no two animals are absolutely alike; Darwin gives many instances of this. The Laplander by long practice knows by sight, and can actually name, each reindeer; the ants of one nest know and recognise one another; the sheep-dog picks out his own sheep unerringly; shepherds have won wagers by recognising each sheep in a flock of a hundred, which they had never seen till a fortnight previously. Voorhelm, an old Dutch florist, kept 1200 kinds of hyacinths, and was hardly ever deceived in recognising each variety by the bulb alone. The whole theory of breeding animals or of cultivating plants and fruits would fall to the ground, and selection would be impossible, unless variations occurred, and there were differences to select from.

Variability, which is a subject attracting much attention at present, is a general rule among animals, and applies to all of them. For example, Carpenter has shown that among the *Foraminifera* the range of variation includes not merely specific characters, but also generic and even ordinal ones. Among anemones great variations are found. Among snails 198 varieties of the common wood-snail have been described. In the case of insects, many of our common English butterflies vary enormously. Among birds, a remarkable series of facts is quoted by Wallace, from a memoir by Mr. Allen[406] on the birds of Florida. Exact measurements were taken of large numbers of these, and all parts were found to vary; not only were variations of 15 to 20 per cent in actual and relative sizes found ordinarily; but the variations affected the length and breadth of the tail and wings; the length, width, depth, and curvature of the bill, the form of the toes, the intensity of colour and nature of the markings. This was therefore not a case of minute or infinitesimal variations, but variations on a large scale, affecting all parts and in all directions.

Variation in habits.—A good illustration of the variation which may occur in habits is found in the *Kea,* a curious parrot inhabiting the mountain ranges of the island of New Zealand, and feeding naturally on the honey of flowers, on insects, fruits and berries, which till quite recently comprised its whole diet. However, since the European occupation of the island, this bird has acquired a taste for carnivorous diet with alarming results. It began by picking the sheep-skins hung out to dry, or the meat in process of being cured. In 1865 it was first observed to attack living sheep, which were frequently found with raw and bleeding wounds in their backs. Since then it is stated that the bird actually burrows into the living sheep, eating its way down to the kidneys, which form its special delicacy. In consequence of this, the bird is being destroyed as rapidly as possible, and one of the rare and curious members of the New Zealand fauna will no doubt shortly cease to exist.[407]

Variation occurs to a large extent also in plants, as seen by the different number of varieties which are described by different observers. Of the bramble, for instance, Bentham described five British species; while Babington, about the same time, described as many as forty-five.

---

406 This is J.A. Allen (not to be confused with Charles Martin Allen, Wallace's assistant on his travels in the Malay archipelago). His morphometry on Florida birds is discussed in detail in Chapter 3 of Wallace's *Darwinism.*

407 It is gratifying to note the Kea (*Nestor notabilis*) remains alive and well amongst the fauna of New Zealand's South Island, although the IUCN lists it as "vulnerable".

Man himself gives as good illustrations of variation as any animal; for instance, in his stature, habits, mental and bodily powers; in such individual details as minute inflections of the voice, or in the shape of the ear. To take the most recent development, Galton's work on finger-prints, the patterns formed by the ridges at the tips of the fingers and thumb are found to be unlike in any two cases, and to retain their peculiarities unchanged throughout life, thus forming one of the most trustworthy modes of identification yet discovered.

We read of the dead body of Jezebel being devoured by the dogs of Jezreel, "so that no man might say—This is Jezebel;" and that the dogs left only her skull, the palms of her hands, and soles of her feet.[408] It is a curious satire that these parts should now be shown to be the very ones by which a corpse could be most surely identified.

The causes of variation are very imperfectly understood, and careful inquiries are now being made in order to elucidate them. Variation is undoubtedly influenced greatly by external conditions, such as nutrition, cold, &c., and these may affect the young, or even the embryo, or the egg before it is laid by the parent. No two animals can ever come into existence under absolutely identical conditions; neither can any two after birth be exposed to absolutely the same conditions.

Variation under domestication is the rule instead of the exception, and occurs more or less in every direction. Consider, for instance, the extraordinary variations in size and mode of growth of the cabbage; the solid heads of foliage utterly unlike any plant in a state of Nature; the curiously wrinkled leaves of the savoy, the purple leaves of the pickling cabbage, the compact heads of flowers of the broccoli and cauliflower, the curious stem of the Kohlrabi, which grows like a turnip. Again, of the apple there are at least a thousand varieties known, all descended from the common crab-apple. In fact, as Wallace says, *"there is hardly an organ or a quality in plants or animals which has not been observed to vary; and further, whenever any of these variations have been useful to man, he has been able to increase them to a marvellous extent by the simple process of always preserving the best varieties to breed from."*

Limits to variation must of course exist, and it is evident that up to some point or other variations must be predetermined on definite lines. The inconstancy of chemical composition or instability is specially characteristic of living things. Variations are spoken of as accidental, not in the sense of their not being all due to natural causes, but inasmuch as they are accidental in relation to the sifting process of natural selection.

E. NATURAL SELECTION, OR THE SURVIVAL OF THE FITTEST.—Those animals which are most in harmony with their surroundings will survive. Just as in the breeding of animals by artificial selection, those animals are selected to survive which have certain favourable peculiarities, usually too slight for any but a practised eye to detect; so under natural conditions the possession of some useful variation, such as a slight increase of speed, or power of endurance or strength, or a keener sense of vision, will determine which shall be the survivors in a large herd of animals.

The action of natural selection is well shown by the following example. Many insects of Madeira have either lost their wings, or had them so much reduced as to be

---

408   Bible, II Kings 9:35.

useless for flight, while their allies in Europe have them well developed. The explanation of this is that Madeira, like other temperate oceanic islands, is much exposed to sudden gales, and the most fertile land being near the coast, the insects if able to fly are liable to be blown out to sea and lost. Year after year the individuals which had the shortest wings, or which used them least, would have an advantage, and so would survive. Hence the survival in the island of the insects with the smallest wings.[409]

In Kerguelen Island, one of the stormiest places on the globe and a place entirely without shelter, all the insects are incapable of flight, and most of them are entirely destitute of wings. These insects include a moth, several flies, and many beetles. Now these are the descendants of winged insects, which must have reached the island by flying, and gradually lost the power of flight, as in the insects of Madeira.

The importance of small variations.—We are apt to overlook the importance that slight variations may have, which is well shown in the artificial breeding of animals. So it is with human affairs, where important points, such as the fate of a Ministry, or even the determination of peace or war between two countries, often depends on side issues. In trade, accidental variations may determine success by attracting attention. The success of a novel, play or oratorio is often impossible to predict, and often depends on a mere caprice. Change for the mere sake of change may involve the misery or even death of thousands, and cause alternating periods of great prosperity and greater distress. This is well seen in the changes of fashion in dress, which in the case of the feathers for ladies' hats, or a particular kind of fur, may mean destruction and wholesale slaughter, even to extermination, of particular animals.

INHERITANCE.—The more favoured ones will not only survive, but will tend to hand down to their descendants their special advantages; and of these descendants some will have these special peculiarities in a less marked degree than their parents, some equally and others more strongly marked. The latter will in the long run survive, if the further development of this special advantage confers further benefit on the individual. The whole theory of the breeding of animals depends on inheritance. For instance, the pedigrees of race-horses are kept with most scrupulous care, and enormous prices are paid for horses for breeding purposes. So it is with pigs, poultry, dogs, and cattle, and in the improvements effected in fruits and flowers. Not only are good characters inherited, but bad ones also, and even diseases and malformations, such as insanity, gout, short-sight, cataract, and colour-blindness, among men.

F. THE INFLUENCE OF ENVIRONMENT.—We have seen that the struggle for existence results in the survival of the fittest. Now, if the conditions remain permanent there is no reason to suppose that the race would alter, for the fittest now would be so a thousand years hence, provided the external conditions did not change in the meantime. Variations would no doubt occur, but as none of these would confer an advantage, they would not be preserved. In a very few cases this is so, but constant change is the rule. For instance, consider the change effected in Australia by the arrival of civilised man with his dogs, horses, &c., resulting in the aboriginal inhabitants,

---

409 AMM treads on dangerous ground here with the phrase *or which used them least*. It gives a distinctly Lamarckian flavour to the argument, although one could imagine a behavioural adaptation which persisted through selection in the sense that the rest of the sentence implies.

human and animal alike, being killed off by competition. Man's influence is no doubt great; but other influences are still more potent.

Here we derive much assistance from the evidence afforded by geology, which tells us that, as regards the earth we live in, things were not always as we find and know them now. The marks on boulders and deposits of glacial mud and clay, show that these boulders have been brought from afar, that their only possible means of transit was by glaciers, and hence that our climate was once much colder than it is at the present time.[410] If, on the other hand, we turn to Greenland, which is now in the glacial condition, we find beneath the ice, beds containing fossil plants, showing the former existence in Greenland of such plants as the chestnut, oak, plane, beech, and poplar; nay even the magnolia, vine, walnut, and plum; proving the former existence not only of a moderate, but of a warm climate.

Geology shows us that the boundaries of land and sea are not constant; for instance, that Britain and France were once united, and that the sea is encroaching on the land on one side, while the land is encroaching on the sea on the other. The crust of the earth is made up chiefly of rocks deposited under water; therefore where there is now dry land, there must once have been open sea. Geology further shows that these changes have not been of a sudden cataclysmal character, but gradual ones, changes which are actually in progress at the present day, and which must always have been going on since the earth began.

The last link in the chain is now complete. Owing to incessant geological change in environment, variations in structure, previously useless or harmful, become advantageous, and their possessors thereby triumph and survive, and hand down their advantages to their descendants; thereby in course of time causing structural modifications of greater or less extent in the race. All Nature is in a condition of more or less unstable equilibrium. The action of environment is *indirect,* and changed conditions of life bring to the front variations previously useless, a slight change often causing great results.

The fittest to survive is not necessarily the one most perfect ideally, but rather the one best adapted to, and most in harmony with, the environment at the time. To be too far ahead of the times is far more fatal, as regards worldly prospects in human society, than to be conspicuously behind them; for in the latter case the individual is pitied and allowed the crumbs of charity; in the former he is regarded with suspicion and starved. This may constitute a consoling thought to those who are temporarily out in the cold, and who see the place they covet occupied by manifestly inferior men.

COMPARISON OF NATURAL AND ARTIFICIAL SELECTION.—In comparing natural with artificial selection we meet with the same principles, and yet domestic races differ from natural ones. The reason of this is that man selects and propagates modifications solely for his own advantage or pleasure, and not for the creature's benefit; he always tends to exaggerate, to go to the extreme point of selecting useful

---

410   Amongst AMM's papers in the Manchester University Archive is a cutting from the *Manchester Guardian,* 2 October 1889, reporting a meeting of the Owens College Court. It mentions the donation of the famous Glacial Boulder which stands in the University Quadrangle, adjacent to Owens College. The boulder, of ancient andesite lava, was discovered during drainage work on Oxford Road. It was evidently carried by glacial action from Borrowdale in the Lake District, a distance of some 80 miles.

or pleasing qualities. Races of animals are thus produced which would be incapable of independent existence; for instance, the prize-pig, which has to be fed with a spoon like a baby, would have a poor chance of existence in the wilderness; and races such as the Italian greyhound, the Fantail pigeon, hornless bulls, or the bull-dog of our dog shows, could not survive unless artificially protected.

Artificially-bred animals and plants are in fact in a condition of unstable equilibrium, and have a tendency to *revert* or slip back to a former and more stable condition, and are kept with difficulty at the stage they have reached. This is well seen in the tendency which crossed pigeons have to revert to the ancestral stage of the blue rock-pigeon. We may illustrate this point by a simple mechanical comparison.

A pack of cards lying on a table are in a condition of stability, and they may be taken to represent the normal or ancestral condition. If now the cards are built up to form a pagoda they are eminently unstable, although forming a more imposing structure, and are liable to collapse with the slightest touch, and revert to their former condition of stability. So the artificially-produced pigeons are much more imposing birds than the blue rock-pigeon, but they are in a condition of great instability, and readily revert to the ancestral condition.

Natural selection, on the other hand, acts, not for the good of man, but for the good of the species,[411] and tends to preserve, develop, and perpetuate all characters which will give the species an advantage in the struggle for existence. There is no known instance of an animal or plant having either structure or instinct developed in order to benefit another species. Every species is for itself and for itself alone.

---

411   See footnote 349, Lecture 13.

LECTURE 16

THE ARGUMENT FROM PALAEONTOLOGY

The title of this lecture and the next may have been inspired by Herbert Spencer. His *Principles of Biology*, Volume 1, Part III *The Evolution of Life* contains the following chapters: (iv) The Arguments from Classification; (v) The Arguments from Embryology; (vi) The Arguments from Morphology; (vii) The Arguments from Distribution.

The final section of the lecture, *Geographical Distribution*, is notable for its logic and for its advocacy of the fossil record as reliable evidence of ancestral species and their whereabouts. Alfred Russel Wallace, amongst others, had thought that *land bridges*, including the Bering Strait, completely explained the presence of similar terrestrial species on discontiguous land masses, believing that the loss of such connections could be solely explained by changes in sea level.

The tectonic movements of the Earth's continental land masses and the associated supercontinents, which were unknown to AMM and of course to Huxley and to Wallace, remove the problem of intercontinental animal migrations with which they and their contemporaries were trying to grapple. It is hard to imagine that they were really content to accept land bridges as the explanation, especially when the regions concerned were as distant as South America, Africa and the Malay Archipelago, yet they would have found it equally hard to believe that the continents were not immobile.

LECTURE 16

THE ARGUMENT FROM PALAEONTOLOGY

IN the first lecture we discussed the theory of Evolution, which claimed that animals now living are the descendants of those that lived formerly. We found the objection to the theory to be that animals which lived formerly were unlike those now living, and therefore that modification was necessary. In the second lecture, we saw how this objection was met by the theory of Natural Selection, and that causes were shown to be in existence not only competent to give rise to modification, but inevitably leading to it.

Let us now test this theory by seeing whether or not it is in accordance with the facts with which it has to deal. There is no possible doubt as to which series of facts we must deal with first. We must unearth these ancestors, put them in the witness-box, examine and cross-examine them, and see whether they support our case or not.

Let us first examine the Crust of the Earth, and its great division into Stratified and Igneous rocks. A headland, for example, consists of stratified rocks; the waves eat away the shore, and the cliffs fall in; streams carry down the mud and sand ultimately into the sea, where it is deposited in a plane over the sea-bottom. The nature of the deposit will depend on the source of the supply. If we ask how and why the cliff is stratified, the cliff itself will tell us. The stratified condition is due to deposition under water. Igneous rocks are intrusive and are caused by the volcanic heat of the deeper parts of the earth.

The crust of the earth is made up of sedimentary or stratified rocks deposited one above another, the most recently formed being at the top. Their position may be subsequently disturbed, yet the general relation can usually be determined. Geologists find the sequence to be much the same, and to show general agreement in all parts of the earth, so that the same names can be employed.

Particular deposits may be thicker or thinner and of variable nature in different localities, or absent altogether. To interpret the crust of the earth we must read it as a history of the earth in successive chapters, like successive centuries or ages in the history of mankind. The chief differences are that it consists of several chapters of unequal length, of which there is no means of determining the absolute age or duration, separated by gaps about which we have no record whatever. The history of these times is revealed by fossils, "imprints on the pages of time," which can be compared to the descriptions and drawings in the written records of man. Bones, teeth, shells, and other hard parts are often found in extraordinarily perfect condition. These records tell us, for example, that but a short time ago, geologically speaking, the lion, bear, rhinoceros, mammoth, and hippopotamus lived in Britain.

This evidence is of great importance, for it consists of real remains of former inhabitants of the earth, who stand in the same relation to the present fauna as our Saxon or Norman ancestors do to ourselves; that is, they are the remains of the actual ancestors of living animals, and must include among them, were our collection complete, such ancestors of all living animals. This is evidence of peculiar value, and there is no gainsaying it. The true nature of fossils was neglected and greatly misunderstood until the early part of the present century; but since the time of Cuvier fossils have been collected diligently, and large numbers have been obtained from all parts of the earth. It is now known that the fossil Mollusca are considerably more numerous than those now living on the earth, and probably fossil mammals are almost as numerous as recent forms. The age of fossils cannot be determined absolutely; their relative age is, however, known, and we are able to draw up tables giving the order and sequence of events, though not the actual dates; sequences, moreover, that will apply not merely to one, but, with certain reservations, to all parts of the globe.

In order to learn the lesson taught by fossils let us take those found in England at different periods.

THE CRUST OF THE EARTH

		FEET
TERTIARY OR KAINOZOIC: 3,100 ft.	Historic	.
	Pre-historic	.50
	Pleistocene	130
	Pliocene	440
	Miocene	300
	Eocene	2,180
SECONDARY OR MESOZOIC: 9,350 ft.	Cretaceous	1,360
	Jurassic	2,340
	Liassic	900
	Triassic	4,750
PRIMARY OR PALAEOZOIC: 111,750 ft.	Permian	550
	Carboniferous	13,000
	Devonian	16,200
	Silurian	32,000
	Cambrian	30,000
	Archaian	20,000

Table of the geological strata, with approximate thickness of the several strata in Britain. Total thickness about 23½ miles.

(1) THE TERTIARY OR KAINOZOIC PERIOD.—The pre-historic cave deposits found in the caves used as dens by wild beasts, who dragged into them the carcases of their prey, prove the occurrence of the *Hyaena, Cave-bear, Rhinoceros, Lion, Reindeer, Bison, Hippopotamus,* and *Beaver*—with man. These forms are now living, but not in England. In the Glacial period the earlier deposits show some of these disappearing, others persisting, and animals appearing that now do not exist anywhere—viz., the *Mammoth, Woolly Rhinoceros,* the *Sabre-toothed Tiger* and the *Irish Elk* or deer. In the Pliocene period we find evidence of the *Mastodon* and *Tapir;* and in the Eocene period, of the *Didelphys, Opossum,* and *Hyracotherium.*

(2) THE SECONDARY OR MESOZOIC PERIOD.—The entire group found at this period is extinct. The Ammonites are characteristic secondary forms, and are very abundant, but none survive into the Tertiary period. The *Ichthyosaurus, Plesiosaurus,* and *Mosasaurus* are extremely characteristic, and confined to the Secondary period.

(3) THE PRIMARY OR PALAEOZOIC PERIOD.—Here we find fossil plants in the Coal measures, the *Lepidodendron, Calamites, Sigillaria,* and *Stigmaria.* Fish are also found abundantly. The *Trilobites* are exclusively Palaeozoic; also the Sea-scorpions, *Pterygotus,* and *Eurypterus.*

The general conclusions we arrive at are:

(1) There is a general advance in organisation from the lower to the higher or more recent deposits, and an increase in the diversity of type.

(2) There is no evidence of sudden breaks or cataclysms; there is no break between the Tertiary period and the present day. Some species die out and others appear, and some persist unchanged. The very evidence which Cuvier relied on to prove Catastrophism disproves it when examined more carefully and with fuller knowledge.

(3) Some forms, known as *persistent types,* remain unchanged for great periods. This constitutes no real difficulty, for Natural Selection does not of necessity involve progression or change of any kind, and is quite consistent with a stationary condition, provided that the environment, or at least all the features of the environment affecting them, remain unchanged. These are examples of the real aristocracy of animals, for they can date back their descent not merely to the time of the appearance of man, but almost to the first appearance of animal life of which we have positive evidence.

As examples of *persistent types* may be cited *Globigerina,* which shares in the formation of chalk, and is found in the Trias, or bottom of the Secondary strata; *Limulus,* the king-crab, which occurs in the Trias, and is found on the American coast at the present day; *Dentalium,* the tusk-shell, an animal in some respects between an oyster and a snail, with a tubular conical shell about two inches long, by which it burrows in the sand, is found in the Devonian and perhaps Silurian strata; the *Pearly Nautilus,* a rare animal, is found in the lower Silurian and at the present day; *Lingula,* an animal with extreme tenacity of life, is found in the lower Cambrian and at the present day, and, so far as we can see, is unchanged. (Figs. 4 and 5.)

Human customs and people have often been referred to for examples of the laws of biology, and there is no need to look beyond them. As an example of a persistent type, cannot we at once call to mind a nation, a homeless nation, the members of

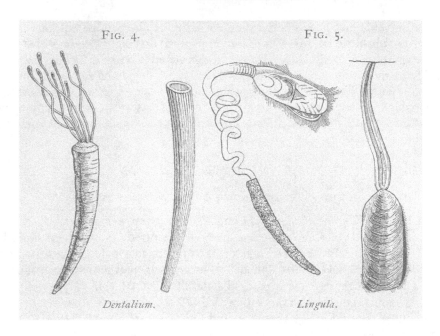

Fig. 4.   Fig. 5.

Dentalium.   Lingula.

FIGS. 4 and 5

which occur in all countries, yet have no country of their own; a nation which, in spite of persecution of unexampled severity, endured not once only, but many times repeated and in most diverse forms, has held its own; a nation which, in spite of various and shifting environments to which it has been exposed, has with singular tenacity retained and preserved its language, traditions, and ceremonial observances in all essential features intact[412].

THE IMPERFECTION OF THE GEOLOGICAL RECORD.

Undoubtedly the most interesting and important point concerning fossils yet remains to be considered. Fossils are the former inhabitants of the globe, and therefore, on the Theory of Evolution, the ancestors of the animals now living. Now, fossil forms are unlike existing ones, therefore modification, and very considerable modification, must have occurred. Do the fossils themselves show evidence of such modification? Can we point to a series of forms showing progressive modification towards the present condition? I admit at once that fossils do not give us all the evidence we could

---

412   AMM's audience would have been in no doubt about his allusion. Despite the widespread emancipation of Jews in Europe in the second half of the 19th century, fears of persecution were rife and widespread migration from Europe to the United States had begun around 1880. Theodor Herzl published *Der Judenstaat* [*The State of the Jews*] in 1896, about ten years after these lectures were given. (British Foreign Secretary, previously Prime Minister, Arthur Balfour, elder brother of AMM's late friend and mentor Francis Maitland Balfour, penned his eponymous declaration advocating a separate Jewish state within Palestine in 1917. The state of Israel was eventually established in 1948.)

wish for; in some instances a fairly complete series can be pointed out, but in most cases we are unable to do this. This is undoubtedly at first sight a serious check, and is one often referred to. By Darwin himself it was stated to be the difficulty that would probably be most widely felt. This difficulty must be considered fully and from two main standpoints: First, of what nature is the record yielded by fossils, and how far is it reasonably complete? Secondly, are we quite certain that we know in what direction to look for intermediate forms, and are we clear that we should recognise such if we found them?

The geological record is imperfect for the following reasons:—

1. Only certain parts of certain animals can be preserved as fossils.

Among PROTOZOA the *Foraminifera* and *Radiolaria* are well preserved, *Infusoria* not at all. In PORIFERA the skeletons are well preserved, and of these there is a long record. Of COELENTERATA such forms as Hydra, jelly-fish, and sea-anemones cannot be preserved, save very exceptionally; but corals are peculiarly suitable, and few classes are so well represented in a fossil state. ECHINODERMATA are well represented, excepting the *Holothurians*. Among VERMES there is no trace of flat or round worms, but of *Annelids* the jaws, tubes, and tracks are found. ARTHROPODA are well preserved, especially the *Crustacea*, and as a general rule aquatic forms are much more completely preserved than terrestrial ones. In MOLLUSCA the shell is well adapted for preservation, but the air-breathing terrestrial forms are rare. Among the VERTEBRATA we find the bones, teeth, and footprints preserved. Mammals, being terrestrial, are at a disadvantage. Birds, whose bodies are light enough to float, are probably devoured as food, and have less chance of preservation; they are therefore rare as fossils.

These parts can, as a rule, only be preserved in a fragmentary condition.

2. Only certain deposits can so preserve them, notably mud.
3. Animals must die at such places and times that they can be preserved; bones, &c., must be carried down regularly to a particular locality.
4. The deposit must extend over a very long time and continuously, if the series is to be complete, or even approximately so.
5. The area must be in subsidence or else it would be filled up.
6. Fossils once imbedded must be raised above the sea.
7. They must escape denudation and be exposed at some workable spot.
8. The intervals between the deposits often represent times of denudation. The denuded parts will be the uppermost—*i.e.*, the most recent, and will break the series effectively by the removal and destruction of fossil records.

All these conditions can very seldom be fulfilled. The difficulty is well illustrated by our state of knowledge with regard to domestic animals. Where are the bones of the intermediate forms between the rock-pigeon and the pouter, or fantail, for example, to be found? Yet we know that these existed but a short time ago. Speaking of this point, Darwin says: "*We shall perhaps best perceive the improbability of our being enabled to connect species by numerous fine intermediate fossil links, by asking ourselves whether, for instance, geologists at some future period will be able to prove that our different breeds of cattle, sheep, horses, and dogs are descended from a single stock or from several aboriginal stocks.... This could be effected by the future geologist only by his discovering in a fossil state numerous intermediate gradations; and such success is improbable in the highest degree."*

It is irrational to demand a perfect gradational series in any number. The actually preserved record is well described as a "chapter of accidents." We have no right to expect, in any particular case we choose to select, that the chain shall be complete, and the links all forthcoming on demand; but we have a right to expect some few well-marked examples of transitional series of forms, and also that none of the facts actually ascertained shall be inconsistent with our theory.

THE GEOLOGICAL EVIDENCES OF EVOLUTION.

The second preliminary point concerns the nature of fossils. Directly intermediate forms between two existing genera must not be expected, and can indeed very rarely have existed. The theory of Evolution requires that distinct genera shall be linked together, not by a direct connection, but by the descent of both from a common ancestor. Wallace says: *"The fantail and pouter pigeons are two very distinct and unlike breeds, which we yet know to have been both derived from the common wild rock-pigeon. Now, if we had every variety of living pigeon before us, or even all those which have lived during the present century, we should find no intermediate types between these two—none combining in any degree the characters of the pouter and fantail. Neither should we ever find such an intermediate form, even had there been preserved a specimen of every breed of pigeon since the ancestral rock-pigeon was first tamed by man."*

This point may be illustrated by an example taken from another department of knowledge—the science of language. Take a word such as *regnum,* and see how the word has become modified in the different languages of European nations, when exposed to different conditions of environment. The following table shows various modifications derived directly or indirectly from the original word *regnum.*

REGNUM.

REINO	*Portuguese.*
REINADO	*Spanish.*
REGNO	*Italian.*
RÈGNE	*French.*
REIGN	*English.*
REGIERUNG	*German.*

The word *regnum* corresponds to the blue rock-pigeon in that it is the parent of all the other forms of words, and represents a fossil form or extinct word, belonging to a dead language.

Again, to return to biology, *Phenacodus* (Fig. 6), one of the most important of recent fossil discoveries, was found in the Eocene of North America, and in several forms, varying in size from that of a small terrier to a leopard. This is a good example of a generalised type, having five clawed digits, a small brain, and complete radius and ulna. In many ways it suggests the ancestral form from which the *Artiodactyla* (deer and sheep) and the *Perissodactyla* (rhinoceros and horse) may have sprung; perhaps, also, it is nearly the parent form of the Carnivora. Among *Birds and Reptiles* a well-known series is known, due to Professor Huxley, connecting the one group with the other. Birds form a very compact and sharply limited group, characterised

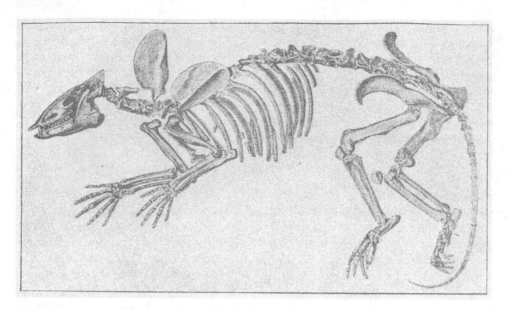

FIG. 6. *Phenacodus.*

by their wings, feathers, and other peculiarities. In the first edition of the "Origin of Species," Darwin said: *"We may thus account even for the distinctness of whole classes from each other—for instance, of birds from all other vertebrated animals, by the belief that many animal forms of life have been utterly lost, through which the early progenitors of birds were formerly connected with the early progenitors of the other vertebrate classes."*

This was a prophecy out of which much capital was made at the time. It appeared an easy way out of the difficulty to suppose extinction and disappearance of all those forms whose existence, at one time or another, it was necessary to assume. At the time of the utterance of this prophecy, in 1859, there was no positive evidence at all. But in 1862, the *Archaeopteryx* (Frontispiece) was shown to be a true bird as regards its feathers and wings, combined with several Reptilian characters, such as the long tail, the nature of the hip-bones, legs, and vertebra. A second specimen was found in 1879, having a skull with numerous teeth, clawed fingers, perfect feathers, and bi-concave vertebra. In 1868 Professor Huxley showed in fossil reptiles *(Dinosaurus)* the nature of the modification in virtue of which the quadrupedal reptile passed into the type of a bipedal bird. Again, in 1875, the discovery of toothed birds in chalk by Marsh completed the series of transitional forms between birds and reptiles. From that time Darwin's prophecy could be replaced by demonstrated facts. There are actual fossils which bridge over the gap between reptiles and birds, in this sense that they enable us to picture to ourselves forms from which both birds and reptiles as we know them could have sprung. The *Pterodactyl* shows how flight is possible to a reptile, and is possibly related to birds, although this point is doubtful. (Fig. 7.)

The most famous instance of geological evidence is found in the Horse, and, although familiar, is so important as to bear repetition. The typical number of toes and fingers is five, as in ourselves. In quadrupeds generally the number is reduced, but the horse, zebra, and ass stand alone in having only one digit on each foot, corresponding to the middle finger and toe. If we compare the foot of the horse with that of man, we find the "hock" of the horse corresponds with man's heel; the "cannon-bone"

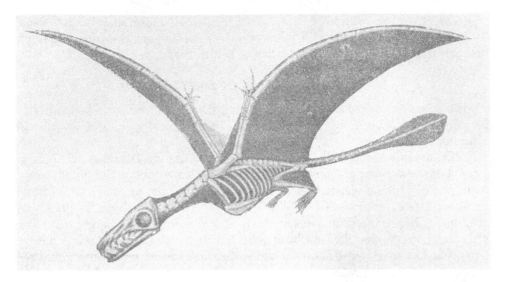

FIG 7. Flying reptile (diagrammatic figure).

is the metacarpal; the "pasterns" form the first two phalanges, and the "coffin-bone" the terminal phalanx of the toe. The "hoof" corresponds to the nail. In the fore-limb the "knee" of the horse is equivalent to the wrist. The "splint-bones" represent the metacarpal bones of the first and third digits.

Now the ancestors of the horse are *Protohippus* or *Hipparion,* which is found in the Pliocene; *Miohippus* and *Mesohippus,* found in the Miocene; *Orohippus*, in the Eocene; and *Eohippus* at the base of the Eocene. In *Protohippus* each foot has three well-formed digits; *Miohippus*, in addition to this, has a rudimentary metacarpal bone of a fourth digit in the fore-foot; in *Mesohippus* this rudimentary metacarpal bone more fully developed; in *Orohippus* there are four well-developed digits in the fore-foot, three in the hind-foot; while in *Eohippus* five digits are present. Thus this series of fossil forms furnishes a complete gradation from the older tertiary forms with four or five toes up to the horse with one toe. These forms differ not only as regards the number of toes, but also in other respects, chiefly in the gradual diminution and loss of independence of the ulna and fibula, and in the gradual elongation of the teeth and increasing complexity of their grinding surfaces.

An excellent series of gradational forms is shown in the case of *Paludina*,[413] of which six or eight disconnected forms were known first, and described as distinct species; later on, connecting forms were discovered, and it was realised that we had a case of progressive modification from the older geologic beds to the newer ones, and all forms are now included as varieties of one species. Over 200 varieties have been discovered in enormous numbers; one characteristic form being found in each horizon. The simpler unornamented shells are from the lowest layers; the most recent forms being identical with a form now living only in North America and the fresh-water lakes of China, which was formerly described as a distinct genus. This evidence was found since the publication of the "Origin of Species" in 1859, and renders the record less incomplete.

---

413   River snails, now called *Viviparus* spp.

Now that men realise the value of Palaeontology, more attention has been directed to the subject; for in all cases positive palaeontological evidence may be implicitly trusted, although negative evidence is worthless.

THE EXTINCTION OF SPECIES.

When a species or group has once disappeared there is no reason to suppose that the same identical form ever reappears—i.e., its existence, so long as it lasts, is continuous.

The influence of the size of animals, and its bearing on Extinction of Species, is of the greatest possible interest and importance. Many zoologists hold the view, in support of which evidence is steadily increasing, that the primitive or ancestral members of each group were of small size.[414] Thus, in the case of birds, on the whole small birds show more primitive conditions of structure than the larger members of the same group, and the first birds were probably smaller than Archaeopteryx. Reptiles and mammals also show in their earlier and smaller types more primitive features than their larger descendants.

Again, in the pedigree of the horse, one of the most striking points is the progressive reduction in size met with as we pass backwards in time. The Pliocene *Hipparion* was smaller than the existing horse; the Miocene *Mesohippus* was about the size of a sheep; while the Eocene *Eohippus* was no larger than a fox. Not only is there good reason for holding that, as a rule, larger animals are descended from ancestors of smaller size, but there is also much evidence to show that increase in size beyond certain limits is disadvantageous, and may lead to destruction rather than to survival. It has happened several times in the history of the world, and in more than one group of animals, that gigantic stature has been attained immediately before extinction of the group, a final and tremendous effort to secure survival, but a despairing and unsuccessful one.[415] The Ichthyosauri, Plesiosauri, and other extinct reptilian groups, the Moas, and the huge extinct Edentates, are well-known examples; to which before long will be added the elephants and the whales.[416]

The same classification applies to both recent and fossil animals; the large divisions are the same, but many minor groups have become extinct. All existing groups are not known to have existed for all time, and many have certainly not done so. Still no one primary division of the animal kingdom is entirely extinct: it is merely the subdivisions that have died out. The earliest origin of all the great groups is driven back to extremely remote times, to the Palaeozoic period, and palaeontology tells us nothing about the mode of origin of the great divisions of animals. Darwin says: "*I look at the geological record as a history of the world imperfectly kept, and written in a changing dialect; of this history we possess the last volume alone, relating only to two or three countries. Of this volume, only here and there a short chapter has been preserved; and of each page, only here and there a few lines. Each word of the slowly changing language, more or*

---

414 See Lecture 8.

415 We should be wary of a logical inevitability here. If it is true that animals get progressively larger as they evolve, the form which immediately precedes extinction (or the contemporary form if not extinct) will be the largest. Extinction does not, therefore, need to be the result of size.

416 See also Lecture 7.

less different in the successive chapters, may represent the forms of life which are entombed in one consecutive formation, and which falsely appear to us to have been abruptly introduced. On this view the difficulties above discussed are greatly diminished, or even disappear."

GEOGRAPHICAL DISTRIBUTION.

The explanation of the distribution of animals on land and in the sea is a subject of great importance, which I propose here to touch upon only as it is affected by palaeontological evidence.

Much information has been collected on this subject by exploration and by systematic observations obtained by dredging expeditions. A gradually growing conviction has arisen that we must not be content with mere facts, but must demand an explanation of these facts, and that this explanation is in our power to find. It is to Wallace that we are especially indebted for our knowledge of the geographical distribution of animals.

The nature of the problems we have to consider is best shown by examples, of which the following will serve our purpose.

A. CAMELIDAE, or Camels, are an exceedingly restricted group, the majority of species now living in domestication.

1. *Camelus* is highly characteristic of hot, parched deserts, and is found in Sahara, Arabia, Persia, Turkestan, and Mongolia, as far as Lake Baikal. There are none now living perfectly wild. Of Camelus there are two kinds: the dromedary, found in Asia Minor and Africa, has one hump; the Bactrian camel, possessing two humps, is confined to Asia, and especially Central Asia, north of the Himalayas.

2. *Auchenia* is of smaller size, with slender legs, and has no hump. It is confined to the mountainous and desert regions of the southern part of South America, and is often found on rugged snow-clad slopes at great elevations. Of this group, *Llama* and *Alpaca* are entirely domesticated—the former being used as a beast of burden in Peru and Bolivia; the latter is cultivated both for its wool and for its flesh. *Vicuna,* the smallest member of the group, is found at elevations or 13,000 feet and upwards, in the Andes of Peru, Ecuador, and Bolivia. *Guanaco,* an animal the size of a fallow-deer, is distributed in the plains of Patagonia and Tierra del Fuego.

Camels are thus distributed over two areas, comprising the mass of two continents, divided by a great ocean; one area being north of the equator, the other south of it, and separated by half the circumference of the globe. They are animals of large size, and it is hardly possible for their existence to have been overlooked. Hence we may assume that their geographical distribution is known correctly.

This is a good example of the difficulties in accounting for geographical distribution, and of the way in which they may be met. Evolution tells us that close anatomical resemblances mean near kinship, and forbids us to contemplate the possibility of animals, agreeing in a number of important points, having come into existence independently.

The anatomical characters of camels are well marked; they have two toes—viz., the third and fourth, and walk on the palmar surface of the middle phalanx, not on hoofs. The sole of the foot is formed by broad integumentary cushions, and the nails are small and flattened. The stomach consists of a paunch with smooth lining, provided with two groups of water-cells with narrow mouths. The cervical vertebrae are peculiar, in that the canal for the vertebral artery pierces the arch of the vertebra, instead of the transverse process.

The theory of gradual modification of animals renders it impossible that identical conditions could have been acquired twice independently. The fact that the two groups of camels agree in a large number of points, in which they differ from all other animals of the same class, must be taken as a proof of near blood-relationship.

Near blood-relationship means common origin—*i.e.*, one of the two groups of camels must be descended from the other, or both groups must be descended from some common ancestors, which were already camels. In other words, either the new-world camels must be descended from the old-world camels, or *vice versa*; or both must be descended from camels that formerly lived elsewhere, but are now extinct.

Our problem is now becoming more clearly defined, and we have to consider the means of migration of mammals. The only means of migration is by walking; for although most of them can swim, it is only for short distances, and it is doubtful whether any land mammal can swim across an arm of sea fifty miles wide. Captain Webb's swim across the Channel has perhaps never been beaten by a land mammal.[417] The practical proof of the efficacy of the sea as a barrier to migration is seen in the fact of the absence in most oceanic islands of indigenous mammals, except bats. To put it in plain words, if mammals are to get from one place to another, they must walk.

The only explanation possible is through fossils, which thus have a new interest, depending on what parts of the earth we find them in. The evidence of fossils with regard to camels is very imperfect, but still points in a fairly definite direction. Fossil camels are found in South America, in Brazil; in North America, in Texas, California, Kansas, and Virginia. In Asia, in the Himalayas, *Merycotherium*, a large fossil camel, is found widely distributed over Siberia, extending to the extreme coast.

Now, there was almost certainly a former land connection between Asia and North America, across the Behring Straits, which are narrow and shallow. Hence the conclusion is, that there is strong reason for holding that camels originated in North America, and thence spread in two directions, southwards to South America, and westwards through Asia; and that their areas of distribution, though now disconnected, were once continuous.

B. MARSUPIALS.—These constitute a large group of animals, of which there is a great variety of forms, the kangaroo and opossum being perhaps the best known examples. Marsupials are a well-marked group, which in their habits, appearance, and structure, especially as regards the skeleton and teeth, curiously simulate the higher divisions of mammals. Carnivorous, insectivorous, and herbivorous forms are all well established and differentiated. The ant-eater or *Myrmecobius*, and fruit-eater or *Phalanger*, are found in this group, which is characterised by low organisation and great tenacity of life.

---

417 Captain Matthew Webb (1848–1883) made the first recorded swim across the English Channel on 25 August 1875. He died trying to swim through rapids below Niagara Falls.

Marsupials occur in two chief regions:

(i.) *Australian region.*—The wombats and Myrmecobius occur only in Australia and Tasmania, kangaroos and phalangers extending northward to New Guinea and adjacent islands; phalangers to Timor, the Moluccas, and Celebes.

(ii.) *American region.*—The opossums are most numerous in the forest region of Brazil, south of the river La Plata; also west of the Andes in Chili. Their distribution extends northwards to Mexico, Texas, and California; and in the States from Florida to the Hudson river, and westwards to the Missouri.

Marsupials form a good example of discontinuous distribution, the explanation of which is yielded by fossils. Opossums are found in the Tertiary deposits of England, France, in other parts of Europe, and in North America. The Cretaceous period shows no trace of them, but in the Jurassic, and in the yet older Triassic, at the base of the Secondary series, large numbers of mammalian remains of small size have been found, which are considered to represent the early phase in marsupial development. Here the starting-point or birth place appears to have been in the Old World, and the group to have migrated southwards.

C. TAPERIDAE, or tapirs, are found in the equatorial forests of South America, in the Andes of Ecuador, in Panama and Guatemala, and also in the Malay Peninsula, Sumatra, and Borneo. Geological evidence shows that during the Miocene and Pliocene times, tapirs abounded over the whole of Europe and Asia, and their remains are found in the Tertiary deposits of France, India, Burmah, and China. In both North and South America fossil remains of tapirs occur only in caves and deposits of the Post-Pliocene age, showing that they are comparatively recent immigrants into that continent, perhaps by means of the Behring Straits again. The climate even now is much milder than on the north-east of America, and perhaps was warm enough in late Pliocene times to allow emigration of tapirs, which were driven south to the swampy forests of the Malay region.

CONCLUSION.—On the whole, then, the evidence afforded us by fossils is not so complete as we should wish, and we have seen that from the necessity of the case this must be so. However, evidence is steadily accumulating, and such evidence as we have is not merely favourable, but in some instances remarkably complete. Indeed, since the date of publication of the "Origin of Species," in 1859, our knowledge has increased, and evidence has accumulated so markedly, that it has been said by a highly competent authority, that if the doctrine of Evolution had not existed, palaeontologists would have been compelled to invent it as the only possible explanation of the facts determined. Again, we are not aware of the existence of paleontological facts which can be demonstrated to be inconsistent with the theory; while the explanation which they afford of new and previously unstudied problems, such as some of the questions of geographical distribution we have touched upon, is evidence of a strong nature in support of the theory. Professor Huxley, with regard to this subject, says[418] *"The primary and direct evidence in favour of Evolution can be furnished only by palaeontology.*

---

418   T.H. Huxley *The Coming of Age of the Origin of Species.* Lecture at the Royal Institution, Friday, March 19, 1880. Published in *Science* 1: 15-20.

The geological record, so soon as it approaches completeness, must, when properly questioned, yield either an affirmative or a negative answer: if Evolution has taken place, there will its mark be left; if it has not taken place, there will lie its refutation."

LECTURE 17

THE ARGUMENT FROM EMBRYOLOGY

Many of the examples in this lecture come from others, especially Lectures 3, 10 and 13, although the material is presented in a more summarized form and intended to be accessible for a general audience. What AMM presents as evidence for recapitulation are really just illustrations of it, for one needs to have accepted the principle in order to understand the interpretation. To that extent, this lecture is somewhat unsatisfactory although it does have the benefit of valuable illustrations.

Recapitulation no longer features in biological understanding as it did in the 19th and early 20th centuries, for reasons discussed elsewhere, yet we should acknowledge the valuable context it provided for understanding animal pedigrees. Current interpretation of embryological events rests on understanding the differentiation of cells and on the multipotency or otherwise of cell lineages. AMM had no concept of this, although his intriguing example of skin and nails may suggest that he was starting to think about things in this way. Despite changes of perspective, embryology remains of the utmost importance in modern biology.

In the final paragraph, AMM moves closer than in previous lectures to the notion that earlier stages of development represent adult forms, in the manner required by Gould's rigid historical definition of recapitulation (see *Interlude*).

LECTURE 17

THE ARGUMENT FROM EMBRYOLOGY

THE last lecture was devoted to the consideration of the evidence afforded by fossils in regard to the theory of Evolution, and may be summed up as follows:—The evidence afforded by fossils is not so complete as we could wish, but we are able to point out the causes which render it difficult or impossible for continuous series to be preserved; fossils give no evidence against Evolution, and some remarkable series have already been unearthed, such as those of the Horse and *Paludina,* which would be unintelligible without Evolution; the evidence is steadily increasing in amount and importance; and the evidence of fossils is a disproof of catastrophism.[419]

We are now concerned with the most recent of biological sciences. Embryology, or the Science of Development, is prominently associated with the names of Von Baer, 1792–1881,[420] and Balfour, 1851–1882. It is utterly vain in one lecture to give any idea of the extraordinary multitude of facts accumulated within the last quarter of a century, or of the numerous and fascinating theories to which these facts have given origin; it is here merely as bearing on the doctrine of Evolution that we have to consider Embryology.

There are two great questions to be considered:—First: Does embryology afford evidence for or against the possibility of the descent of animals from unlike ancestors? Secondly: If it gives evidence in favour of such descent, does it afford us any clue in regard to the actual line of descent in a given case, and will it help us to reconstruct the pedigrees or past histories of animals?

The answer to the first question is found in the extraordinary changes which an animal may undergo in its own person, during development, within the space of a few days or weeks, thus showing the possibility of such descent with modification; for instance, the changes which occur during the metamorphosis of the butterfly, and the change of the water-breathing tadpole into the air-breathing frog. This suggests further that such enormous periods of time as are usually demanded to bring about such changes may not really be necessary. A further reply to the question is found in the fact that groups of animals, the relations of which were previously unknown, have had their true zoological positions determined by the study of the changes undergone during their development.

Thus many animals, when adult, present little or no resemblance to other members of the groups to which they really belong. The Ascidians are a well-known example. So long as their adult condition alone was known, zoologists were entirely in the dark as to their real affinities, and by most writers they were grouped with the *Brachiopoda* and *Polyzoa,* as a subdivision of *Mollusca.* As soon as their development was worked out, it was found that they were really members of the *Vertebrata,*

---

419   See Lecture 1.
420   Von Baer died in 1876.

inasmuch as when young they show a curious resemblance to tadpoles—not merely in form and appearance, but in all essential points of structure; while the mode of formation of their nervous system, skeleton, alimentary canal, and other parts, is exactly that obtaining among other Vertebrates, and entirely unlike that of all Invertebrate groups. (Fig. 8.)

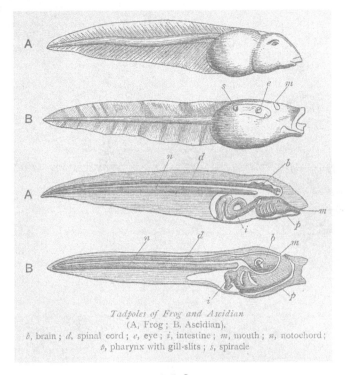

Tadpoles of Frog and Ascidian
(A, Frog ; B. Ascidian).
b, brain ; *d*, spinal cord ; *e*, eye ; *i*, intestine ; *m*, mouth ; *n*, notochord ;
p, pharynx with gill-slits ; *s*, spiracle

FIG. 8.

Again we may take animals such as a prawn, a barnacle, and one of those curious sac-like parasites of the genus *Sacculina*, which are found not uncommonly adhering to the under surface of the rudimentary tail of crabs. These three animals are, when adult, very unlike one another. The prawn is a free-swimming form. The barnacle is fixed firmly to rock, usually between tide-marks; it forms a hard protective shell, has no eyes, no locomotor organs, and is hermaphrodite. Sacculina is altogether unlike the other two animals: it has a soft unjointed body, no trace of limbs, no mouth or alimentary canal, and no sense-organs: it is, in fact, merely a soft-walled bag of eggs, attached to the crab's tail by a number of branching root-like processes penetrating the crab's skin and spreading out in its body, from which they absorb the nutriment on which the parasite lives and grows.

Yet when we turn to the development of these animals, we find that, utterly unlike as the adult forms are, the young of all three genera hatch in the form known as a *Nauplius*. This Nauplius larva has a short unsegmented body, three pairs of appendages, used for locomotion, and a single median eye; and although the Nauplii are not identical one with another, yet they agree very closely in all essential features. Embryology tells us that this means that the three animals must really be members of the same group, and allied to one another; and thereby gives us a clue to the real

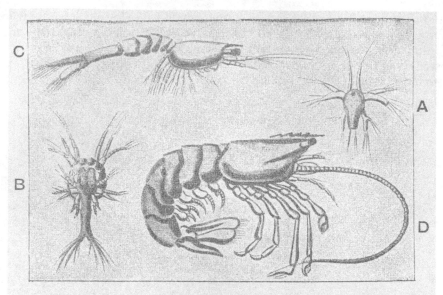

Stages in Development of the Prawn (Peneus).

A, *Nauplius* stage; B, *Zoëa* stage, in which the larva resembles an
dult Copepod; C, *Schizopod* stage, where it corresponds in structure to
the adult *Schizopoda;* D, Adult Peneus.

FIG. 9.

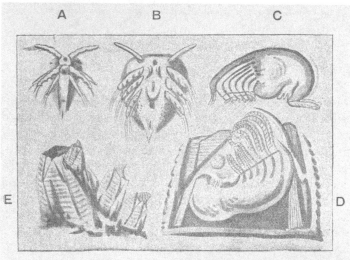

Stages in the Development of the Barnacle (Balanus).

A, *Nauplius* stage; B, Second stage, in which the first pair of
swimming appendages of the Nauplius are converted into antennæ and the
rudiments of the six pairs of cirri appear; C, Pupa stage—in this stage
the animal is free-swimming and has six pairs of legs, antennules, two large
compound eyes, and imperfectly developed masticatory appendages. The
pupa becomes attached by its antennules and develops into D, the adult
Barnacle. E, Group of Barnacle shells.

FIG. 10.

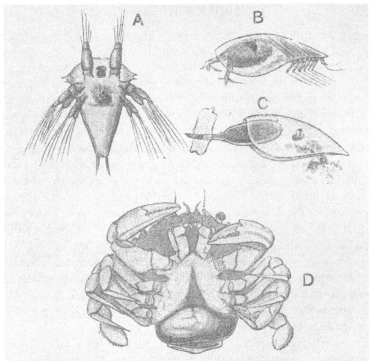

Young and adult specimens of Sacculina, *to illustrate the Degeneration or Retrograde Metamorphosis which the parasite undergoes in the course of its development.*

A.—The Nauplius stage, in which the young *Sacculina* hatches. The three pairs of appendages correspond to the antennules, antennæ, and mandibles of a Crab or Crayfish. The black spot between the antennules is the eye, and the small patch immediately behind the eye and between the two hinder pairs of appendages is the ovary, which is already present at this very early stage. Unlike the Nauplius stage of other groups of Crustacea the *Sacculina* Nauplius has no mouth or alimentary canal.   × 90.

B.—The *Cypris* or pupa stage. It is at this stage that the young *Sacculina*, hitherto a free-swimming animal, attaches itself to a Crab and becomes parasitic. The pupa is characterised by the bivalved carapace ; the stout antennules by which it fixes itself to the Crab ; and the six pairs of locomotor appendages. The black spot is the Nauplius eye, and the mass immediately below it is the ovary.   × 90.

C.—The *Sacculina* three days after fixing itself to the Crab. The six pairs of swimming legs have rotted away and fallen off ; the bivalved carapace is being detached, and is carrying with it the eye and certain *débris* from the body. The sole parts remaining, out of which the adult *Sacculina* will be formed, are the antennules, now modified into a tube, which is represented projecting through a piece of the skin of the Crab ; and the head, which forms a bottle-shaped mass attached to the tube. and containing the ovary.   × 90.

D.—Adult *Sacculina* attached to the ventral surface of the tail of a Crab (*Portunus*). The *Sacculina* is the large dark-coloured bag in the lower part of the figure ; it is attached to the Crab by a short fleshy stalk. not seen in the figure, which, penetrating the skin spreads out into a tuft of branching roots in the Crab's body.   × $\frac{1}{2}$.

FIG. 11.

affinities of barnacles and Sacculina that we could hardly get in any other way. (Figs. 9, 10, 11.)

In parasitic animals generally the shape and structure are liable to be so profoundly modified, in consequence of the special conditions of parasitic existence, that, but for the aid afforded by development, we should often be absolutely unable to determine to which division of the animal kingdom they really belong.

This leads us to our second question; If Embryology gives the clue to the relationship of animals, may it not do more and reveal their ancestry?

THE RECAPITULATION THEORY.[421]

A further explanation is afforded us by what is known as the *Recapitulation Theory*, which states that not merely have existing animals descended from ancestors which are often unlike them, but that each animal bears the mark of its own ancestry and reveals its parentage in its own development. Evolution tells us that each animal has had a pedigree in the past; Embryology reveals to us this ancestry because every animal has an inherited tendency during its own development to repeat its own ancestral history; or, to put it in other words, to climb up its own genealogical tree.

A good example of recapitulation is afforded by flat fish such as the sole, flounder, turbot, and plaice, which are distinguished not merely by the remarkable flattening of the body from side to side, but by the further facts—(1) that the two sides, right and left, of the fish are never coloured alike, one being nearly white, and the other dark-coloured; and (2) that the two eyes, instead of being situated, as in other animals, one on each side of the head, are both on the same side—*i.e.*, the darkly-coloured one. On watching these flat fish in an aquarium, we note that they habitually lie on the bottom on the paler coloured side, and we are at once led to associate the remarkable condition of the eyes with this habit; for it is clear that when so resting, if the eyes were placed in the usual positions, one at each side of the head, the eye on the paler surface—*i.e.*, the surface on which the fish lies—would not only be perfectly useless, but would be liable to injury from contact with the sea-bottom.

On turning to the development of the flat fish we find this supposition confirmed. A sole on hatching, and for some time afterwards, has its eyes one on each side of the head, just like any ordinary fish; furthermore, it swims, like other fish, with the body vertical, and has its two sides coloured alike. It is only after it has attained some size that it gradually adopts the habits of the adult, and takes to resting on its left side on the sea-bottom and swimming with the body horizontal instead of vertical. At the same time, the right side of the body gradually becomes coloured differently to the left, and in such a way as to resemble the sea-bottom closely, and so enable the fish to escape the notice of its enemies; and the left eye, no longer of use in its original position, is gradually displaced upwards on to the top of the head, and then shifts over to the right side; the change in position of the eye being accompanied by very considerable twisting and distortion of the skull. (Fig. 12.)

Inasmuch as flat fish in all other respects agree with more ordinarily-constituted fish, and as their special peculiarities—*i.e.*, the lateral compression of the body, the difference of colouring on the two sides, and the singular position of the eyes—may all be readily and naturally explained by their habits, which again would be clearly advantageous to the fish in aiding them to escape from enemies, it becomes in the highest degree probable that flat fish are descended from normally-formed fish, which first acquired the habit of lying on one side for the sake of protection, and then underwent structural changes in consequence of this habit.[422]

---

421   See Lectures 10 and 13.

422   We may hypothesize, with modern hindsight, that things happened the other way around: suppose that distorted development in the adult skull, appearing as the fish matures, is the result of a gene mutation in the germline; then suppose further that such a distorted fish survived predation by adopting a sessile habit and reproduced successfully. Thus would be founded a line of flatfish which might subsequently

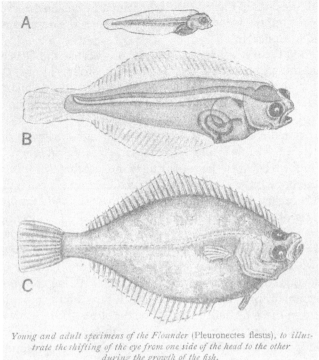

Young and adult specimens of the Flounder (Pleuronectes flesus), *to illustrate the shifting of the eye from one side of the head to the other during the growth of the fish.*

FIG. 12.

If this be correct, then the developmental history of the flat fish becomes intelligible by assuming that each individual has an inherited tendency to repeat in its own development the history of the species; every flat fish during its own growth passing through the same series of changes by which we have supposed the whole race of flat fish to have acquired their special peculiarities.

The case with regard to the sole is really a very strong one: for the only alternative view is that flat fish are not descended from normally-shaped fish, but have sprung into existence independently; and not only is this view absolutely contradicted by what we know of other animals, but it would render the development of the flat fish an incomprehensible mystery. The one view gives a complete and intelligible

---

diversify into a range of similarly distorted cousin species. It is known that some flatfish, at both species and individual levels, are *right-sided* and others *left-sided*; thus it is easy to imagine that the mutation may affect either side of the developing skull. This oversimplified explanation turns out not to be too far from the truth (Why the flounder is flat. *Science Daily*, 5 December 2016, www.sciencedaily.com/releases/2016/12/161205110908.htm) even if the genetics is a little more complex than imagined (see footnote 232, Lecture 10 for references). A fascinating recent finding is that the hormones which determine the asymmetrical development also cause the discrepant coloration between the fish's two sides, upon which AMM also remarked. Two paragraphs later AMM discounts the notion that flatfish *sprung into existence independently*, yet in a sense that is what they did.

explanation of all the facts of the case; the other not only has no direct evidence in its favour, but is totally opposed to all experience, and leaves the developmental features unexplained and inexplicable.

I have selected the sole for special description because the facts of the case are well known, and the argument is a simple and easily-followed one. Other animals, however, would serve the purpose equally well, and would afford illustrations quite as striking of the aid given us by the Recapitulation theory in unravelling embryological problems.

Thus, a crab and a lobster are animals closely agreeing with one another in essential structure, and clearly belonging to the same zoological group. The characteristic difference in form between the two is due to the fact that, while in the lobster the hinder part of the body, or "tail," is well developed, forming about half the length of the animal, and being used as a swimming organ; this "tail" is in the crab very greatly reduced in size, is of no use for swimming, and instead of projecting horizontally backwards, is carried bent forwards under the anterior part of the body, to which it is so closely fitted as to escape notice at first sight. Here again, as in the case of the flat fish, the structural differences may clearly be traced to difference in habit. Lobsters not merely walk about on the sea-bottom on their legs, but are able to swim freely in a backward direction, by powerful jerks of the tail. Crabs, on the other hand, walk but do not swim; and in them the tail, being no longer of use, has greatly diminished in size, and become rudimentary. Crabs, however, in their early stages of development, are free-swimming animals, and have tails quite as well developed and fully as large, relatively to the whole animal, as lobsters; and it is only after they have reached a certain size that they abandon their free-swimming habits, sink to the bottom, and henceforth move by walking only. The case is exactly parallel to that of the flat fish; and the Recapitulation theory explains the developmental history of a crab by saying that it is a repetition of the ancestral history of crabs in general: that crabs are descended from animals essentially similar to lobsters—i.e., from *Macrurous* ancestors, and that each crab passes through a lobster stage in its development, because of the inherited tendency that all animals have to climb up their own genealogical trees. (Fig. 13.)

The evidence of the descent of crabs from Macrurous ancestors involves, moreover, the supposition that they came into existence later than the *Macrura*. This supposition is supported by the evidence of palaeontology, for the *Macrura* are found as fossils in the Devonian and Carboniferous periods, and abundantly so in the Jurassic and Cretaceous, but are comparatively scanty in the Tertiary period; the *Brachyura*, or crabs, on the other hand, are very abundant in the Eocene and numerous in the Cretaceous, but doubtfully represented in the earlier periods.

Good examples of recapitulation are found in Molluscs. The typical Gasteropod has a large spirally-coiled shell; the Limpet, however, has a conical shell, which in the adult animal shows no sign of twisting, although the structure of the animal shows its affinity to forms with spiral shells. However, in its early stages of development the Limpet has a spiral shell, which is lost on the formation of the conical shell of the adult.

Recapitulation is not confined to the higher groups of animals, and good examples are found among the Protozoa. One of the best instances is that of *Orbitolites*, one of the most complex of the Foraminifera, which, during its own growth and development, passes through the series of changes by which the discoidal type of shell is derived from the simpler spiral shell. This forms an instructive example, for, owing

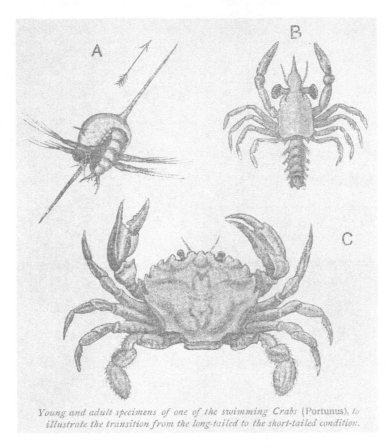

Young and adult specimens of one of the swimming Crabs (Portunus), *to illustrate the transition from the long-tailed to the short-tailed condition.*

FIG. 13.

to the mode of growth by addition of new shelly matter, the older parts are retained often unaltered, and in favourable examples all stages can be determined by simple inspection of the adult shell. (Fig. 14.)

The mode of growth of shells is important, since it gives an opportunity for *comparing the palaeontological and embryological records.* In such a shell as that of the *Nautilus* the central chamber is the oldest and first formed one, to which the other chambers are added in succession. If then the development of the shell is a recapitulation of ancestral history, the central chamber should represent the palaeontologically oldest form, and the remaining chambers in succession forms of more and more recent origin.

In the shells of *Ammonites* it has been shown that such a correspondence between historic and embryonic development really exists. In the middle Jurassic deposits the older Ammonites are flattened and disc-like, with numerous ribs; in later forms the shell bears a row of tubercles near the outer side of the spiral, and later still a second inner row of tubercles as well, while the ribs gradually become less conspicuous, and ultimately disappear. In more recent forms the outer row of tubercles disappears, then the inner row, the shell becoming smooth, swollen, and almost spherical. On taking-one of these smooth spherical shells, such as *Aspidoceras cyclotum,* and breaking away the outer turns of the spiral so as to expose the more

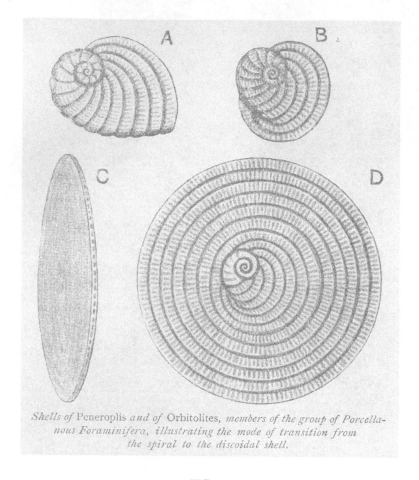

Shells of Peneroplis *and of* Orbitolites, *members of the group of Porcella-*
nous Foraminifera, illustrating the mode of transition from
the spiral to the discoidal shell.

FIG. 14.

central and older turns, first an inner and then an outer row of tubercles appear, which nearer the centre disappear, and in the oldest part of the shell are replaced by the ribs, characteristic of the earlier, and presumably ancestral forms.

Another illustration of the parallelism between the palaeontological and the developmental series is afforded by the antlers of deer, which are shed annually, and grow again of increased size and complexity in each succeeding year. In the case of the red-deer *(Cervus elaphus)*, the antlers are shed in the spring, usually between the months of February and April; during the summer the new antlers sprout out, and growing rapidly, attain their full size at the pairing season in August or September; they persist through the winter, and are shed in the following spring. The antlers of the first year are small and unbranched; those of the second year are larger and branched; in the antlers of the third year three tynes or points are present; in the fourth year four points, and so on until the full size of the antler and the full number of points are attained.

The geological history of antlers is of great interest. In the Lower Miocene and earlier deposits no antlers have been found. In the genus *Procervulus,* from the Middle Miocene, a pair of small, erect, branched, but non-deciduous antlers were present, intermediate in many respects between the antlers of deer and the horns of antelopes. From slightly later deposits a stag *(Cervus dicrocerus)* has been found with forked deciduous antlers, which, however, do not appear to have had more than two points. In upper Miocene times antlers were more abundant, larger, and more complex; while from Pliocene deposits very numerous fossils have been obtained, showing a gradual increase in the size of the antlers and the number of their branches, down to the present time.[423]

Antlers are therefore, geologically considered, very recent acquisitions: at their first appearance they were small, and either simple or branched once only; while in succeeding ages they gradually increased in size and in complexity. The palaeontological series thus agrees with the developmental series of stages through which the antlers of a stag pass at the present day, before attaining their full dimensions.

The Recapitulation Theory, if valid, must apply not merely in a general way to the development of the animal body, but must also hold good with regard to the formation of each organ and system, and with regard to the later as well as the earlier phases of development. Take for example the mode of renewal of nails and of the epidermis generally in the human skin. Each cell begins its career in an indifferent condition, and gradually acquires the adult peculiarities as it approaches the surface, through removal of the cells lying above it.

RUDIMENTARY ORGANS OR VESTIGES.

The Recapitulation Theory also affords most valuable assistance in determining the meaning of special points in the structure of animals which otherwise would be hard to understand. More especially is this the case with regard to what are spoken of as "rudimentary organs." These are structures, such as the eye of the mole, or the rudimentary teeth present in whalebone whales at an early developmental stage, but got rid of before birth: structures which are constantly present in all members of the species, but which are of no use to their possessors. They cannot be nascent structures—i.e., ones which are in process of formation, but which have not yet become functionally active, for the law of natural selection requires that no structure can be developed and retained by a species unless it either is of direct use to its possessor, or else has been of use to some of its ancestors. Their presence would be a complete enigma, but for the Recapitulation Theory, which explains them as structures which were formerly of use to the ancestors of the existing animal, and which appear in the latter because of the inherited tendency of all animals to repeat their ancestral history in their own development. Eyes are developed in the mole, although quite useless to it, because moles are descended from mammals in which the eyes were functionally active. Through the burrowing habits of the mole, the eyes have become degenerate and rudimentary; but owing to the law of Recapitulation, they are still developed. So with whalebone whales, which are toothless when adult, the presence of teeth at an

---

423   This conclusion requires greater caution than AMM indicates, for it could be greatly influenced by sampling error resulting from the paucity of the fossil record. However, the increase in antler size over time presumably resulted from sexual selection in successful species. The parallel made in the next paragraph with maturational development in individuals, illustrates recapitulation if one accepts recapitulation, but is not evidence for it.

early stage of development can only be explained by the descent of whalebone whales from toothed ancestors.

Rudimentary organs are of exceedingly common occurrence; indeed there are probably few if any of the higher animals in which some may not be found. Thus the splint-bones of a horse's leg are rudiments of the metacarpals or metatarsals of the second and fourth digits of the manus or pes, which were fully developed in the extinct *Hipparion*, and in other more remote ancestors of the horse. Almost all parasitic animals undergo, as we have seen, retrogressive or degenerative change in certain parts or the whole of their structure, and it commonly happens that vestiges of these lost organs linger on as rudiments, whose presence would be inexplicable but for the history of their formation.

Man himself is no exception to the rule. The muscles of the ear, whereby it can be pulled upwards or twitched forwards or backwards, are in a degenerate condition, and comparatively few men have any real power over them; while other smaller muscles which run from one part of the ear to another, are in such a completely rudimentary state, that but for anatomy their presence would never be suspected.

There are other parts of man's bodily structure—his teeth, for example—that show degeneration quite as clearly as do these ear-muscles; while most excellent examples are furnished by his laws, habits, clothing, and speech. In such a word as "reign," the letter "g" is mute and rudimentary; no attempt is made to sound it when pronouncing the word, and its presence can only be explained in accordance with the laws of rudimentary organs generally. Turn to the past history of the word, refer to its ancestors, and you find in the Latin "regnum" a word in which the "g" has full value, and from which we know that our own "reign" has been derived. The "b" in "doubt," the "n" in "solemn" are other examples; while in the "lf" of "halfpenny" we have a case in which degeneration is in the act of taking place at the present time.

Disturbing Causes hindering Recapitulation.

We must now turn to another side of the question. Although it is undoubtedly true that development is to be regarded as a recapitulation of ancestral phases and that the embryonic history of an animal presents to us a record of the race history; yet it is also an undoubted fact, that the record so obtained is neither complete nor straightforward.

It is indeed a history, but a history of which entire chapters are lost, while in those that remain many pages are misplaced and others are so blurred as to be illegible; words, sentences, or entire paragraphs are omitted, and, worse still, alterations or spurious additions have been freely introduced by later hands, and at times so cunningly as to defy detection.

The chief disturbing cause arises from the necessity of supplying the embryo with nutriment. This acts in two ways. If the amount of nutritive material in the egg is small, then the young animal must hatch early, and in a condition in which it is able to obtain food for itself. In such cases there is of necessity a long period of larval life during which natural selection may act so as to introduce modifications of the ancestral history, or spurious additions to the text. If, on the other hand, the egg contains a considerable quantity of nutrient matter, then the period of hatching can be postponed until this has been used up. The consequence is that the embryo hatches at a much later stage in its development, and if the amount of food material, or food yolk as it is called, is enough, may even leave the egg in the parent form.

This varying condition, as regards amount of food-yolk, affects recapitulation—i.e., the tendency of the embryo to pass through the ancestral stages—in two principal directions. If there is very much food-yolk, there is a tendency for the embryo to shorten its development by the *omission of certain of the ancestral stages,* and especially by the suppression of characters which, though functional in the ancestors, are of no use in the adult state of the animal itself. Thus tadpoles, after hatching, breathe for a time by gills: this gill-breathing condition being an ancestral one for all Vertebrates. In the West Indies there is a little frog *(Hylodes)* which lays its eggs, not in water, but on the leaves of plants. These eggs are larger than those of the common frog—*i.e.,* contain more food-yolk—and the young embryo is thereby enabled, just like the lobster or crayfish, to develop to a later stage before hatching: it passes through the tadpole stage within the egg, and hatches, like the crayfish, in the form of its parent. Although it passes through a gill-cleft stage no gills are developed; being of no use to the embryo, it would be a sheer waste of time to form them, and so they have dropped out of the ontogeny or individual development. (Figs. 15, 16.)

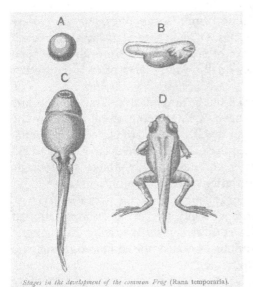

Stages in the development of the common Frog (Rana temporaria).

A.—The egg.   × 3
B.—The Tadpole at the time of hatching. The mouth is not yet formed, the Tadpole being still dependent on the food-yolk present in its body. Two pairs of external gills are present as branched finger-like processes at the sides of the neck. Below and in front of these is the horse-shoe shaped sucker by which the Tadpole fixes itself, and at the sides of the front of the head the rudiments of the nose and eye are seen.   × 3
C.—A Tadpole shortly after the time of appearance of the limbs; the hind limbs are seen at the junction of the body and tail : the fore limbs are present, but are concealed by the opercular folds covering the gills. At this stage the Tadpole breathes by both gills and lungs.   × 1
D.—A young Frog with the tail only partially absorbed.   × 1

Stages in the development of the West Indian Frog (Hylodes), illustrating the effect of increased amount of food-yolk in causing the omission of ancestral stages. The free living Tadpole stage of the common Frog is entirely suppressed, and no gills are ever formed. The entire development from the laying of the eggs to the hatching of the Frogs occupies from a fortnight to three weeks.

A.—The larva at the end of the first week. The head, eyes, stumps of the limbs, and the long tail are well shown. The food-yolk is contained within the large yolk-sac in the middle of the figure.   × 3
B.—The young Hylodes shortly before hatching. There is still a very large tail present, which is believed to be used as a respiratory organ.   × 3
C.—Young Hylodes at the moment of emerging from the egg : a short stump of a tail is still present.   × 3
D.—Young Hylodes at the end of the first day. The tail is completely absorbed, and the Frog has already the form of an adult.   × 3

FIG. 15.   FIG. 16.

Exactly the same thing has happened in reptiles, birds, and mammals, in which gill-clefts are found, but gills are not developed.[424] A similar tendency to the omission or blotting out of useless characters is seen in the development of all forms which have sufficient food-yolk to carry them over the stages at which these characters would be of functional value before the time of hatching.

In the case of embryos developed from small eggs there is a *tendency to distortion of the ancestral history* of a very different nature. Such embryos, owing to the small supply of food-yolk in the egg, have to hatch very quickly—*i.e.*, not merely of small size, but in a condition representing a very remote ancestral stage. The intervening stages between the early condition and the adult one have to be repeated while the larva is enjoying a free existence: this process will necessarily be slow, for the larva has not merely to develop, but to obtain for itself food, at the expense of which the further development may be effected. Hence such larvae may take weeks, months, or even years in recapitulating these later stages, which their larger-egged allies get through in a few days. Moreover, during the whole of this time they are exposed to competition, both amongst themselves and with other animals; they have to obtain food for themselves, and they are liable to be themselves devoured as food by other animals.

Owing to this competition, and to the length of time during which it lasts, these *larvae are liable to acquire, through natural selection, characters which are connected with their existence as larvae, but which form no part of the ancestral history;* characters which will aid them in obtaining food, or in escaping from their enemies, but which were not found in any of the ancestors of the species. Of such secondary larval characters the long spines with which the *Pluteus* larva of sea-urchins are provided are good examples; so also are the enormous spines on the young larva of crabs and other crustacea.

Other excellent illustrations are afforded by the developmental history of many fresh-water forms, in which, from the danger of their being swept down by the currents of the rivers or streams in which they dwell, or to obtain protection from the cold of winter, special characters are often acquired. The *glochidium* larva of the fresh-water mussel is a good instance of the former, and the specially-protected statoblasts or winter buds of *Polyza* and sponges, of the latter specially-acquired character.

It is not easy to distinguish between these later acquired, or larval characters, and the features that are really due to inheritance; while the correct discrimination of them is one of the greatest problems which an embryologist has to solve. As a general rule, secondary characters will be more variable, because the different groups will probably have acquired them independently. Again, secondarily acquired characters must always be useful, must confer some advantage on their possessors, or otherwise they would never have been preserved; while on the other hand, structures such as rudimentary organs, which are of no practical use, must be inherited, for in no other way can their presence be explained.

One other cause of falsification of the ancestral history in actual development may be briefly alluded to. It happens not uncommonly that the larvae and adults have entirely different habits, and in such cases the transition from one to the other is not always a gradual one, but may be effected by an abrupt, almost violent metamorphosis.

---

424   But gill clefts develop into other organs, including ears, so they are not useless.

Take the case of a butterfly or moth. From the egg emerges a caterpillar, a soft-bodied vermiform animal with short fleshy legs adapted for crawling along the branches and leaves of plants, with jaws adapted for biting these leaves, and an alimentary canal fitted to digest them as food. The caterpillar feeds and grows rapidly, but retains its shape, and, except in size, makes no appreciable approach towards the adult condition. Having reached its full size, it changes into a chrysalis or pupa, and during this state, which lasts for weeks or months, it takes no food, having indeed no mouth, but lives at the expense of nutriment which it has accumulated in its body during the caterpillar stage. After a time, longer or shorter in different cases, the pupa skin is cast off and the full-blown butterfly or moth appears: an animal altogether different to the caterpillar; provided with two pairs of wings, with three pairs of long jointed legs, with much more perfect sense-organs, and with its jaws modified so as to form a long tubular proboscis, by which it can suck up the juices of flowers on which it now feeds.

This is a typical case of abrupt metamorphosis, and it is obvious that the developmental history cannot here be a true recapitulation. It is quite impossible that the pupa, for instance, should ever have been an adult condition, and the abrupt character of the changes from caterpillar to pupa, and from pupa to imago, cannot be ancestral. Without entering at length into the origin of metamorphoses such as these, it may be pointed out that they only occur amongst insects, in forms in which the nature of the food, and therefore the structure of the jaws, is very different in the caterpillar and in the adult condition respectively; and that in such cases a gradual transition from one to the other would be quite impossible, for a mouth intermediate in its characters between the masticatory mouth of the caterpillar and the suctorial one of the butterfly would manifestly be incapable of either biting leaves or sucking the juices of flowers.

In the case of other insects, such as the locust, cricket, or grasshopper, the developmental history is one of gradual progression, and not one of abrupt metamorphosis, each step being a step onwards towards the adult insect. In the cockroach, again, the process of development is gradual, and parts not present in the larva, such as wings, appear not suddenly, but step by step and progressively. The differences in these cases between the young and the adult are not great.

There is no doubt that gradual transformation is the simpler and more primitive condition, and that the action of Natural Selection may cause the larval and adult forms to move apart and constitute periods of growth and reproduction respectively. Natural Selection may also cause such specialisation of the adult condition as to make it differ widely from the larva in habits and the nature of its food as well as in structure, and the larva and adult may move so far apart as to render a period of quiescence necessary in order to allow the change of organs into those required for new work. Again, Natural Selection may act on the larva as well, fitting it better for its own life as distinct from that of the imago; thus leading to still further divergence, and resulting in the larva acquiring characters that are no part of its ancestral history.

A similar explanation applies to the process whereby the young sea-urchin is formed within the larva. The larva is adapted for a free-swimming existence, the sea-urchin for crawling on the sea bottom. A gradual transformation from one to the other would be undesirable, for the intermediate conditions would be imperfectly adapted to either mode of existence; hence we find the external form of an early ancestral stage preserved, while internally the larva is passing through the later stages and gradually working its way up to the adult form and structure. A similar process occurs in many other animals from various groups, the explanation in all cases being found in considerations such as the above.

TESTS OF RECAPITULATION.

An important consideration is that, if the developmental changes are to be inter-preted as a correct record of ancestral history, then, first, the several stages must be all possible ones, the history must be one that could actually have occurred i.e., the several steps of the history as reconstructed must form a series, all the stages of which are practicable ones. Secondly, each stage must be an advance of the preceding one, otherwise it would not have been retained, for it must constitute an advance so distinct as to confer on its possessor an appreciable advantage in the struggle for existence. It is not enough that the ultimate stage should be more advantageous than the initial ones, but each intermediate stage must also be a distinct advance. Intermediate stages, which are not and could not be functional, can form no part of an ancestral series.

A good example of an embryological series fulfilling these conditions is afforded by the development of the eye in the higher *Cephalopoda*. First let us consider the evolution of eyes in the Mollusca. In *Solen* we find the simplest condition of the molluscan eye, merely a slightly depressed and slightly modified patch of skin, which can only distinguish light from darkness, and in which the sensitive cells are protected

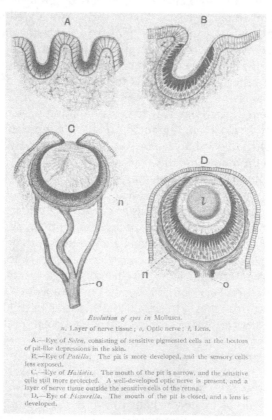

Evolution of eyes in Mollusca.

n, Layer of nerve tissue ; *o,* Optic nerve ; *l,* Lens.

A.—Eye of *Solen,* consisting of sensitive pigmented cells at the bottom of pit-like depressions in the skin.

B.—Eye of *Patella.* The pit is more developed, and the sensory cells less exposed.

C.—Eye of *Haliotis.* The mouth of the pit is narrow, and the sensitive cells still more protected. A well-developed optic nerve is present, and a layer of nerve tissue outside the sensitive cells of the retina.

D.—Eye of *Fissurella.* The mouth of the pit is closed, and a lens is developed.

FIG. 17.

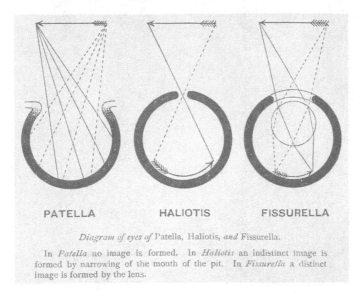

PATELLA   HALIOTIS   FISSURELLA

Diagram of eyes of Patella, Haliotis, *and* Fissurella.

In *Patella* no image is formed. In *Haliotis* an indistinct image is
formed by narrowing of the mouth of the pit. In *Fissurella* a distinct
image is formed by the lens.

FIG. 18.

by being situated at the bottom of the fold of skin. In *Patella* the next stage is found, where the eye forms a pit with a widely open mouth. This is a distinct advance on the preceding form, for owing to the increased depth of the pit, the sensory cells are less exposed to accidental injury. The next stage is found in *Haliotis,* and consists in the narrowing of the mouth of the pit. This is a simple change, but a very important step forwards, for in consequence of the smallness of the aperture, light from any one part of an object can only fall on one particular part of the pit or retina, and so an image, though a dim one, is formed. The next step consists in the formation of a lens at the mouth of the pit, by a deposit of cuticle: this form of eye is found in *Fissurella*. (Fig. 17.) The gain here is twofold—viz., increased protection and increased brightness of the image, for the lens will focus the rays of light more sharply on the retina, and will allow a greater quantity of light, a larger pencil of rays from each part of the object, to reach the corresponding part of the retina. (Fig. 18.) Finally, the formation of the folds of skin known as the iris and eyelids provides for the better protection of the eye, and is a distinct advance on the somewhat clumsy method of withdrawal seen in the snail. This is found in the Cephalopoda, such as *Loligo*.

If now we study the actual development of the eye of a cuttle-fish, we find that the eye, although a complicated one, yet passes in its own development through all the above series of stages, from the slight depression of skin, through the stages of a pit with large and small mouth; lens and finally eyelids being developed. (Fig. 19.)

The important point here is that we are able to show that the series fulfils our conditions, that all stages are possible ones, and that each is a distinct step onwards and an improvement on its predecessor; and furthermore, that each stage is retained as the actual permanent condition in some actually living mollusc.

It is not always possible to point out so clearly as in the above instance the particular advantage gained at each step, even when a complete developmental series is known to us; but in such cases our difficulties may be largely ascribed to ignorance of the particular conditions that confer advantage in the struggle for existence.

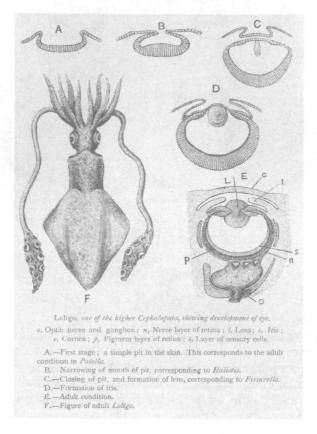

Loligo, *one of the higher Cephalopoda, showing development of eye.*

o, Optic nerve and ganglion; *n,* Nerve layer of retina; *l,* Lens; *i,* Iris; *c,* Cornea; *p,* Pigment layer of retina; *s,* Layer of sensory cells.

A.—First stage; a simple pit in the skin. This corresponds to the adult condition in *Patella.*
B. Narrowing of mouth of pit, corresponding to *Haliotis.*
C.—Closing of pit, and formation of lens, corresponding to *Fissurella.*
D.—Formation of iris.
E.—Adult condition.
F.—Figure of adult *Loligo.*

FIG. 19.

EMBRYONIC STAGES VIEWED AS ANCESTORS.

Early larval stages are of much interest, as possibly indicating the forms of the earliest ancestors. The most important and fundamental point is the fact that all the higher animals arise from eggs, and that the bodies of the higher animals are built up of cells or units, as a wall is built of bricks.

The lowest animals, or *Protozoa,* are single units or cells, and the egg of the higher animals is also a single cell. Therefore each of the higher animals begins its life as a single cell—*i.e.,* in a Protozoon stage. Does not this indicate the descent of *Metazoa* or multicellular animals from *Protozoa* or unicellular ones? If there is a blood-relationship between the highest and lowest animals, and if the higher are descended from the lower, is it not reasonable to look for the origin of Metazoa in the Protozoa? May not this be the explanation of the origin of all Metazoa in their actual development from single cells, and may not the egg represent the Protozoon stage in the ancestry?

Further, if animals really recapitulate, and if the reason why all Metazoa begin life as single cells—which is the most remarkable fact in the whole of embryology—is that they are descended from Protozoa, may not we hope to find from the study of early stages of development some hint as to the mode of origin of Metazoa in the first instance?

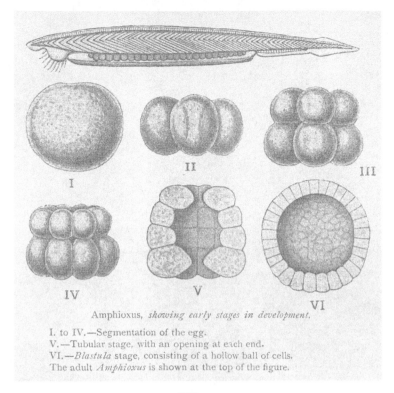

Amphioxus, *showing early stages in development.*
I. to IV.—Segmentation of the egg.
V.—Tubular stage, with an opening at each end.
VI.—*Blastula* stage, consisting of a hollow ball of cells.
The adult *Amphioxus* is shown at the top of the figure.

FIG. 20.

Let us consider, for example, the actual development of *Amphioxus*. The egg divides into a number of cells, which, instead of separating, remain together and continue to divide again and again, giving rise to the *morula* stage. The next stage is the *tubular* condition, where the cells are arranged regularly round a central cavity with an aperture at each end. This is followed by the *blastula* stage, which consists of a hollow ball, the outer cells of which are furnished with cilia enabling the embryo to swim freely. During later stages foldings take place, caused by outgrowths in some places and depressions in others, whereby the shape is gradually altered. (Fig. 20.)

Our present point is to ascertain whether these earliest stages are possible ones; whether there are organisms which remain permanently in one of these conditions— viz., (1) a single cell; (2) a heap of similar cells; (3) a hollow tube; (4) a ciliated hollow ball.

As examples of the first condition, or that of a single cell, the *Monads* may be taken. These are among the most minute and the simplest of living organisms, having an oval body, a nucleus, and a flagellum.

They are found in infusions of animal and vegetable matter. An example of the second condition, a heap of similar cells, is found in *Pandorina,* a colony of similar flagellate cells, all alike and living together in a common capsule. The third condition is found in *Salinella,* one of the most recent discoveries, and one of the most remarkable animals known, which is found in water containing 2 per cent. of salt. This organism consists of a tube, open at both ends, the wall of which is formed of a single layer of cells. The fourth condition, that of a hollow ball or blastula, is represented by

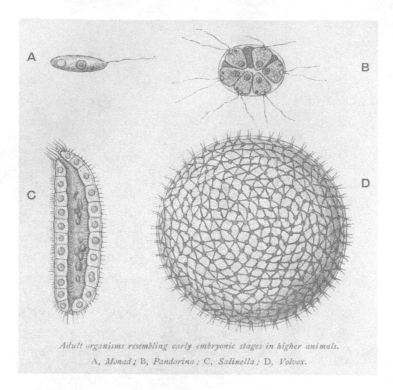

Adult organisms resembling early embryonic stages in higher animals.
A, *Monad*; B, *Pandorina*; C, *Salinella*; D, *Volvox*.

FIG. 21.

Volvox, the well-known fresh-water organism, which consists of a hollow ball formed by a single layer of flagellate cells like monads. (Fig. 21.)

It must be noted that these examples do not form a continuous series in the sense of being derived from one another. They are mentioned merely with the object of showing that the very earliest stages of development, like the later ones, may be recapitulatory; and that these early larval stages represent possible forms of adult beings (whether animals or plants it is impossible to say) capable of independent existence; and they illustrate very forcibly the interest which attaches to embryology in the light of the Recapitulation theory, coupled with the theory of Evolution with modification.

LECTURE 18

THE COLOURS OF ANIMALS AND PLANTS

AMM gave a lecture with this title in the New Islington Hall, Ancoats, on December 6, 1885, as the 14th lecture of the fourth series of the Recreation in Ancoats *Sunday Afternoon Lectures to Men and Women.* It was followed by a concert by Manchester Philharmonic String Band. He gave it again, this time *With Diagrams*, on November 7 the following year as one of the first series of *Sunday Lectures for the People* in Temperance Hall, Leeds Road, Bradford. CFM duly records that it was a Victoria University Extension Lecture.

Alfred Russel Wallace published a pair of papers entitled "The Colours of Animals and Plants" in *Macmillan's Magazine* in 1877 (**36**(215): 384–408 and 464–471) and discussed the topic in a great many other publications, some of which are quoted by AMM. The whole topic of colour, camouflage and survival was a favourite of late 19th-century biologists.

A key issue, which AMM addresses here, was the extent to which the environment determines the reactive colours and patterns of individuals as distinct from changes in the frequency of coloration within populations brought about by selection. The interpretation of experiments by Weismann and others was confounded by this problem, at the exact period in biological history when the question of whether acquired characteristics could or could not be inherited was being resolved.

The matter continued to be discussed at length well into the 20th century. The ability to apply strict mathematical analysis, for example to the classic melanism of the peppered moth, was used to defend the contention that Mendelian genetics was a proper science, in contradistinction to the ideological Lamarckism promulgated by Lysenko in the USSR (see Julian Huxley, *Soviet Genetic and World Science*, London, Chatto & Windus, 1949).

The lecture ends with a timely warning about the human perception of beauty in Nature and the interpretation of colour. Once again, Darwin rides to the rescue. A similar sentiment is also expressed, even more succinctly, in the next lecture.

THE COLOURS OF ANIMALS AND PLANTS

S O universal and universally acknowledged are the power and charm of colour, and so accustomed are we to associate bright colours with health, happiness, and merriment, and gloomy colours with misfortune, evil, or with death, that we are apt tacitly or explicitly to assume that the existence of colour is sufficiently explained by the pleasure it gives us; that the exquisite and varied colours of butterflies, birds, and flowers, for instance, are developed and acquired for our special enjoyment. But this is not so, and the poet has warned us that this explanation is not enough, for *"full many a flower is born to blush unseen, and waste its sweetness on the desert air."*[425] These lines man, in his supreme conceit, usually interprets as an expression of a half-contemptuous pity for the unfortunate flower which fails to meet his lordly eye, and so wastes alike its beauty and its sweetness. The poet Gray is, however, quite right, and there is something to be explained, as a moment's consideration will show us.

Birds and insects are the most gorgeously-coloured animals, and many of the most brightly-coloured are inhabitants of parts of South and Central America, the Malay Archipelago, &c., where man is almost unknown and wholly unwelcome. Again, man is a comparatively recent arrival on the earth, and the exquisite shapes and mouldings of many fossil shells show us that at any rate beauty of form existed in rare perfection long before his advent. This is a subject worth inquiring into, and one which has attracted the attention of many men, especially Wallace. The result of these inquiries is to show that colour is no mere accidental attribute of animals and plants, but has a very definite reason for its existence, and that in various ways and for divers reasons it may contribute materially to the welfare of its possessor.

The colours of animals in a state of nature are constant, or nearly so. Tame rabbits vary greatly, but wild ones are all very much alike; each kind of animal having a particular colouring, which does not vary very greatly, save in exceptional cases. Not merely in the individuals of a given species, but in genera or even in entire families a certain constancy of colouring may be noticed. For instance, the "blue" family of butterflies are characterised, not only by their blue colour, but more markedly still by the eye-spots on the under-surface of the wings. The mottled under-surface of the wings of Vanessidae, and the silvering of the wings of Fritillaries, are other examples.

Colour cannot be explained as due to the direct action of light or heat; for although it is true that there is an immensely greater number of richly-coloured birds and insects in tropical than in temperate and cold countries, yet the majority of tropical birds are dull-coloured, and in some groups the most brightly-coloured members are not tropical: for instance, the Arctic ducks and divers are more handsome than the tropical ones. The hummingbirds found in the Andes form another instance, for here they are confined to lofty mountains, sometimes to a particular mountain, *"just*

---

425 Thomas Gray, *Elegy Written in a Country Churchyard* (1751).

beneath the line of perpetual snow, at an elevation of some 16,000 feet, dwelling in a world of almost constant hail, sleet, and rain." Again, most tropical and brightly-coloured birds are denizens of the forests, and shaded from the direct action of the sun; they abound also near the equator, where cloudy skies are very prevalent. In the case of flowers, Wallace remarks that *"in proportion to the whole number of species of plants, those having gaily-coloured flowers are actually more abundant in the temperate zones than between the tropics."*[426]

NON-SIGNIFICANT COLOURS.

Many colours are the incidental results of chemical and physical structure, for the same reason that sulphate of copper is blue. The red colour of blood, the white colour of fat, the silvery colour of the bladder of many fish, the pigmented condition of the frog's peritoneum, and the green tint of the bones of many fish, are examples. The brilliant and varied colours of deep-sea animals are probably devoid of any significa-tion, and the green colour of grass and the blue colour of the sky, for all we know, are non-significant.

The red colour of *Tubifex*, for example, is associated with the physiological activity of haemoglobin; the red colour here is probably disadvantageous as such, but is counterbalanced by the physiological utility of the pigment for respiratory purposes. A similar explanation holds with chlorophyll, the green colouring matter of plants.

It is important to note that red is only red in the presence of light, and that a red animal if put in a dark place ceases to be red; or if put in a green light, which it is incapable of reflecting. Non-significant colours *"form the material out of which natural or sexual selection can form significant colours,"* and *"all animal colours must have been originally non-significant."*[427]

THE DIRECT ACTION OF ENVIRONMENT.

Distinct colour varieties occur locally among the Lepidoptera, and a great prevalence of green is shown by the fauna of Ceylon, not only by terrestrial forms, but by echino-derms, corals, and other animals. That differences in food have an effect on colour has been shown by feeding the larvae of various kinds of Lepidoptera on different plants: the larvae of the eyed-hawk moth, the brimstone moth, and the peppered moth show changes of this kind. This effect is not due to the colour showing through the skin, but must be effected through the nervous system, the particular pigment being actually built up by the caterpillar.

Pupae assume in many cases the colour of the objects to which they are attached; for instance, it was shown that the pupa of the small tortoise-shell butterfly, when placed on a dark background, became itself very dark; but when placed on a white background it became light-coloured. This susceptibility to change of colour is

---

426   These quotations from Wallace come from *Tropical Nature and other Essays* (1878) and can also be found in periodical contributions. They also appear in *Natural Selection and Tropical Nature: Essays on Descriptive and Theoretical Biology* (1891), published after AMM's lecture.

427   Source not known.

greatest at the stage when the larva first fixes itself before changing to the pupa. The effect is produced through the skin generally, and not through the eyes.

Again, trout change colour according to that of the bottom of the stream they inhabit; so also do minnows; hence the interior of a minnow-can is painted white, so that the bait may be light-coloured, and more conspicuous to a pike or perch. Changes of colour depend on the eye in the case of fish, and a blind trout remains dark, the pigment cells relaxing and becoming flattened, thereby exposing their maximum amount of surface area. Cave animals, on the other hand, become pale, because the pigment which is now useless degenerates and disappears.

The effect of cold in causing change of colour indirectly through the nervous system, has been demonstrated by suddenly exposing animals to cold which had previously been protected from it for some time, the result being distinct blanching. Melanism, or dark coloration, is common on oceanic islands, and humidity of the atmosphere is as a rule associated with the darkening of colours. So it is with increased elevation, and it is possible that the object of this is to increase the absorption of heat. Brilliant colours are not dependent on or proportionate to the amount of light.

The effect of environment in causing changes of colour is well shown in cases of what is known as *seasonal dimorphism*, where animals produce two broods in each year, each of different appearance with regard to colouring, and each capable of producing the other. A good example of this is found in the two continental butterflies *Vanessa prorsa* and *Vanessa levana*.[428]

Vanessa levana, the spring form, has a red ground colour, with black spots and dashes, and a row of blue spots round the margin of the hind wings; *Vanessa prorsa*, the summer form, is deep black, with a broad yellowish-white band across both wings, and with no blue spots. These were formerly called distinct species, but have recently been shown to be varieties of one and the same species. That these differences are due to the direct action of cold and heat has been shown by keeping the pupas of *levana* at a low temperature. These, which would ordinarily have produced the summer form *prorsa*, hatched, under the altered conditions of temperature, partly as *levana* and partly as an intermediate form.

Again, very young Canaries change their colour to orange when given Cayenne pepper, and certain parrots have been shown to change their colour when fed on the fat of a particular fish[429].

It is very difficult to draw the line between the direct action of environment through the nervous system, and the action of Natural Selection; for to which can we attribute the whitening of Arctic animals?

---

428   See also Lectures 2 and 5. These references to experiments of *Vanessa* relate directly to the work of Weismann on seasonal dimorphism and the emergence of species, carried out on *Vanessa* and other butterflies from about 1865 and continuing on and off for the rest of his life.

429   Examples such as these—the pink colour of flamingos makes another—are known to be due to the assimilation of colour pigments from food into the skin or feathers. It is not clear whether AMM was intending to imply this, but the next sentence suggests not.

We have now to consider the great mass of cases illustrating the preservation and accentuation of colour through the agency of Natural Selection—one of the most striking of the later developments of the theory.

SIGNIFICANT COLOURS.

These are colours which are of direct advantage to their possessor as colour, and not merely because they are associated with other properties which are useful, as in the case of haemoglobin and chlorophyll. The classification of these colours is a matter of some difficulty, for cross-relations occur which are difficult to express.

There are three chief classes or groups:—

1. *Apatetic;* the purpose of which, or rather the object gained by which, is to hinder recognition by other animals,

2. *Sematic,* or signalling colours; the purpose of which is to facilitate or aid recognition by animals of the same or of other kinds.

3. *Epigamic;* which include those cases in which differences occur between the male and female sex, as in the peacock and pea-hen, the duck and drake, &c. This is a special and important group.

APATETIC COLOURS are again divided into

a. *Protective resemblances,* aiding escape from enemies, as in those cases where animals resemble sticks or plants, and so escape notice.

b. *Aggressive resemblances;* the purpose of which is to aid the approach to prey; for example the resemblance of the colour of the lion to that of the desert.

c. *Alluring resemblances;* constituting a small group of cases, in which an animal acts as a bait by taking on the form of something attractive to its prey.

SEMATIC COLOURS have two subdivisions:

a. *Warning colours.* These constitute a curious group of cases, in which animals have bright conspicuous colours, for the purpose of warning other animals off them, and which are signs of inedibility or of the possession of dangerous powers of attack.

b. *Recognition colours.* These are for the purpose of easy recognition by animals of the same kind; and are best seen in the cases of gregarious animals, such as deer, whose safety largely depends on association and mutual defence.

PROTECTIVE RESEMBLANCES.

Such forms of protective colouring as aid the escape from enemies by hindering recognition may be one of two kinds:

(1) *General;* in which the colouring is such as to assimilate the animal to its environment, and so render it less conspicuous, as in the case of the whiteness of Arctic animals, such as the Polar bear; and the sandy colour of desert animals, or the transparent blueness of pelagic forms.

(2) *Special;* where the resemblance is to some particular object, and where the animal escapes, not through being concealed from view and so overlooked, but through being mistaken for something else. Of these cases some extraordinary instances are known. The resemblance may be to another animal or to a plant, flower, or leaf, or to inorganic substances.

Furthermore, the protective colouring may be either constant or variable; a good example of variable protective colouring being shown by the *Octopus* and *Chameleon*. Again, in animals such as insects, which undergo metamorphosis, and in which the form, structure, and habits are widely different in the larval and adult stages respectively, both these stages may be protectively coloured, but the resemblance will be to entirely different objects.

Let us take examples from the different groups of animals, and we shall see that the reality of protective colouring is impossible to doubt.

MAMMALS.—The whiteness of Arctic animals has already been referred to. The American polar bear is white all the year round; the ermine or stoat changes to white in the winter, and the Arctic fox usually does this also. The Alpine hare always becomes white in the winter in Scandinavia, and usually in Scotland, although rarely so in Ireland. This change consists in an actual blanching of the hairs from the tips inwards, with a new growth of additional white hairs. The general tawny colour of deer is also protective; the protection afforded by spots is seen by their resemblance to the circular spots of light caused by sunlight passing through the leaves of a wood, while stripes facilitate escape in long grass or reeds.

To fully appreciate the protective value of colours it is necessary to see the animals in their native haunts. Thus, speaking of the Zebra, Francis Galton says that although *"no more conspicuous animal can well be conceived,"* yet the proportion of the black and white stripes *"is such as exactly to match the pale tint which arid ground possesses when seen by moonlight."*[430] With regard to the Giraffe, Sir S. Baker graphically describes the way in which, when seen at a distance, it resembles a dead tree stem. Again, the green colour of the Sloth is due to parasitic algae, which cause it to resemble a lichen-covered branch; an oval buff-coloured mark on the back giving the impression of the broken end of the stump.

BIRDS.—The summer plumage of the Ptarmigan conceals it among the heather very effectively; in winter it becomes white. The Heron again, is almost impossible to find among the rushes, where it stands in an absolutely vertical position, with the tip of its beak tilted up. The ventral surface having pale yellow stripes, closely resembles the surrounding rushes. Moreover, it turns slowly round so as always to present the ventral surface to view, while the striped back and broad dark-coloured sides are never presented to the observer. In the case of birds which build open nests, the female is protectively coloured: this is well seen in the pheasant. When both sexes are brilliantly coloured, such as the kingfisher, parrots, &c., the nest is of such a nature as to conceal the sitting bird.

INSECTS.—These afford the best examples of protective colouring both in the larval and adult states.

I. *Larval Insects.*—Lepidopteran larvae, or caterpillars are in the great majority of cases absolutely defenceless: their bodies *are* soft and their very shape is due to their containing fluid under pressure, therefore a slight wound involves much loss

---

430   *Narrative of an Explorer in Tropical South Africa* (1853).

of fluid or blood; hence their great need for protection. The great purpose in life of a caterpillar, next to feeding, is not to be seen, or rather not to be recognised.

General protective resemblances are found in their green colour, which is the most usual, and harmonises with that of the food-plant. This green colour is partly due directly to food and partly to metachlorophyll, a special pigment in the blood, and a slightly altered derivative of the chlorophyll of the food[431]. Those caterpillars which have the habit of feeding either on grass, or on low-growing plants among the grass, are protected by longitudinal striping. In the larger caterpillars, such as those of the Privet hawk moth, which are striped transversely or obliquely, the colour is usually that of the flower of the food-plant, and the stripes serve to break up the surface of the body. These larvae turn brown at the time of descending to the earth to change into pupae.

Special protective resemblances are best seen in the larvae of the geometer moths, or "stick caterpillars" as they are called. These are very common, and are rarely seen, or rather detected, owing to their resemblance to twigs. They have only two pairs of legs or claspers, a long thin and cylindrical body, which stands out at an acute angle with the stem upon which they fix themselves, and upon which they sit motionless for hours; this absence of movement being very important in order to increase the deception. The head is modified in shape to increase the resemblance to twigs, and a silk thread is spun, attached to the twig, to relieve the tension involved by remaining in the same position for so long a time. They feed at night, when there is less need for protective devices. The protection here is so real that a green lizard will generally fail to detect a stick caterpillar in its position of rest, though it will seize and greedily devour it directly it moves.

2. *Adult Insects.*—In butterflies the under surface of the wings is coloured protectively, the upper surface attractively; and the sudden change when they fold their wings over their backs is often enough to defy detection. The most perfect examples are found in the leaf butterflies, such as *Kallima*, found in the Malay Archipelago, India, and Africa. This is a very common and showy butterfly, with orange and purple colouring on the upper surface of the wings. It is a rapid flier, and frequents dry forests, always settling where there is dead and decaying foliage. The colouring on the under surface of the wings bears a remarkable resemblance to that of a dead leaf, and when the wings are turned up, with the head and body hidden between them, it is often very difficult to distinguish it from dead leaves, the resemblance being rendered even more close by the short tail, which looks like the stalk of a leaf, and by the markings on the under surface, which closely imitate the mid-rib and veins of a leaf. Speaking of this insect, Mr. Wallace says: "*The colour is very remarkable for its extreme amount of variability, from deep reddish-brown to olive or pale yellow, hardly two specimens being exactly alike, but all coming within the range of leaves in various stages of decay. Still more curious is the fact that the paler wings, which imitate leaves most decayed, are usually covered with small black dots, exactly resembling the minute fungi on decaying leaves.*" (Fig. 22.) This is an extreme case of what is really a general law among

---

431   This conclusion appears to be based on the studies of Poulton EB, The essential nature of the colouring of phytophagous larvae (and their pupae); with an account of some experiments upon the relation between the colour of such larvae and that of their food-plants. *Proceedings of the Royal Society, London* 1884; **38**: 269–315.

Kallima, *showing under surface of wing.*

FIG. 22.

butterflies. The mode in which it is acquired is as follows: At first there is a more or less accidental resemblance. Large numbers of butterflies are killed by birds, lizards, and other animals, and any whose markings and habits of perching render them less easy to detect will have a better chance of escaping, and so of laying eggs and transmitting their peculiarities to their offspring. This protection becomes, through selection, better from generation to generation, and the imperfectly protected forms are weeded out and eaten.

Similar examples are found in the Herald and Angle shade moths,[432] which resemble decayed and crumpled leaves; in the Buff-tip moth, which resembles a broken piece of decayed and lichen-covered stick; and in the Lappet moth which has the appearance of dried leaves. But the best example perhaps is that of *Phyllium*,[433] the leaf insect found in the East Indies, which has in the adult condition a most extraordinary resemblance to a leaf, or rather a bunch of leaves. The colour is green, and the wing-cases are marked with lines like the veins of a leaf: some of the joints of the limbs are flattened and expanded. Andrew Murray relates how an Indian species exhibited in the Botanical Garden at Edinburgh deceived everybody by its resemblance to the plant upon which it lived. The deception was ultimately the cause of its

---

432   Family *Noctuidae.*

433   Now *Phylliium*, family *Phlyliidae.*

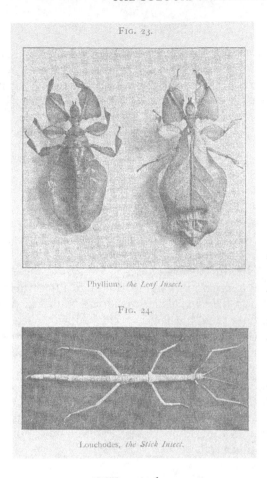

FIG. 23.

Phyllium, the Leaf Insect.

FIG. 24.

Lonchodes, the Stick Insect.

FIGS. 23 and 24.

death; for the visitors, sceptical as to its animal nature, insisted on touching it before they would be convinced. (Fig. 23.)

VARIABLE PROTECTIVE RESEMBLANCES.

Instances of this we have already mentioned in the case of the Trout, which changes colour according to that of the bottom of the stream. This effect is always produced through the eye, and is consequently absent in blind animals. How far it is under the control of the will is difficult to determine.

Another excellent illustration is afforded by the Chameleon and many lizards which possess the power of changing colour. This power depends on the presence of chromatophores or pigmented bodies found in the skin, which have the power of changing shape. These are of different colours, and are arranged in layers, some near the surface, some deeper; the light yellow cells being most superficial, then the red and brown, and the black deepest of all. If the superficial ones contract, the deeper ones will become more apparent, and *vice versâ*. These changes of colour always bear a relation to the surface on which the animal is placed at the time, and are therefore supposed with good reason to be protective.

AGGRESSIVE RESEMBLANCES.

These aid the approach of their possessor to its prey, through a superficial resemblance to its general environment or to other objects. The Lion and Tiger are good examples of general aggressive resemblances, the lion resembling in colour the desert in which it lives, and the stripes on the tiger rendering it easy to conceal itself among the long grass. So with the Polar bear, which not only is of the same colour as the snow, but also has a curious shapeless outline when lying down, not unlike a heap of snow.

ALLURING RESEMBLANCES.

These attract or entice the approach of prey. *Lophius piscatorius*, the angler-fish, which conceals itself in the mud at the bottom of the water, is provided with long tentacles, which are mistaken by small fish for worms writhing in the muddy water. The body of the angler, being concealed in the mud, is not seen, and the prey, deceived by what they imagine to be worms, are themselves eaten by the angler-fish. Again, certain deep-sea fish have a luminous phosphorescent organ at the end of the foremost tentacle, suspended as a lure in front of the mouth, to attract their prey.

One of the lizards, *Phrynocephalus mystaceus*, found in Asia, not only in its general colour resembles sand, but is also furnished with a red fold of skin at each angle of the mouth, which is produced into a flower-like shape resembling a little red flower which grows in the sand. Insects, attracted by what they believe to be flowers, approach the mouth of the lizard, and fall victims to its snare. The Indian Mantis, *Hymenopus bicornis*, feeds on other insects, which it attracts by its flower-like shape and pink colour, which resemble an orchid, the shape being due to flattening of the proximal joints of the limbs, which radiate from the body like the petals of a flower. *Thomisus decipiens*, a spider found by Forbes, and seen twice only, resembled exactly the droppings of birds on a leaf.[434]

ADVENTITIOUS COLOURING.

This may possibly be either aggressive or protective. The Caddis worms, for instance, make cases of sand, shells, twigs, and any objects found at the bottom of streams. *Xenophora*, a genus of Gasteropods, build pieces of dead shells, rocks, and corals into the edge of their growing shells. Some crabs, again *(Stenorhynchus)*, have a habit of fastening pieces of sea-weed, &c., on to their bodies or limbs. Mr. Bateson describes how these crabs tear the weed in pieces with their claws, and after chewing the pieces in their mouths in order to soften them, rub them on their head or legs till they are caught by the hairs which cover them. *"The whole proceeding is most human and purposeful."* These processes are gone through both by night and clay, and a blinded Stenorhynchus, if cleaned of its weed, *"will immediately begin to clothe itself again with the same care and precision as before."* It shows no disposition to take up a position among similarly coloured objects; and some individuals which have taken up stations among weeds do not dress themselves at all.[435]

---

434   The Bird Dropping Spider, now *Phrynarachne decipiens*, was discovered in the Malay Archipelago in 1883 by Henry Ogg Forbes (1851–1932).

435   Bateson W, Notes on the senses and habits of some Crustacea. *Journal of the Marine Biological Association of the United Kingdom, New Series* 1889; **1**: 211–214. The inclusion

WARNING COLOURS.

If an animal belonging to a group liable to be eaten by others, is possessed of a nauseous taste, or if an animal, such as a wasp, is specially armed and venomous, it is to its advantage that it should be recognised quickly, and so avoided by animals that might be disposed to take it as food.

Hence arises *warning coloration,* the explanation of which is due to Wallace. Darwin, who was unable to explain the reason for the gaudy coloration of some caterpillars, applied to Wallace, stated his difficulty, and asked for suggestions. Wallace thought the matter over, considered all known cases, and then ventured to predict that birds and other enemies would be found to refuse such caterpillars if offered to them. This explanation, first applied to caterpillars, soon extended to adult forms, not only of insects, but of other groups as well.

An excellent example is afforded by the Skunk (*Mephitis mephitica*), a small black and white animal possessing a very offensive secretion, which it ejects over its enemies, and which protects it from their attacks. Owing to this offensive weapon the skunk is seldom attacked by other animals, and its black and white coloration easily distinguishes it from unprotected animals.

Insects afford many admirable examples of warning colours, and many well-known instances are found among butterflies. The best examples among these are found in three great families of butterflies—the *Heliconidae,* found in South America, the *Danaidae,* found in Asia and tropical regions generally, and the *Acraeidae* of Africa. These have large but rather weak wings, and fly slowly. They are always very abundant, all have conspicuous colours or markings, and often a peculiar form of flight; characters by which they can be recognised at a glance. The colours are nearly always the same on both upper and under surfaces of the wings; they never try to conceal themselves, but rest on the upper surfaces of leaves and flowers. Moreover, they all have juices which exhale a powerful scent; so that if they are killed by pinching the body, a liquid exudes which stains the fingers yellow, and leaves an odour which can only be removed by repeated washing. This odour is not very offensive to man, but has been shown by experiment to be so to birds and other insect-eating animals.

Warning colours are advertisements, often highly-coloured advertisements, of unsuitability as food. Insects are of two kinds—those which are extremely difficult to find, and those which are rendered prominent through startling colours and conspicuous attitudes. Warning colours may usually be distinguished by being conspicuously exposed when the animal is at rest. Crude patterns and startling contrasts in colour are characteristically warning, and these colours and patterns often resemble each other; black combined with white, yellow, or red, are the commonest combinations, and the patterns usually consist of rings, stripes, or spots.

Other examples are found in the bright colours of some sponges which have been proved to be nauseous to fish; in the Anemones, Ascidians, and many brightly-coloured Nudibranchs.

One of the best-known instances is that of the frog found by Mr. Belt in Nicaragua, a small animal, gorgeously coloured with red and blue, which never hides itself; whereas most frogs are coloured green or brown, and hide during the day-time,

---

of this quotation from Bateson shows that the lecture was rewritten at least four years after its first presentation as part of the Ancoats series.

to avoid being eaten by snakes and birds. Suspecting this animal to be uneatable, Mr. Belt offered it to ducks and fowl, all of which refused to touch it, except one young duck, which took the frog in its mouth, but dropped it directly, "and went about jerking its head, as though trying to throw off some unpleasant taste."

Very numerous examples are found among caterpillars such as that of the Cinnabar moth, which is coloured black and yellow, and rejected even by a toad. The Magpie-moth caterpillar, which is cream-coloured, with orange and black markings, and extremely conspicuous, is either refused altogether by birds, lizards, frogs, and spiders, or else causes them to exhibit signs of most intense disgust after eating it. The caterpillars of the Tiger-moth, the Burnet, and the Buff-tip are all brightly coloured and nauseous.

It has been objected, first, that the protection afforded by warning colours is only imperfect, that it must not be overdone, and is only available to a few; that the likes and dislikes of insect-eating animals are purely relative, and hunger will drive them to extremes and overcome taste, hence giving rise to contradictory results of experiments. But this is the very essence of natural selection, which preserves advantageous characters, but does not lead to perfection. Secondly, it has been stated that "tasting is quite as dangerous to the caterpillar as swallowing outright," and hence it is argued that there is no advantage. But this, again, is a misconception, for although the individual may suffer, the species will benefit through the lesson learnt by its death.

Eisig's theory of warning colours[436] states that the pigment itself is the cause of the distastefulness, and that it is very probably excretory in nature. According to this, the brilliant colours—i.e., the abundant secretion of pigment—have caused the inedibility of the species, rather than that the inedibility has necessitated the production of bright colours as an advertisement. Brilliant colouring is the normal condition in caterpillars, and the advent of birds led to protective modifications, except when combined with inedibility.

MIMICRY.

Examples of mimicry are really cases of protective colouring, but are only intelligible through knowledge and appreciation of the value of the warning colour. Alluring colours, such as those of the *Mantis*, which simulates an orchid, may aptly be described as cases of "wolves in sheep's clothing," while the cases of mimicry may be considered as "asses in lions' skins."

Warning colours are conspicuous advertisements of inedibility, and certain colours and groupings of colours are usual, so that the lesson may be more easily learnt. Black, white, red, and yellow, in startling and striking contrasts, form the

---

436   Hugo Eisig put forward his theory in Part C, Section V of his major treatise on polychaete worms, *Die Capitelliden*, published in *Naples Monographs*, the journal of the Naples Zoological Station, in 1887. It came to prominence when Frank Beddard mentioned it in a letter to *Nature* (26 November 1891) and discussed it in his book *Animal Coloration* published in 1892 (London: Swan Sonnenschein). Beddard thought the idea worthy but doubted if it was applicable in every case. EB Poulton, author of *Colours of Animals* (London: Kegan Paul, 1890), also took issue with it: *Nature* 1891; **45**: 174–175.

usual types of warning colours. If these are successful—i.e., generally recognised as signs of inedibility—it is clear that other and different animals, which resemble them sufficiently closely to be mistaken for them, might benefit by the mistake, and escape.[437]

Certain butterflies, the best examples of which are found among the *Heliconidae, Danaidae*, and *Acraeidae*, are nauseous, slow-flying, gaudily-coloured insects, having an unpleasant smell, and taking no pains to conceal themselves. Alongside these occur edible forms belonging to totally different genera and families, each of which shows a striking resemblance to one particular species of the protected butterfly; this being in many cases confined to the female, which has greater need for protection. A large number of cases of this mimicry are now known; for instance, *Leptalis*, a form allied to the common garden white, mimics *Methona*, one of the protected Heliconidae, in the shape of its body and wings, in its colour and even in its habits and mode of flight; so much so that they are difficult to distinguish from one another. (Fig. 25.)

Methona (protected).

Leptalis (mimetic).

FIG. 25.

Other examples of mimicry are found in the Beehawk moth, which mimics the humble-bee; in the Clearwing moths and Hornet moths, in which protection is obtained by mimicking insects possessing stings, and some of these are even said to writhe the abdomen when captured, as though pretending to stings; others are said to have the characteristic odour of the hornet. The "devil's coach-horse," the beetle with

---

437   Batesian mimicry.

the habit of turning its tail over its back, pretending to have a sting, certainly deceives children, and perhaps grown-up people as well.

There is no doubt as to the success of this deception, which may deceive expert naturalists, or even the insects themselves; and Fritz Müller says; *"I have repeatedly seen the male pursuing the mimicked species, till after closely approaching and becoming aware of his error, he suddenly returned."*[438]

Perhaps the most remarkable instance known is that of *Papilio merope*, one of the South African swallow-tail butterflies. The female of this species alone mimics; and of the female three forms are known, each of which mimics a different species of *Danais* prevalent in its own district. In Madagascar, which in so many other instances furnishes us with a glimpse of what the ancestral African fauna must have been, the female *Papilio merope* closely resembles the male, and is not mimetic.

In butterflies the female is more in need of protection, because it is of slower flight and liable to be exposed to attack while laying eggs. All an adult butterfly has to do is to pair and lay eggs; and as these events take up a very short time, protective colouring is not very useful, but warning colours are far more so.

Among birds, the Cuckoo is undoubtedly protected by its similarity to a hawk in appearance and in mode of flight. In the snakes, again, the harmless ones are said to mimic the venomous.

The conditions necessary in order to effect mimicry are given by Wallace as follows:—

(1) The two species, the imitating and the imitated, must occur in the same locality.

(2) The imitating species must be the more defenceless.

(3) The imitating species must be less numerous than the imitated, in individuals.

(4) The imitating species must differ from the bulk of its allies.

(5) The imitation, however minute, is external only, never extending to internal characters or to such as do not affect external appearance.

The mode of acquisition of mimicry is by the gradual action of natural selection, and must have been accidental in the first instance.

RECOGNITION MARKINGS

These are closely allied to warning colours, and their purpose is to facilitate recognition, not by enemies, but by friends. They are seen especially in gregarious animals, and specific markings and colours are very probably in many cases not protective, but for recognition. A good instance of this class of colouring is seen in the upturned white tail of the Rabbit, which, although making it conspicuous to its enemies as well as friends, is probably a signal of danger to the other rabbits; and when feeding together, in accordance with their social habits, soon after sunset or on moonlight nights, the upturned tails of those in front serve as guides to those behind to run home on the appearance of an enemy. Many birds, antelopes, and other animals, have markings believed to serve a similar purpose, and probably the principle of distinctive colouring for recognition has something to do with the great diversity of colour met with in butterflies.

---

438   In a letter to Darwin, 14 June 1871, https://www.darwinproject.ac.uk/letter/DCP-LETT-7820.xml.

EPIGAMIC[439] COLORATION.

This is seen in mature animals, especially in butterflies and birds, where the two sexes differ markedly as regards colour. As a general rule, the male is of the same hue as the female, but of a deeper and more intensified colour; for instance, in thrushes, hawks, and in the Emperor moth. Sometimes patches of colour found in the males are absent in the females, as in the Orange-tip butterfly. In some cases there are more extreme differences, as in the drake, peacock, cocks and hens, pheasants and Bird of Paradise; gay colours being the special privilege of the male. (Fig. 26.)

FIG. 26. Bird of paradise, male and female.

It is curious to note how with man the conditions are reversed, for the female butterfly or bird is as a rule larger and plainer than her mate. So it is with the organs of voice; the male cricket or grasshopper can alone produce sound, and many female birds have no song. The power of talking was originally the exclusive possession of the males—a privilege that with us they have long had to surrender.

The origin of the bright colours of the male may possibly be due to the selective action of the sexes on one another. As among deer now, and men in several countries, the best fighters carry off the prizes; so among birds and butterflies the most gaily or tastefully arrayed males are the most highly favoured, and have the best chance of securing the most eligible mates. In the case of the Argus pheasant, the wing feathers are enormously elongated and of marvellously beautiful colouring; while

---

439 Concerned with sexual attraction.

during courtship the male erects his tail like a fan, displaying his glories to their best advantage. Similar examples are afforded by humming-birds.

There is no possible doubt as to the appreciation of colour by animals, and the chief differences between men and butterflies would appear to lie in the infinitely better taste displayed by the latter in the selection and combination of colour, both as regards marvellously delicate gradations of tint, and as regards daring combinations of strongly-contrasted colours. It is in fact as rare for a bird or butterfly to offend against good taste in matters of colour, as it is for a man to conform to it.

The theory of sexual selection, which was proposed by Darwin as supplementary to natural selection, is disputed by Wallace, who holds that brighter colour is the physical equivalent of greater vigour.[440]

THE COLOURS OF FLOWERS AND FRUIT.

The essential parts of a flower are the ovary, in which the ovules are produced, and the anthers, in which the pollen is contained; and in order that the ovule may give rise to a seed—i.e., to something capable of growing into a new plant—it must be fertilised by the pollen. There is a great advantage as regards the number of seeds produced, and the vigour of the offspring, if the ovules are fertilised by pollen, not from the same flower, but from a different flower or plant. This *cross fertilisation* is always highly beneficial and often absolutely essential. It is effected mainly by the agency of insects, especially bees, flies, and butterflies. These are induced to visit the flowers by bribes of honey secreted by the flower in such a position that, in order to reach it, the insect must brush against the anthers and get dusted with the pollen, by which, on visiting a second flower, fertilisation is effected.

The purpose of the coloured part of the flower is to form a conspicuous advertisement to insects of places where honey is to be found; and more detailed markings in the flower direct the insect towards the store of honey. Curiously ingenious contrivances are found in order to prevent self-fertilisation, and to ensure that the insect shall effect its work properly.

A familiar instance is that of *Orchis mascula*, the spotted orchid, which is abundant in meadows and in damp places in open woods. This consists of a spike of flowers, the calyx of which is formed by three coloured sepals, and the corolla by three petals. One of the petals, called the *labellum*, is larger than the others, and forms a sort of landing-stage. This is prolonged backwards into a spur-like nectary with spongy walls. The male organs consist of one anther with two cells, each of which contains a pollen mass. The ovary has three pistils, united together and twisted, ending above in two almost confluent stigmas. The third stigma forms the *rostellum*, a rounded projection overhanging the other stigmas and the entrance to the nectary. The pollinia, which lie in the anther cells, are club-shaped, the head of the club consisting of a number of packets of pollen grains united by thin elastic threads: the stalk of the club ends in a disc with a ball of very viscid matter on its under side, lying in the rostellum. The anther cells open when ripe, exposing the pollinia; the rostellum is very delicate, and is ruptured by the slightest touch, exposing the viscid balls.

---

440 A common interpretation today does not require this separation between colour and vigour: the magnificence of the male peacock's tail reflects the health of its possessor and has evolved as the wisest choice of the female for the promulgation of her genes.

The problem is to transfer the pollinia from one spike of flowers to another. The manner in which this problem is solved through the agency of insects is as follows.

The insect, alighting on the labellum, pushes its head into the flower in order to reach the spur with its proboscis. In doing this it knocks against the rostellum, displacing its covering membrane and exposing the viscid balls to which the pollen masses are attached. On withdrawing its head the pollen masses come away firmly cemented to it and standing erect. In about thirty seconds the viscid disc contracts, causing the pollen mass to bend forwards through an angle of 90°, so as to become horizontal. By this contraction the pollen mass will be in a position to be applied directly to the stigma, when the insect visits the next flower. This manoeuvre can be imitated by pushing the point of a pencil into a flower as shown in the figure, when the pollen masses will come away fixed to the pencil. (Fig. 27.)

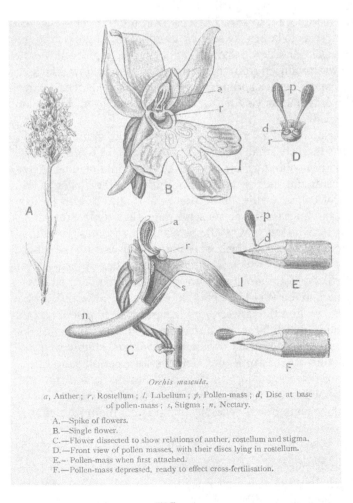

Orchis mascula.

a, Anther; r, Rostellum; l, Labellum; p, Pollen-mass; d, Disc at base of pollen-mass; s, Stigma; n, Nectary.

A.—Spike of flowers.
B.—Single flower.
C.—Flower dissected to show relations of anther, rostellum and stigma.
D.—Front view of pollen masses, with their discs lying in rostellum.
E.—Pollen-mass when first attached.
F.—Pollen-mass depressed, ready to effect cross-fertilisation.

FIG. 27.

In this way an insect flying from flower to flower effects cross-fertilisation regularly, and humble bees were actually watched in the act of fertilising by Hermann Müller.[441] He saw them insert their heads into the flower and emerge with the pollinia attached, visit other flowers on the same spike, where they tried, more or less ineffectually, to rub off the pollinia, and finally fly off to other plants. Out of 97 humble bees which he caught, 32 bore the pollen masses of orchids. He proved that the bees visit the flower to obtain the fluid in the nectary, the walls of which they pierce with their maxillae. Moreover, he timed the bees and found that they spent three or four seconds at each flower; two or three seconds being sufficient to fix the pollinia. The average time spent at a given spike of flowers was twenty to twenty-two seconds, the bees then flying to another spike. In twenty-five to thirty seconds the pollinia were depressed and cross-fertilisation ensured.

The beauty and odour of flowers and the storage of honey are thus due to the existence of insects, and in a large number of cases the actual insects are known which effect cross fertilisation. Such is the case with regard to all conspicuous flowers: honey is secreted in order to attract insects, and the flowers are large and conspicuously coloured, so as to be readily seen by them. A striking illustration of this is seen in the common Clover. Darwin showed that by protecting 100 flowers with a net, not a single seed was produced from them; whereas the 100 flowers which were outside the net were visited by bees and produced 2720 seeds. Hence, but for humble bees, which are the only insects visiting the common red clover, there would soon be no clover.

Large conspicuous flowers are visited much more frequently and by many more kinds of insects than are small inconspicuous ones. The long tubular corolla of many flowers is acquired so that certain insects alone should be able to get at the honey, these insects being the ones best suited for fertilising the flower.

The bright colour of the whole flower is to attract insects at a distance; the coloured dots and lines on the petals serve to guide it to the store of honey. This fact was proved by Darwin, who cut off the petals of *Lobelia,* and found that these flowers were then neglected by bees, which were perpetually visiting the other flowers.

In the case of flowers which are fertilised by means of the wind, such as grasses and trees, the flowers are small and not gaily coloured, and possess an enormous amount of pollen and a very large stigma. Moreover, in localities where insects are few in number, we find the flowers very insignificant in colour: for example, in the Galapagos Islands, which have only one butterfly and no bees.

White flowers are fertilised by nocturnal insects, chiefly moths. These flowers are always odorous, the jasmine and clematis for example, and often odorous only at night. Alpine flowers, again, are peculiarly beautiful, and the size of individual flowers is increased owing to the comparative scarcity of insects in the places where they grow, and the consequent necessity of attracting them from afar.

Fruits.

Fruits consist of seeds with surrounding envelopes of various kinds, and require to disperse their seeds so as to reach places favourable for growth and germination. Dispersion of the seeds is effected in some cases, such as the dandelion, by means

---

441   See also Lecture 21.

of the wind; in the edible fruits, on the other hand, it is effected by the fruit being swallowed by animals as food. Fruits are divided into two great groups—*attractive fruits* and *protective fruits*.

ATTRACTIVE FRUITS are soft, pulpy, and agreeable to the taste—such as the cherry, grape, strawberry, &c.—and are devoured by birds or mammals. In these the seeds themselves are hard, and pass through the animal unchanged. It is probable that every brightly-coloured pulpy fruit serves as food for some species of bird or mammal.

PROTECTIVE FRUITS, such as nuts. In these the part that would be eaten by an animal is the seed itself, and this is protectively coloured, being green while on the tree, and turning brown as it ripens and falls to the ground. Many seeds are specially protected, such as the chestnut by its prickly coat, and the walnut by its nauseous covering.

It thus appears that we owe the existence of many flowers to insects, and of many fruits to birds and mammals. This is an excellent example of the interest imparted to everyday life by the theory of Natural Selection, which tells us that we have merely to watch closely, to note carefully what is going on every day before our eyes, in order to obtain the clue to problems of extraordinary and widely spread interest. On the other hand, the support this theory receives through being able to offer a ready and complete explanation of so many and such divers facts is very great indeed, and becomes all the more significant when we reflect that the facts themselves only came to light, or received serious attention, some time after the promulgation of the theory.

The conclusion we have arrived at is, that the colours of animals and plants are no mere accidents, and are not created for our special benefit, but are directly useful to their possessors, and have been acquired because they are useful. Were any additional argument necessary, it would be easy to find it in the fact that men are so far from being in agreement as to what is and what is not beautiful, that the ideal of one nation may be the horror of another; that a picture which an Art Committee may select as beautiful, may appear to the public, for whom it is purchased, as entirely destitute of beauty; that the various devices which savage races practise in order to render themselves, as they consider, beautiful, appear disfigurements to other nations. So then it appears, on the one hand, that not only is there no general agreement among mankind as to what is beautiful, but that different nations, or the same nation at different times, absolutely contradict each other. Some other explanation is necessary of the beautiful colours of animals and plants, and we see what that explanation is in the great law of Utility, expounded by the doctrine of Evolution, for the full enunciation of which we are indebted to Darwin.

LECTURE 19

OBJECTIONS TO THE DARWINIAN THEORY

Lecture 19 addresses the main criticisms that were levelled at Darwin's theory by its detractors, from when the *Origin* was published right through to the end of the 19th century (and even into the 20th). AMM tackles these objections head on in his typically robust but respectful style, often using examples that have appeared in earlier lectures. In reading his analysis, one might wish at times for a different use of language to describe the course of development and the attainment of complexity but his intended meaning becomes clear.

To this day, it is not uncommon to find a newly discovered fossil described in the popular press as a "missing link", the convenience of the shorthand carrying implications of which the ill-educated writer is blissfully ignorant. Many non-biologists remain perplexed by what they see as the impossibility of an eye emerging from nothing by natural selection, and it is regrettably common, even in better-informed literature, to see contemporary cousin species referred to as "ancestral" or "primitive".

Recalling that this is a lecture for a non-specialist audience (also part of the 1893 series of Victoria University Extension Lectures), any teacher will admire the way AMM deals with the challenge of combining scientific detail and intellectual integrity with accessible explanation.

LECTURE 19

OBJECTIONS TO THE DARWINIAN THEORY

T HE best possible mode of testing a theory is to consider the objections which have been raised against it. It is impossible for us to deal here with all the objections which have been put forward, but I propose to select those which appear the most important—i.e., those which have been urged with the greatest force and persistency by men specially competent to deal with the subject. It is interesting to note that, in spite of the fierce storm of criticism to which the theory has been exposed, and the considerable amount of literature written on the subject—no small number of books having been written for the express purpose of "smashing Darwin"—yet nowhere are the objections and difficulties more clearly stated than by Darwin himself. Very few of any real importance have been added to the list given by Darwin, while he himself has indicated others that had escaped the notice of his opponents.

It seems strange to have to claim credit for candour; yet candour so striking as Darwin's does demand special and cordial recognition. Whether he was right or wrong in his conclusions, Darwin simply sought to determine the truth, and was always ready to discuss and consider in detail even the most trivial and thoughtless objections.

MISSING LINKS

The most popular objection, and in many ways the most famous, is that of the so-called Missing Links. If the present existing animals are descended from ancestors which were unlike them in former geological times, and if all animals are really akin or cousins, where are the intermediate forms, the missing links? These must on the theory of Evolution have existed. Can we produce them? or if not, can we give any reasonable explanation of our failure? We must at once admit that the demand is absolutely fair, and one which must be met. This question, although dealt with incidentally in former lectures, it is well to reconsider more directly. We have really two distinct problems to deal with, two kinds of links to be sought for—viz., (1) *Links between existing animals,* which must occur if the animals are akin to one another; (2) *Links between existing and extinct animals.* Let us consider these separately.

Links between the several kinds of existing animals. —Failure to find these has often resulted from misdirected efforts, from looking in the wrong direction. A straight line being the shortest distance between two points, it is commonly and not unnaturally assumed that the link must lie in the line connecting the two forms directly. True links are, however, not directly intermediate, the real relation being that of descent from a common ancestor, or branches of one stem.

Take for example the domestic pigeons. The blue rock is the common ancestor of all our domestic races, and is not in any sense intermediate between existing forms, such as a pouter and a fantail. Such intermediate forms have not existed at any time. So it is with ourselves: the real bond of union is through descent; two brothers are related, not directly, but through the parents; and with cousins the grandfather forms the real link.

These examples show that the actual links are not direct, but indirect; and that, unless careful, we may spend much time in looking for links where they could not exist, and overlook the real ones which are before our eyes all the time. This may be rendered deceptive by the occurrence of actual intermediate forms which are not true links, and against which we must be on our guard. An example will make this point clear.

The horse and the donkey are closely allied animals: midway between them is the mule, which shares the characters of both its parents. Yet this is clearly no true link, but an artificial unnatural creation that could not possibly have existed prior to either the horse or the donkey.

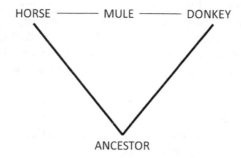

HORSE ———— MULE ———— DONKEY

ANCESTOR

Developmental evidence we have found to be of the utmost value, for the early stages in the development of animals are themselves the very links we want; sometimes distorted or modified, but usually recognisable with sufficient care. Take, for example, the fact of the prawn and the barnacle both commencing life in the same form—i.e., the Nauplius. This is evidence of a most cogent character in support of their descent from a common ancestor, and here the Nauplius is the ancestral form, or link from which both were derived.

Rudiments or vestiges also give most valuable evidence. For instance, the short tail of the crab; the splint bones of the horse's leg; the mute letters in our words; points which were fully considered in a previous lecture.[442]

Links between the Present and the Past.—This was Cuvier's difficulty, and formed the basis of an objection which was raised with fatal force. If existing animals are descended from extinct forms or fossils, why do the gaps appear so marked, and where are the intermediate stages which should exist?

This question we have already dealt with at length in a former lecture[443] where we saw that the objection could be met in its chief part by the *imperfection of the geological record.* We saw the extreme improbability that a continuous series of transitional forms could be preserved, owing to the fact that only certain parts of

---

442   {M} Lecture 16, "Rudimentary Organs or Vestiges".
443   {M} Lecture 16, "The Imperfection of the Geological Record".

animals can, in the ordinary course of events, have any reasonable chance of being preserved as fossils; and furthermore, that only certain deposits, such as mud, are capable of preserving these remains uninjured. *"The crust of the earth,"* to quote from Darwin, *"with its embedded remains, must not be looked at as a well-filled museum, but as a poor collection made at hazard and at rare intervals."* We also, however, saw that in spite of this imperfection of the record, several series of fossil links have been obtained; notably, in the cases of the horse and *Paludina;* and that the *Archaeopteryx* is one of the most important links known.

A further consideration in regard to fossil links is that of the unlikeness between the dominant forms of one age and those of succeeding ones: for instance, between the reptiles of the secondary period[444] and the later forms. Undoubtedly, if we regard these alone, there are great and apparently abrupt gaps; but we must not suppose fossil forms all to stand in the direct line of ancestry of living animals.

As it has been with animals, so it has been with men. As we find in human history the Egyptian, Greek, and Roman nations, each in turn dominant, so also do we note that they are not lineal descendants one of another, but collateral branches of the great human family or tree, which in succession attained maturity and gained dominion. Dominant races of animals in successive geological ages have often died out and left no descendants—great reptiles such as the *Dinosaurus* for example. It will be these dominant races which will be the most likely to leave fossil remains— as with the buildings and records of man—and, especially if of large size, to give character to the age. Yet these do not usually stand in the direct line of ancestry of living forms, which are descended from collateral branches, which at those times were insignificant. This consideration explains at once the apparently sudden gaps, and the difficulty in obtaining evidence of actual ancestors. Dominant types often die out, and are succeeded by others whose early history is difficult to unravel, because overshadowed by the then dominant forms.

On the whole, therefore, the search for links is not so hopeless as it appeared at first sight, when we once realise what a link really is, and what we have to look for. Additions are yearly, or almost monthly, made to our knowledge of the former history of the earth, in the discovery of fossils, now that their true importance is recognised.

Before concluding the subject of missing links it will be well to refer briefly to that aspect of the problem known familiarly as the "monkey question." The Darwinian

---

444 The secondary period of the Palaeozoic Era, now called the Devonian (416–359 mya), when fish and then tetrapods first appear in the fossil record. Although this is not explained, many of AMM's audience will have understood what was meant, either because they had attended the *Argument from Palaeontology* lecture (Lecture 16) or because of a popular best-seller of the time: *Vestiges of the Natural History of Creation* by Robert Chambers (1802–1871). Chambers' book, which was published anonymously, first appeared in 1844 and reached its 12th edition in 1884, discusses the appearance of reptiles in the fossil record during that period of the era. It was influential in bringing to the general public the idea of the transmutation of species on which the Darwin–Wallace theory (1854) relies. Wallace was greatly inspired by it, and it influenced his decision to travel, with Bates, to South America to collect insects. Most notably in the present context, *Vestiges* linked the appearance of fossil forms to stratigraphical markers of geological age.

theory does most undoubtedly imply that there is a blood-relationship between man and the lower animals; and it is also a most undoubted fact that, of these animals, the anthropoid apes—the orang, the chimpanzee, and the gorilla—are those which are most closely allied to man. Such being the case, much ingenuity has been exercised in the search for "missing-links" which will bridge over the gap between man and monkey, and very bold statements have been based on the failure of these efforts.

Such ingenuity is, however, misdirected, and such efforts predestined to fail; for if what we have said above as to the real nature of links be true, it follows that the links between man and monkey will most certainly not be directly intermediate forms, any more than the link between the horse and donkey is a mule. From time to time we are provided with such alleged direct links, and it is a source of very legitimate amusement to a Darwinian to note the extreme anxiety people always show to prove that these waifs are really monkeys and not men, and the readiness with which they are able to accomplish this to their own satisfaction. And yet in all cases they have proved to be really human beings after all. Thus, of late years we have had a couple of idiot children spoken of as Aztecs, and a Siamese child called Krao, who differed from the rest of the family in nothing except unusual hairiness; not to mention other celebrities, all of whom were pronounced positively to be monkeys on their first appearance.[445] Indeed, the only conclusion to be drawn from such cases, appears to be how very easy it is to persuade people to believe that human beings are monkeys.

PERSISTENT TYPES.

PERSISTENT TYPES, a group of cases which we have already considered,[446] have been brought forward as an objection to the Darwinian theory, on the ground that, as the fittest survive, there should be a continual improvement in the race; whereas, we have seen forms such as *Lingula* and *Nautilus* persisting apparently unchanged for periods of enormous duration; or even through the whole time of which we have any evidence in regard to life on the earth. We must bear in mind, however, that natural selection does not necessarily imply advance, and is perfectly consistent with a stationary condition; for the Fittest now may be the fittest many years or ages afterwards. It is quite practicable for particular forms to remain stationary for enormously long periods, provided their environment, or at least all features of the environment affecting them, remain constant.

Persistent types are usually marine, for there the conditions are more constant. They are very commonly found on sandy shores, in which they often burrow; their tenacity of life and power of withstanding injury are very great; they are usually also unpalatable, and hence not eaten as food by other animals. Their tenacity of life is

---

445   Exhibitions of freaks were popular in Victorian Britain. Barnum & Bailey's Circus, "The Greatest Show on Earth", visited London in 1889–90 and included a freak show amongst the static entertainments for visitors on their way to their seats. Amongst the delights on view were an Aztec man and wife (Pettit FY, 2012, *Freaks in Late Nineteenth-Century British Media and Medicine*, PhD thesis, University of Exeter). Krao Farini was a girl with hypertrichosis, and possibly other unusual features, who was taken from Laos (part of Siam; the adjective is used here with its correct meaning) and exhibited in Europe, England and the US until her death in 1926 aged 50.

446   {M} Lecture 16, "persistent types".

well shown in the case of *Lingula,* which was carried by Morse in his pocket during a three days' journey across America without coming to any harm; and *Balanoglossus* has been known to live in a bucket of unaerated water in a hot climate, with the hinder part of its body completely macerated, and its branchial skeleton exposed.

The persistence of lowly organised forms alongside more highly organised ones is often felt as a difficulty; for if the higher animals are descended from more lowly constituted ancestors, as we believe, and if these advances, these steps forward have been preserved because they were improvements, and because they gave advantage in the struggle for existence, how is it that we find the lowly organised forms still living alongside the higher and improved ones?

The answer is that these lowly organised forms occupy places which cannot be filled by the higher forms; as Wallace says: *"There is no motive power to destroy or seriously modify them, and they have thus probably persisted, under slightly varying forms, through all geologic time."* Again, if we compare human history, we see that advance in civilisation does not involve the advance of all the members. For instance, some shops become larger and more pretentious, yet there are still plenty of places where small shops can survive, while there is no room for large ones. Again, some nations still exist which use bows and arrows, or slings and clubs, which are easily replaced and more suitable for their purposes than more modern implements of warfare.[447]

DEGENERATION, OR RETROGRADE DEVELOPMENT.

It is very commonly assumed that, as in the struggle for existence, the fittest survive; therefore each generation must be rather more highly or more perfectly organised, and fitter to survive than the preceding; and that in all cases there must be a steady and continuous, though slow, progress upwards. It is then asked, how is it that even at the present day we find numerous representatives of the simplest groups of animals living? And how is it that we find many cases of degeneration—i.e., of animals which, in the early stage of their existence —representing ancestral phases—are more highly organised than in the adult condition?

Now, an animal may be placed under conditions in which organs useful to its ancestors, and inherited from them, may be no longer of service. Such organs tend to become degenerate, persisting for a time as vestigial structures, and ultimately

---

447   These three paragraphs reveal a possible flaw in AMM's understanding to which we have alluded previously. That there should even be a concept of *advancement* or *improvement*, and their counterpart *constancy*, implies a notion of developmental progress towards some ideal state or condition of perfection. The next section, too, uses related terminology suggesting regression, although it is redeemed by explaining perfection in terms of fitness. When evolution is interpreted strictly according to the Darwin–Wallace theory, phenotypic changes are seen as selective responses to environmental pressures, and to that alone. No form is more advanced than another, other than by chronology, and forms are only improved inasmuch as they are adapted to new conditions. As before, we can give AMM the benefit of the doubt if we read him as being forced to adopt the terminology of his time, but it remains an unsatisfactory position. It is not until we read the final paragraph of the next section that our anxieties are relieved.

perhaps disappearing altogether. Of such cases of degeneration we meet with numerous examples, of which the following are the most important:

(a) An animal fixed in the adult state, but free when young: such as sponges, hydroids, corals, polyzoa, oysters, and barnacles. This involves loss, or modification, of the locomotive organs, and often of the sense-organs as well.

(b) Parasites which live on or in other animals, and of which *Sacculina* is a good example. In these animals the whole body often becomes degenerate, the conditions of life rendering locomotor, digestive, sensory, and other organs entirely useless. In such cases, those forms which avoid the waste of energy resulting from the formation and maintenance of these organs will be most in harmony with their surroundings. Parasitic worms, molluscs, &c., show similar wholesale degeneration, and live immersed in the body fluids of their victims. The explanation of the extreme degeneration of parasites is that special food is required to meet the drain at the time of ripening of the eggs. For instance, in *Copepoda*, the female is alone parasitic, and that only at the time of laying eggs. The new phase intercalated in the life-history involves the necessity of laying more eggs, and there is greater difficulty in completing the ancestral history in individual development. This reacts on the parasitic stage, rendering it more important; more food is required, and hence further modification ensues.

(c) Special organs show signs of degeneration even in the highest animals, and give evidence of a former more perfect condition in their ancestral forms. This is seen in the eyes of the mole, and in many cave animals; in the splint bones of the horse, and in all the examples of rudiments or vestiges mentioned in a former lecture[448]

In a sense, all the higher animals are degenerate; that is, they can be shown to possess certain organs in a less developed condition than their ancestors, or even in a rudimentary state. Thus, a crab, as compared with a lobster, is degenerate in regard to its tail; a horse, as compared with Hipparion, in regard to its outer toes. It is a mistake, however, to speak of a crab as a degenerate animal in comparison with a lobster, for an animal should only be spoken of as degenerate when the retrograde development has affected, not one or two organs only, but the totality of its organisation.

No animal is at the top of the tree in all respects; man himself being primitive in retaining the full number of toes, and degenerate as regards his ear muscles. Care must also be taken not to speak of an animal as degenerate merely because it possesses organs less fully developed than allied animals. An organ is not degenerate unless its present possessor has it in a less perfect condition than its ancestors had. A man is not degenerate in the matter of the length of his neck as compared with a giraffe, nor as compared with an elephant in respect of the size of his front teeth, for neither elephant nor giraffe enters into the pedigree of man. A man is, however, degenerate, whoever his ancestors may have been, in regard to his ear muscles, for he possesses them in a rudimentary and functionless condition, which can only be explained by descent from. some better equipped progenitor.

The theory of Natural Selection does not say that the ideally best survive, but those most in harmony with their surroundings for the time being. If these are of such a kind as to render certain organs useless, such as the eyes of cave dwellers,

---

448   {M} Lecture 16, "Rudimentary Organs or Vestiges".

their possession is no longer an advantage, and the energy previously devoted to their production can be better utilised in other directions. Hence, though it is quite true that on the whole there has been a progress towards greater specialisation, and that differences between extreme groups are greater now than ever, yet there are many individual exceptions, and natural selection actually requires that there should be such exceptions.

THE ALLEGED USELESSNESS OF SMALL VARIATIONS.

This is an objection which has often been put forward. Admitting that no two animals are absolutely identical, it is urged that the differences are in most cases too small and too trivial to have the effect assigned to them; namely, to determine between survival or destruction. This is, however, a misapprehension, for if four out of five are to die, a very small matter may determine success or failure.

In a race for which there is but one prize, a victory by a short head is, so far as securing the prize is concerned, as conclusive as a win in a canter. Again, the whole theory and practice of the breeding of animals and plants afford absolute proof of the importance of attention to minute details, so slight as to escape the notice of all but the most skilful observers.

Think how in commerce a very small and subordinate point may determine survival; such as, for instance, the use of bye-products resulting from certain chemical manufactures, which had previously been neglected and regarded as waste products. Think what small events have decided the fate of battles and of nations. *"The death of a man at a critical juncture, his disgust, his retreat, his disgrace, have brought innumerable calamities on a whole nation. A common soldier, a child, a girl at the door of an inn have changed the face of fortune and almost of nature."*[449]

THE DIFFICULTY WITH REGARD TO THE EARLIEST COMMENCEMENT OF ORGANS.

This difficulty is a very serious one, for Natural Selection can only act on an organ after it has already attained sufficient size to be of practical importance and utility. Natural Selection accounts for any amount of modification in an organ when once established; modification in any direction, either of increase or decrease, but does not offer any explanation of the first appearance of such an organ. This is best understood by a few examples, showing the continuous preservation of a series of very minute variations.

I. *The wing of the Bat,* a flying mammal, is clearly a modified arm with great elongation of the fingers and webbing of the skin, which also extends from the side of the body and involves the hind-legs and tail (Fig. 28). It is easy to see that, when once established as a flying organ, Natural Selection would cause survival of those with the best wings, and so lead to gradual improvement and perfection of the wing. But how does the wing first commence?

Bats are a specialised group of mammals which must have been descended from non-flying ancestors. If the first commencement of the wing was a slight accidental

---

449   Edmund Burke (1729–1797) *Letters on a Regicide Peace* (1795/6).

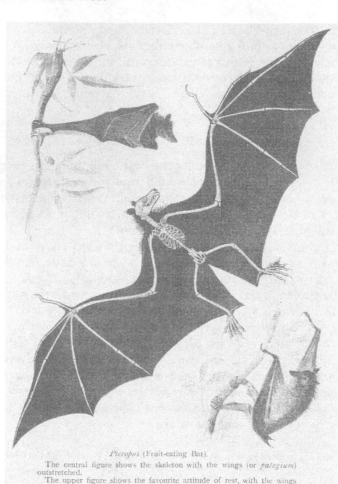

Pteropus (Fruit-eating Bat).

The central figure shows the skeleton with the wings (or *palagium*) outstretched.
The upper figure shows the favourite attitude of rest, with the wings folded up.
The lower figure shows the mode of progression along a branch, by means of the claws on the thumbs.
[N.B.—In this figure, the left-hand side represents the top, the right-hand side the bottom of the figure.]

FIG. 28.

elongation of the fingers, and a slight increase in the webbing, this would not give the power of flight, and would be of no use as a wing until it had attained a considerable size. In other words, *such an organ as a wing would in its earliest stages be useless for the purpose which it ultimately fulfils.*

This very important objection applies to a great number of cases, of which the origin of the wing is a typical one.

2. *The Origin of the Lung* is another example of the difficulty we are considering. The lung develops as a small saccular outgrowth from the throat, and it is quite unintelligible that a slight depression at the back of the mouth should have been preserved because it was useful for breathing air directly.

The explanations in these cases are good instances of many which are afforded by an important theory.

THE THEORY OF CHANGE OF FUNCTION.—This theory was suggested by Darwin, and afterwards developed more fully by Dohrn, as affording a possible solution of difficulties such as those we are considering. The principle is, that an organ may lose its original function, and yet persist because it is useful for another purpose—i.e., that an organ may be used for two or more different purposes, one predominating at one time, another at another time; and, further, that structural modifications may ensue fitting the organ better for its adopted function.

This theory offers an explanation of the first commencement of the lung, which is shown to have arisen from the *swimming- bladder* of fishes through change of function, and a series of forms is known to exist connecting the air bladder of fishes with the lung of the higher Vertebrates, which is undoubtedly the same organ.

The swimming bladder of most fish—the sturgeon for example—is a closed sac lying beneath the vertebral column, and is used for the purpose of flotation, to keep the back uppermost[450]. In many fish it acquires a connection with some part of the alimentary canal, and then becomes an accessory breathing organ.

The mud-fish, *Ceratodus* of Queensland and *Protopterus* of Africa, inhabit rivers which during the dry seasons are apt to become dried up. These animals lie buried in the mud for months, and can live for a long time out of water, owing to the fact that the swimming bladder is used as a lung, a slight change in the circulation causing aerated blood to be returned from it to the heart.

In *Menobranchus*[451], found in North America, both lungs and gills are present throughout life, and it is equally at home in water and on land. The lung is better developed than in the Protopterus and Ceratodus.

In *Amphiuma*[452], found in North American swamps, the lungs are still more perfect. The gills are lost but the gill-slits remain.

From this we reach the condition met with in the newt and frog, which possess gills in the tadpole stage, but lose them in the adult or lung-breathing state. (See Fig. 29.)

We have thus a series of animals, all now living, showing the actual transition from the swimming bladder to the lung, and from the gill-breathing to the lung-breathing condition; and we further see that the frog *actually repeats this history in its own development.*

The explanation of the first commencement of the bat's wing is more difficult, and there is still some uncertainty about it. Let us consider another group of Mammals, the SQUIRRELS, in which the finest gradation is known from animals with their tails slightly flattened, the hind parts of their bodies wide, and the skin of the flanks full, to the "flying squirrels," in which the limbs and even the base of the tail are united by a broad expanse of skin; this fold of skin, acting like a parachute, enables them to glide through the air for a great distance from tree to tree, and so escape their enemies. Here each step is useful, and similar modifications are met with in other animals.

In *Galeopithecus*, the flying lemur of Borneo, the skin fold extends from the neck to the hand, thence to the foot and from this to the tail, and includes the limbs

---

450   And to facilitate buoyancy adjustment at different depths. Swim bladders are found in bony fish but not in cartilaginous fish.

451   Now *Necturus*, for example *N. maculosus*, the Mudpuppy, a type of salamander.

452   The aquatic "conger" salamander.

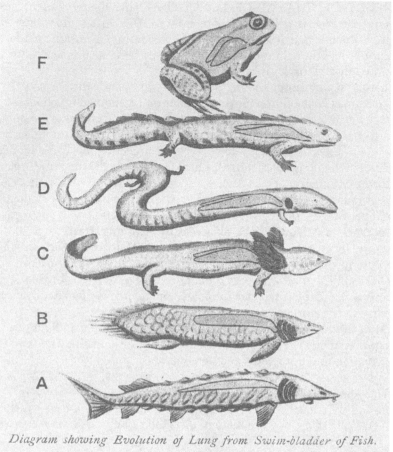

Diagram showing Evolution of Lung from Swim-bladder of Fish.

A, Sturgeon; B, *Ceratodus*; C, *Menobranchus*; D, *Amphiuma*;
E, Newt; F, Frog.

FIG. 29.

with the elongated fingers. From this Darwin suggested that the bat's wing could be derived by elongation of the fingers.

Further illustrations of the utility of imperfect wings to arboreal animals or fish are found. The flying frog of Borneo is a tree-frog with very long and fully webbed toes, which enable it to take long leaps in the air. The flying lizard *(Draco volans)* has the skin of the flanks supported by ribs. In both these cases there is no true power of flight, the action being that of a parachute. Of flying fish there are two chief groups, *Dactylopterus* (the gurnard), and *Exocoetus* (the flying herring). In both animals the pectoral fins are largely developed, and in the gurnard are almost certainly moved like wings.

There is no difficulty in understanding these cases as being acquired by Natural Selection, and the bat's wing may be not such a serious difficulty after all.

Other Examples of Change of Function.—A good instance of change of function is that of the *Hyo-mandibular* gill-cleft. The presence of gill-slits in the early stages of development of the higher Vertebrates is very interesting, and the only possible explanation of their presence is that their possessors are descended from gill-breathing ancestors. These gill-slits never bear gills, and all close up early with the exception of one, the hyo-mandibular cleft, which is preserved because its function is changed. This cleft passes close to the ear, which is buried in the side of the skull, and by remaining open becomes advantageous for the purpose of hearing. So what was once a gill-cleft has by change of function become part of the organ of hearing.

The *electric organs* of some fish are good examples of change of function. These are always formed by modification of muscular tissue, and in all muscular contractions electric changes occur. Perhaps this is an instance of a secondary function becoming primary. In *Gymnotus*, the electric eel, the electric organs lie just underneath the skin along the sides of the tail, In *Malapterurus*, the electric cat-fish, they extend over the whole body between the skin and muscles, being especially developed at the sides. In *Torpedo* they are used for stunning the animals on which it preys, and also for purposes of defence.

THE INSUFFICIENCY OF TIME.

Natural Selection is a slow process depending on the gradual accumulation of small variations, for the acquirement of which we have no actual standard of time. Palaeontology as yet tells us nothing as to the origin of life, or even the origin of the large groups of animals. In the earliest fossil-bearing rocks we find the great groups typically represented, and in some cases, such as *Nautilus*, *Chiton*, and *Lingula*, by genera now living. We seem driven to require an amount of time behind the Silurian period vastly greater than that which has elapsed since; how much no one can say.

On the other hand, physicists say we can only have a certain amount of time; for the earth is cooling, and it is a matter of calculation how long it has been cool enough for life to be possible. If evolution has really occurred, there must have been time, and the question for the biologist is whether there is evidence of evolution or not. Embryology suggests that the rate of change may be more rapid than is commonly suspected.

EVIDENCE OF DESIGN AND FORETHOUGHT.

This is a subject a little difficult to touch upon without trenching on matters which I wish to avoid.[453] The evidence of adaptation of means to ends is especially manifest when we find contrivance or beauty. That there is harmony everywhere between

---

453   Exactly a century after AMM wrote these words, the term *Intelligent Design* came to replace *Creationism* in popular discourse, acquiring currency amongst those who wished to attribute a purpose to life and its evolution without necessarily appealing to any particular scriptural or religious authority. Of course, most scientists rejected the notion outright and pointed to its illogicality, just as their predecessors had done since the time of Darwin, Paley and others. Scientists and others have lobbied strongly, if not entirely successfully in all countries, for its exclusion from educational syllabuses. We know nothing at all of AMM's religious persuasions, if any, but his hesitancy here

animals and plants and their environment is undoubted; yet it appears to have escaped the notice of the objectors that this is the *very essence of the theory of Natural Selection.* That there is evidence that any animals or plants are specially designed to satisfy the wants or to delight the senses of man is most absolutely denied; and could such cases be proved, they would be fatal to the whole theory. In Nature those characters alone are preserved which are advantageous to the species.

INSTINCT.

Two objections have been raised with regard to instinct: (i) that it could not have been acquired through Natural Selection; (2) that it does not benefit its possessors, and therefore that its preservation is unintelligible.

Let us consider some examples of instinct. The eggs of butterflies and other insects are laid in places as safe as possible and near to their future food-supply. The butterfly never sees her young, and, feeding on the juices of flowers, can have no idea from her own experience as to the respective merits of different leaves; yet she makes no mistake. Again, certain wasps sting the larva of beetles, so as to paralyse without killing them; they then lay a single egg on the paralysed victim and leave it to its fate. The grub emerges and devours its prey, passing the winter in the pupa stage and emerging in the spring with the instincts of its parent. Here the individual wasp derives no advantage, but the gain to the species is enormous.

Preservation of habit, or instinct, is due to the fact that those individuals which take the greatest care to make provision for their young, will be most likely to give rise to offspring which will survive in the struggle for existence. Natural Selection will tend to preserve the instinct because it is *advantageous to the species, although of no benefit to the individual.*

---

shows that he was well aware of the dangers of straying into that territory within a scientific lecture. This paragraph, and indeed those on Instinct which end the Lecture, would be far from out of place today in any presentation of bio-philosophical thought.

LECTURE 20

THE ORIGIN OF VERTEBRATED ANIMALS

I n this Victoria University Extension lecture from the 1893 series, AMM gives what is perhaps the broadest and deepest account of his understanding of the evolutionary process by which modern animals came into being. Starting with an overview of the Craniota [sic], he makes links to fossil vertebrates, including the recently discovered Archaeopteryx, before broadening out to animals with which his audience would have been less familiar but which he sees as crucial to a proper interpretation of probable relationships. The undoubted goal of the lecture is to locate, albeit in general terms, the evolutionary position man, and he feels it necessary to remind his listeners that we are animals too.

He insists on a clear appreciation of the fact that contemporary organisms arose from common ancestors, not by ancestry from currently existing forms. His preoccupation with this point, especially when talking about man in the later parts of the essay, suggests that he felt the need to correct misunderstandings as well as to inform. It is remarkable that to this day it is possible to find lazy writers referring to other primates as if they were our ancestors rather than our cousins, equating their supposed *primitivity* with earlier stages in pedigree. (See also Lecture 19.)

The language example, which we have met in different forms in earlier lectures, serves neatly as both a metaphor for pedigree and an example of mankind's extreme evolutionary development.

AMM uses the *Conclusion* to define the limits of his discourse, boldly excluding the irrational as he did towards the end of the previous lecture. These limits are not restrictive—on the contrary, there is an explicit, Darwin-inspired statement of the power of reason. The great value and fertility of Darwin's theory emerge again at the beginning and end of the final lecture.

THE ORIGIN OF VERTEBRATED ANIMALS

I N order to illustrate the conclusions arrived at in previous lectures, and to test the validity of the Darwinian theory, I propose to apply it, in a more detailed manner than we have yet considered, to one group of animals. The group selected is that of Vertebrates, as being a well-defined group, consisting of forms which are familiar and usually of comparatively large size; and also for the reasons that the fossil forms in this group are numerous and characteristic, and the embryology of the group has been worked out with more detail than in the Invertebrates. Moreover, Vertebrates have a special interest from the fact that it is to this group that man himself belongs.

The problems we have before us are, first, to determine the mutual affinities of different groups now living; secondly, to determine the relations of Vertebrates to other animals; this being the less important of the two problems at the present time.

The evidence available is of three chief kinds

(1) *Comparison of Structure.*—For example, a cat and a dog are clearly more closely allied to one another than is either of them to a fish or a bird.

(2) *Development.*—Here we obtain evidence from the Recapitulation theory, or the tendency of animals to repeat their past history in actual development.

(3) *Fossils.*—These afford the most valuable of all evidence, because it is the most direct and convincing, although at the same time the most fragmentary and incomplete.

Vertebrates form a good group for our purpose, inasmuch as in them evidence of all three kinds is available. The main characteristics of a typical Vertebrate are *the tubular nervous system*, forming the brain and spinal cord; the *notochord;* forming the main skeleton or backbone, and situated between the nervous system and the alimentary canal; the *myelonic eye*, or eye developed from the brain.

CLASSIFICATION OF VERTEBRATES.

I. CRANIOTA.—In these the skull and brain are present, and limbs nearly always so, or when absent have been clearly lost. The heart, liver, and other organs are well developed.

a. Pisces, or fish, are aquatic Vertebrates possessing gills, and provided with fins instead of limbs. They are rarely able to leave the water.

b. Amphibia, such as frogs, newts, and toads, are fresh-water or terrestrial. In early life they are aquatic, breathing by gills. Later on in life they may lose their gills and take to land life.

c. Reptilia. —These never have gills. They possess fore and hind limbs, furnished with fingers and toes, except in cases where they have been lost, as in the snakes. Among reptiles we find lizards, crocodiles, turtles, and snakes; and many extinct groups are known.

d. *Aves,* or birds, are characterised by the possession of wings and feathers, and have special modifications of the skeleton to aid in flight.

e. *Mammalia.*—These are typically terrestrial animals. Their main characteristics are the possession of hair, the presence of two pairs of limbs, furnished with claws or hoofs, and the fact that they do not lay eggs, but give birth to young which are suckled by milk glands.

II. ACRANIA.—This group contains a number of lowly organised forms, possessing no skull, no distinct brain, and no limbs. Their sense-organs and other parts, such as the heart and liver, are in a very primitive and simple condition. They are interesting as telling us in which direction to look for ancestors, for they show us that these may be destitute of limbs, eyes, ears, or backbone, or even without any skeleton. I propose to leave this group for a time, and first consider in more detail the several groups of *Craniota.*

PISCES.—The characteristic mode of breathing in fish is by gills. Gill-clefts have no meaning apart from their respiratory function, yet they are present in the early stages of development of all Vertebrates without exception. They are one of the most characteristic features of Vertebrates, and have a constant relation to the heart, blood-vessels, nerves, muscles, and skeleton. In fish and some Amphibia they are preserved in a functional state throughout life.

The inevitable conclusion is that, of the five groups of Craniate Vertebrates, fish are the most primitive; and that the other four groups are descended, if not from fish, at any rate from gill-breathing forms, aquatic and presumably fish-like. The evidence afforded by Palaeontology does not help us much on this point, the oldest known Vertebrates being fishes from the lower Silurian deposits, Geology would favour an aquatic origin for Vertebrates as for other forms, the instability of the land in comparison with the sea being well known.

AMPHIBIA.—The chief point of interest in Amphibia is the well-known transitional series from the gill-breathing to the lung-breathing condition, which we have already dealt with at length when discussing the theory of change of function. Almost all Amphibia commence life as gill-breathers; but in one or two cases, such as *Hylodes,* where the eggs are not laid in water, owing to abundance of food-yolk the early stages are passed through before hatching, and, as in the higher Vertebrates, the gill-breathing stage is dropped out, though gill-clefts are developed (see Fig. 16).

The pedigree of Amphibia may be regarded as established, and is as follows: Amphibia are descended from fish, which migrated into rivers to avoid enemies and to obtain food. Owing to drought the air bladder became converted into a lung, and this at once conferred the power of going on land (see Fig. 29). This led to the conversion of a fin into a definite pentadactyle limb, a change the true nature of which is uncertain. The point of interest is, that every Amphibian repeats this history in its own development. Intermediate stages are not merely possible, but are known to actually occur at the present day, a most important point. The *Axolotl* is one of these, and is provided with large gills and also lungs; the *Siren* is similar, and is found in the swamps of the Southern States of North America; the mud-fish *Lepidosiren,* of Brazil, *Protopterus* of Africa, and *Ceratodus* of Queensland are other examples.[454]

The origin of terrestrial Vertebrates is probably revealed to us in this way, and it must be considered that the fresh-water forms are descended from marine ancestors and the terrestrial forms from the fresh water.

REPTILIA.—Reptiles are an exceedingly abundant group, both absolutely and relatively, in the secondary period; a great number of families and orders are extinct. Although their actual origin is uncertain, their descent from gill-breathing ancestors is proved by their passing through the gill-cleft stage in their development. No reptiles, however, have gills.

It is uncertain whether reptiles are derived from Amphibia, or descended from gill-breathing ancestors directly. The gigantic *Ichthyosaurus* (Fig. 30) and *Plesiosaurus* (Fig. 31), which have more than five digits in the hand and foot, have been supposed to indicate an independent origin from fishes, but are more probably a reversion to the paddle-like form in an aquatic group. Moreover, the oldest reptiles yet obtained as fossils are pentadactyl, and not provided with paddles. Of existing reptiles the

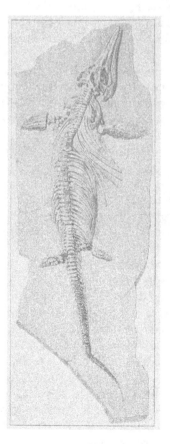

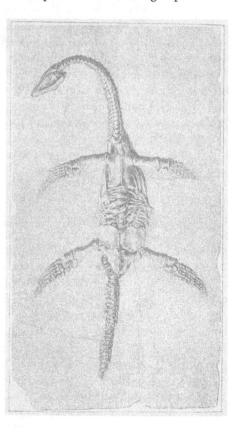

FIG. 30. *Ichthyosaurus.*   FIG. 31. *Plesiosaurus.*

454   Sirenidae are a family of eel-shaped salamanders lacking hind limbs.

lizards are the most primitive. Snakes are comparatively recent tertiary descendants of lizards, characterised by the absence or vestigial condition of the limbs. Crocodiles and tortoises both date back to a very remote period.

AVES. – Birds are a very special group, characterised by the possession of feathers, and the conversion of the fore-limb into a wing. Their bones are very light, with a marked tendency to fusion; as seen in the metacarpal and metatarsal bones. The greater number of their distinguishing points are obviously correlated with the power of flight. Birds are warm-blooded and really terrestrial, and form a compact, well-developed group, at first sight very independent, and having no obvious kinship with any one group of Vertebrates rather than another. The affinities of birds we have already discussed, and Palaeontology speaks very decidedly on the point. There are no forms directly intermediate between birds and reptiles, but such forms, we have seen, should not exist.

We find reptiles which gradually lose the distinctive reptilian characters, and birds which, as we proceed backwards in time, gradually show less and less the special features separating birds from reptiles.

This is shown in the case of the pelvic girdle; the pubes in the crocodile pointing downwards and forwards, in the bird downwards and backwards; while in the *Dinosaurus* both processes are present, and one is preserved in birds, the other in reptiles. The *Dinosaurus* therefore forms a true link between birds and reptiles; and this is supported by the evidence given by development, indications of both processes being present in the development of the pelvic girdle in birds. Again, fossil birds, such as the *Archaeropteryx,* are known which possess teeth and biconcave vertebrae, a long and many-jointed tail, and free metacarpal bones; thus showing the convergence between birds and reptiles as we proceed backwards in time.

The actual origin of wings is still doubtful. The fore-limbs seem first to have become small, as is the case among the Dinosaurs, and then in some unknown way to have changed their function and become wings. Ostriches are said sometimes to give a possible clue, but more probably they are forms in which the power of flight has been lost, for wings are not essential to the bird type. The anatomy of birds supports the contention, for in most reptiles, although both left and right aortic arches persist, yet the right is much the larger. The presence of one condyle to the skull, the quadrate bone, and the absence of vertebral epiphyses are other points in favour of this view.

Evidence is also afforded by embryology. Birds pass through the gill-cleft stage, but have no gills; therefore it is probably either to mammals or reptiles that we must look for allies. The large eggs of birds, the presence of egg-shells, and the details of their early development point to reptiles, as also does the development of the optic lobes of the brain, which though laterally placed in the adult bird, are dorsal in the embryo, and exactly like those of the lizard. The development of the metacarpal bones and of the coccyx is also similar to that of reptiles; and the development of feathers is comparable with that of the scales of a reptile and very unlike the hairs of a mammal.

MAMMALIA.—Mammals are characterised by being warm-blooded animals with the following distinctive points:—A heart with four cavities; circular blood corpuscles; a diaphragm; a left aortic arch; a more perfect brain, especially as regards commissures, than the other groups; no quadrate bone; three special ear bones; two occipital condyles; mammary glands and hairs. The anterior limbs are always present, though the posterior may be absent. There are almost invariably seven cervical vertebrae. Great difference in size is met with among mammals, from the harvest mouse to the whale. They form by far the most important group of all animals from an economic standpoint, some of them serving as food, some furnishing clothing, and others being used for transport.

The geological history of mammals commences with small forms found in the Trias, but comparatively scanty remains are found till the tertiary period. From tertiary times onwards, they are found in three great areas, very distinct from one another: (1) Australia; (2) South America; (3) Europe, Asia, Africa, and North America.

The present distribution of Mammals is divided into three great groups:

(1) *Monotremata*: a very small group of lowly-organised animals, confined to the Australian region.

(2) *Marsupials*: formerly wide-spread, but now confined to the Australian region and America; the kangaroo and opossum are examples of this group.

(3) *Placental mammals*, such as the horse, deer, lion, whale, bat, and monkey.

MONOTREMATA.—These are in many respects the most primitive, and it is therefore well to concentrate our attention on them.

The *Ornithorhynchus* (Fig. 32) is one of the most remarkable of animals, and was first described by Shaw in 1799. The animal is from 18 to 20 inches in length, and is covered with short soft hairs. It has a horny beak, something like that of a duck, and

Ornithorhynchus.

FIG. 32.

possesses teeth when young, although these are lost in the adult state. It has small eyes, and no external ears. The limbs are short and strong, and in the fore foot the web extends considerably beyond the fingers. In the hind foot the nails are long and curved, and in the male the heel has a spur an inch long, at the apex of which opens a "poison gland".

The Ornithorhynchus is an aquatic animal living in lakes and streams; it swims and dives well, and forms burrows, sometimes fifty feet long, on the banks of the water in which it lives. These burrows end in a dilated chamber which has two openings, one below, the other above the water level; in this chamber the animal rolls itself up into a ball when going to sleep. It is found in Tasmania and the southern and eastern parts of Australia.

Echidna, the spiny ant-eater (Fig. 33), is another member of the group, and of this there are two species found in Australia, Tasmania, and New Guinea. Its length is from a foot to 18 inches or more. It is covered with fur intermixed with strong sharp-pointed spines like a hedgehog. It has an elongated tapering snout, a long tongue used for licking up ants, and no teeth. The feet are provided with long and strong claws. Echidna burrows very rapidly, and is mainly nocturnal in habit. It can endure long fasts, and even exist for a month or more without food.

Embryological Evidence.—Other Mammals bring forth their young alive; but they are really developed in the same manner as other Vertebrates, such as the frog or bird, from a single cell or ovum; and the embryo is retained within the body instead of being laid.

The eggs of mammals are very small, that of the rabbit being 0.116 mm. (1/250 inch) in diameter. The curious point is, that the eggs of mammals develop after the fashion of large eggs, and not in the manner of small eggs. The difference in size of eggs depends mainly on the amount of food yolk contained in them; a small egg, such as that of the frog, develops entirely and directly into the embryo; a large egg,

Echidna aculeata and *Proechidna Brujnii.*

FIG. 33.

that of the chick for instance, is hindered by the presence of food yolk, and becomes constricted into two parts, the embryo and the yolk sac. The egg of the rabbit develops as though it had a large amount of food yolk, and forms a yolk-sac. This fact is only intelligible on the view that mammals are descended from ancestors having large-yolked eggs. (Fig. 34.)

Monotremata are of special interest in this respect. It was long believed that their eggs were laid as by birds, and in 1829, Professor Grant described, on the authority of a Mr. Holmes,[455] the eggs of Ornithorhynchus as being ovoid in shape, equal at the two ends, 1½ inches long and 3/8 inch in diameter, with a thin, whitish, transparent, calcareous shell. Of these eggs, originally nine in number, four came to England and two were deposited in the Manchester Museum, and labelled "eggs of duck-billed Platypus."[456]

In 1884 Monotremes were shown to be oviparous by Mr. Caldwell, who was sent to Australia[457] to study the subject. The Ornithorhynchus lays two eggs at a time.[458]

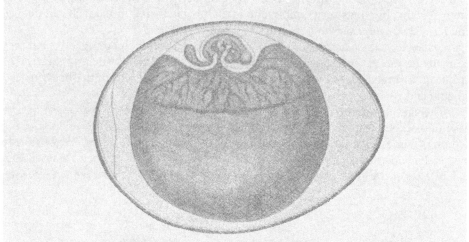

Embryo Chick at the end of the fifth day of incubation, showing relation of embryo to yolk-sac. Typical example of a large-yolked egg.

FIG. 34.

---

455   Robert Edmond Grant in a letter of 24 September, 1829, to Etienne Geoffroy St Hilaire (held at the National Library of Australia). The letter contains the detailed description of the eggs, inserted here by AMM, except that it describes them as being *1⅜ inch in length, and nearly 6/8 inch in breadth (English measure)*. According to the letter, Mr Holmes was a hunter and collector who found them in a platypus nest near the Hawksburgh River in New Holland (now Australia).

456   The University Museum catalogue currently lists a single duck-billed platypus egg but without indicating provenance or acquisition date.

457   William Hay Caldwell (1859–1941), Scottish naturalist and first recipient of a scholarship in memory of F. M. Balfour, under whom he had studied.

458   According to the Grant letter, Holmes found four eggs in the nest.

inch by inch in size, enclosed in a strong flexible white shell; *Echidna* lays only one egg. The details of these eggs are not yet forthcoming, but they strongly suggest affinities to reptiles.

Osteological Evidence.—The vertebrae have no epiphyses; the sternal ribs are well ossified; the mandible is devoid of an ascending ramus; the humerus is very reptilian in character. There are three distinct cartilage bones in the shoulder girdle—viz., coracoid, scapula, and pre-coracoid—agreeing with the Anomodont reptiles.[459] Some of these extinct reptiles from the Permian (the top of the Palaeozoic) and the Triassic deposits—*Anomodontia* or *Theromorpha*[460]—in the character of the shoulder and hip girdles; in the perforation of the columella; in the relation and number of the bones of the hind foot, especially the astragalus and calcaneum; in the nature of the ribs and in other points, are in many ways intermediate between mammals and reptiles. They may possibly be the common ancestors of amphibia, reptiles, and mammals. This is a point of importance as showing a convergence towards a common ancestor.[461]

We now enter on a second stage in our inquiry. We have found evidence in embryology of the descent of terrestrial Vertebrates from aquatic ancestors, in the presence of gill-slits, and in the actual transition shown us by Amphibians, such as the frog and newt. The geological evidence of the instability of land compared with the sea leads us to think of this as of general application, and not confined to Vertebrates, and that the origin of all the great groups of animals is to be looked for in the sea.

So far, we are led to regard fish as the most primitive of the five great groups of Vertebrates, and to view the remaining four as descended from fish or fish-like ancestors. Now, can we get any further? What is there behind fish, and can we trace the pedigree of Vertebrates further back?

This brings us to the consideration of the lower group of Vertebrates—ACRANIA.

Amphioxus (see Fig. 20).—This is a small fish-like animal from one and a half to two inches in length, living in the sand, and found at Messina, Naples, on our own shore, in Australia, and elsewhere. It possesses a notochord, a large number of gill-slits, and a central nervous system along its dorsal surface. Amphioxus is an animal of very primitive character, differing from other Vertebrates in the following points: The skeleton is of extreme simplicity, consisting of an elastic rod, the notochord. It has no limbs, no skull, no ribs, no brain, and no ears. The eye is a single pigmented spot at the anterior end of the nervous system. The alimentary canal is straight, and there are a hundred or more gill-slits (eight being the greatest number in fish). The heart is a straight tube, and the liver a simple diverticulum from the alimentary canal.

Yet Amphioxus is obviously a Vertebrate, for although it differs in almost all the above points from adult Vertebrates, it resembles the embryos of Vertebrates, and may be said to halt at an early stage in development. All Vertebrates pass through stages in which there are no limbs; in which the notochord is the only skeletal structure, and the brain merely the anterior end of the neural tube; in which paired eyes and ears

---

459   A diverse group of extinct herbivores.
460   Now Theriodonta: extinct therapsids with mammal-like teeth.
461   That is, backwards in time, or stemming from a common ancestor.

are not yet formed; in which the heart is a simple tube, the alimentary canal straight, and the liver a simple outgrowth from it.

The position of Amphioxus with regard to Vertebrates is thus closely analogous to that of the tadpole with regard to the frog a stage through which the frog passes, but at which it does not halt. The inference in the case of the frog is that frogs are descended from tadpole-like—i.e., fish-like—ancestors. Now, if this conclusion is rightly founded, we may conclude that Vertebrates are descended from ancestors like Amphioxus, and that of all living animals Amphioxus most nearly represents the common ancestor of Vertebrates. The only alternative, and one that is urged by Dohrn, Lankester, and others, is that Amphioxus is degenerate. I think this alternative wrong, and that there is no real retrograde development, but that Amphioxus merely stops at what is an early stage in the development of the higher forms. To my mind, Amphioxus is one of the most important of Vertebrates, to be regarded as shifting back the origin of Vertebrates to an extraordinary extent, which is best realised by saying that the differences between Amphioxus and a fish are zoologically of far greater importance than those between a fish and a mammal.

Ascidians, or Sea-squirts. Some of these animals are solitary and fixed, some in colonies, and others free-swimming. They have a covering or "test" of cellulose with inhalant and exhalent apertures. The pharynx has numerous gill slits. Water entering the inhalant aperture passes into the pharynx, where the food-matter it contains is filtered from it, and passing through the slits in the pharynx escapes into the atrial cavity, and so out at the exhalent aperture. In Ascidians the nervous system is represented merely by a single ganglion, and there are no sense organs. The notochord is absent. In fact, there is nothing in the adult Ascidian to indicate Vertebrate affinities. (Fig. 35.)

In their development, however, we find that they pass through the stage of a free-swimming larva, like a tadpole, which possesses a swimming tail, a nerve cord, a notochord, and an eye and ear. (See Fig. 8.) This larva after a time becomes fixed; the tail shrivels up and is absorbed; the nervous system becomes reduced to a single ganglion; the ear and eye become aborted, and the pharynx enlarges. In fact, Ascidians are degenerate animals, which in their larval stage stand at about the same level as Amphioxus—if anything somewhat higher. Some Ascidians, such as *Appendicularia*, never get beyond the larval stage, and are found as minute free-swimming pelagic animals, with only one pair of gill clefts.

Concerning the zoological position of Ascidians, it is accepted by every one that they are Vertebrates, which is proved by their development. Mistakes have, however, been made in assuming that they must be ancestral, and this mistake was made by Darwin himself. If ancestral, they should stop at what is a transitional stage in the development of higher animals, but they do not do this. The resemblance is not between the adult Ascidian and the embryo Vertebrate, but between the embryo Ascidian and the embryo Vertebrate. This indicates descent from some common ancestral stock, but not the descent of one group directly from the other.

Amphioxus and *Appendicularia* are the simplest existing forms with the characteristic vertebrate structure, and represent the nearest approach yet made towards the determination of the ancestry of Vertebrates. This suggests that the Vertebrate stem arose very early, and must not be derived from groups such as annelids, lobsters,

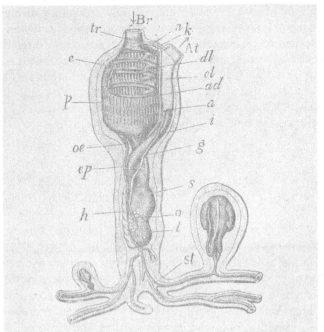

Clavelina, a compound Ascidian, showing the anatomy of an adult member of the colony, and two members in early stages of development (produced asexually). The members of the colony are connected together by the stolon, which consists of an outer tube of ectoderm continuous with the outer layer of the individuals ; and an inner tube of endoderm, which is a direct prolongation of the so-called "epicardium."

a, Anus ; ad, Dorsal nerve cord ; at, Atrial aperture ; br, Branchial aperture ; cl, Cloaca ; dl, Dorsal languets ; e, Endostyle ; ep, "Epicardium"; g, Genital ducts ; h, Heart ; i, Intestine ; k, Sub-neural gland ; n, Nerve ganglion ; o, Ovary ; oe, Œsophagus ; p, Pharynx, with numerous gill-slits ; s, Stomach ; st, Stolon ; t, Testis.

(Clavelina is more primitive than most Ascidians in retaining the dorsal nerve cord in the adult condition.)

FIG. 35.

spiders, and sea-urchins,[462] each of which has been claimed as the ancestor of Vertebrates in recent years. Any attempt to get further back than this is mere speculation. The only available evidence is that of embryology, and that this is trustworthy we have abundant proof as regards the later stages of development. The Ascidian tadpole is shown by *Appendicularia* to be a possible adult, and indeed to correspond very closely to an existing adult animal. With regard to the earlier stages of development we have seen in a former lecture[463] that this will also apply, and that adult organisms are known corresponding to the several stages in the early development of Amphioxus which may possibly represent ancestral forms.

---

462 By modern classification, based on the mechanism of gut formation during early embryo development, the first three of these are protostomes while the fourth are deuterostomes. Vertebrates are deuterostomes. See also Lectures 8 and 13.

463 {M} Lecture 17, Figure 21 and final paragraph.

Thus, by the aid of anatomy, paleontology, and embryology we are able to define fairly clearly the broad lines of Vertebrate ancestry; and the conclusion we arrive at is the usual one, viz., to emphasize the uselessness of search for directly intermediate links between existing forms, and to drive back the origin of this, the highest group of animals, to times earlier than those of the oldest fossiliferous rocks; to show that our only hope of obtaining information concerning these first progenitors depends on the extent to which their existing descendants have preserved the record in their own development, and on our skill in deciphering this record.

THE DESCENT OF MAN.

And now we turn to the last stage in our inquiry, the zoological position of man.

Man is distinctly an animal—i.e., neither a plant nor a mineral—requiring organised food, for which he is dependent on other animals or on plants. He is distinctly a vertebrate, as proved by his backbone, the relations of his nervous system, brain, heart, and sense-organs. Further, he clearly belongs to mammals—the presence of hairs instead of scales or feathers would alone be sufficient to show this; but he also possesses all the other characteristics of the group—viz., two condyles to the skull, seven cervical vertebrae, and a left aortic arch. Of the different groups of mammals, it is allowed on all hands that he is most closely allied to that of monkeys. The general shape of his body; the form of his limbs; the number and nature of his fingers and toes; the power of pronation and supination of the fore-arm; the shape of his head; the structure and size of his brain, and the form of his teeth, all prove this incontestably.

Further inquiry shows this correspondence to be a very close one. It is seen in every detail of structure of the human body, bone for bone, muscle for muscle, nerve for nerve, and even tooth for tooth. Man and monkey can be compared, and the most exact correspondence pointed out. This correspondence is so close that it is almost impracticable to find any constant points of difference of any value whatever. It has indeed been shown by Professor Huxley that the anatomical differences between man and the higher monkeys[464] are markedly less than those between the higher and lower monkeys. Again, as to development, man, like all other animals, commences as a single cell, and passes through a gill-cleft stage indistinguishable from that of other Vertebrates. He has, like other mammals, at one time two aortic arches, a right and a left, but during development he loses the right arch. His brain, eyes, and limbs are all formed in the same manner as in other mammals.

Rudimentary organs are also present in man, and are of the utmost value, because they are only explicable on the supposition that man is descended from some ancestor in which these organs were in a functional condition. For instance, the intrinsic muscles of the ear are present in man in an incompletely developed form, and in a condition in which they can be of no use to their possessors. So also with the *platysma* muscle, which, while extensively developed in some Vertebrates, such as horses, is in a comparatively rudimentary condition in man, and of scarcely any functional value. Again, the "wisdom teeth" form another example of rudimentary

---

464 We would now identify man and monkeys as belonging to the Primate order. Man is a member of the Ape class along with gorilla, orang-utan, and the chimpanzees. Apes are distinguished from the monkey class by the absence of a free tail. Presumably *higher* and *lower* monkeys refer to apes and monkeys, respectively.

organs, since they are always cut long after the others, and sometimes never pierce the gums at all.

The tendency to *reversion* is also met with in man, and is seen in the more or less complete presence of two aortic arches in some cases, and in the occasional development of the muscles of the ear.

Like all higher animals man is primitive in some respects, for he is pentadactyl and plantigrade (*i.e.*, five-fingered and walks on the flat of the foot), and retains the power of supination and pronation of the fore-arm; these conditions being more primitive than those met with in other Vertebrates which are on the whole of lower organisation.

In fact, unless man wishes to continue going about the world stamped with living and palpable proof of his kinship with lower animals, he had better stop up his ears, or, still better, cut them off altogether; for so long as he bears at the side of his head those tell-tale flaps with their aborted and rudimentary muscles, so long as he hears by means of that slit, once a gill-cleft, now by change of function become an accessory organ of hearing, so long does he carry about in sight of all men sure proof of his relationship with lower, even with water-breathing animals.

Yet one can hardly recommend the operation, for if you were to remove one by one the various parts of your body which proclaim this kinship, you would get rid in succession of skin, muscles, nerves, bones, &c., and all that would be left in the end as man's special and peculiar possessions would be: (1) certain parts of his brain, and these only doubtfully; (2) the *extensor primi internodii pollicis* muscle, which straightens the first joint of the thumb; (3) the *peroneus tertius,* a small muscle in front of the lower part of the leg and ankle, inserted into the base of the little toe; (4) certain portions of other muscles.

Again, if we turn from bodily structure to the other characteristics of man, we find the same tendency to over-population, resulting in the same struggle for existence and the same survival of the fittest. Indeed, it was from the study of Malthus' "Essay on Population"[465] that Darwin was led to the theory of Natural Selection. So it is with the history of the rise and fall of nations, with the evolution of human speech, customs, and clothing. All alike conform to the same laws as those regulating the structure and habits of other animals. And so with the influence of man on other animals; the advent of man has simply been the arrival of another animal, better equipped and more cunning, more cruel than any other; acting with supreme selfishness; tolerating the existence of other animals only when they can be made subservient to his own wants or pleasures; ruthlessly exterminating all that offend or thwart him. His very kindness is merely a nominal exception, for if perchance he appear kindly disposed to certain animals, it is only to satisfy his own selfish ends, that he may fleece them of their coats or pluck them of their feathers to adorn himself; or to fatten them, that they may acquire a flavour more acceptable to his palate.

Application of the Darwinian Theory to the Language of Man.—Language has been said to be *"the one great difference between man and brutes,"* and an *"insurmountable obstacle*

---

465   First published anonymously in 1798 and revised several times up to 1828. According to his *Autobiography*, Darwin first "happened to read [it] for amusement" in October 1838.

to the theory of alliance by descent." This has been urged even by those who would accept the theory as applying to all other animals.

But has not language a history, has it not been evolved gradually, and is it not constantly, even daily, undergoing change? Is not this evolution, are not these changes of a nature precisely similar to those which have governed the animal kingdom in other branches, and have made it what it is at the present day? Did the English language suddenly appear, was it specially created, or was it gradually evolved by slow modification of other tongues, such as Latin and Saxon? Why do modern languages—Italian, French, Spanish and Portuguese—have so many words closely similar or identical? Is it not owing to descent from a common Latin ancestor?

Again, Latin and Greek have many likenesses to one another, and with Celtic, Teutonic, Slavonic, and the ancient dialects of India and Persia, may be all regarded as descended from one common Indo-European or Aryan stock. So Hebrew, Arabic, and Syriac form a Semitic group. Moreover, *"suppose,"* says Max Müller,[466] *"we had no remnant of Latin; suppose the very existence of Rome and of Latin were unknown to us, we might still prove on the evidence of the six Romance dialects that there must have been a time when these dialects formed the language of a small settlement; nay, by collecting the words which all these dialects share in common, we might to a certain extent reconstruct the original language, and draw a sketch of the state of civilisation as reflected by these common words."*

Moreover, evidence of recapitulation is shown by the way in which a child learns to speak its own language, and "we know for certain that an English child, if left to itself, would never begin to speak English." We also see examples of rudimentary organs well illustrated by the silent letters in words such as doubt, reign, feign, debt, and answer. Persistent types in language are also met with. *"The language which the Norwegian refugees brought to Iceland has remained almost the same for seven centuries; whereas, on its native soil and surrounded by local dialects it has grown into two distinct languages—Swedish and Danish."*[467]

Constant change is found in words at different periods. We can read Milton, Bacon, Shakespeare, and Hooker, though conscious of unfamiliar words and obsolete expressions; we can make out Wycliffe and Chaucer; but when we come to the English of the thirteenth century we can but guess its meaning. A Bible glossary shows that since the year 1611, three hundred and eighty-eight words, or one-fifteenth of the whole number used, have become obsolete: and, on the other hand, new words are constantly being added.

CONCLUSION.

This is the precise position which I have endeavoured to establish: that there are causes which will account for what we find—for the structure, language, and habits of man; causes which have been in existence ever since life began, and causes which must have tended in this direction.

---

466 Max Müller: *Lectures on the Science of Language* delivered at the Royal Institution, 1861. AMM may also have had Müller in mind for the statements in quotation marks at the start of this section on language, although no direct references have been located.

467 Also Max Müller.

Whether there is anything further than this, whether man has other attributes, either peculiar to himself or held by him in common with other animals; whether there are attributes that cannot be explained by these laws, is a question with which we have no concern here. Science has nothing to do with such matters, and has nothing to say either for or against them.

Such is the doctrine of Evolution as applied to man, and I would ask you, Is there anything humiliating in this? Surely it cannot be more degrading to have risen than to have fallen. Surely the true interest of life lies in the future rather than the past; in the possibility of further achievements; in there being work ahead for us to do. It is in the consciousness that we now possess the key that will compel the past to yield up its secrets, and that opens to us unbounded possibilities in the future; it is in the conviction that there is a reason in and for everything, and that it is within our powers to determine that reason, that we find the great charm and attractiveness of the Darwinian Theory.

LECTURE 21

THE LIFE AND WORK OF DARWIN

O f all the lectures, this last one requires the fewest introductory comments.
The origin of the quotations from Darwin in this lecture can be found in
Darwin Online http://darwin-online.org.uk/. Many of them also appear in two
works by Francis Darwin (1848–1925; Charles' third son and seventh child): *The Life
and Letters of Charles Darwin* and *Charles Darwin: His Life told in an Autobiographical
Chapter, and in a Selected Series of his Published Letters* (both appeared in several
editions). CFM, in his Preface to the second volume of lectures, thanks Francis Darwin
for permission to use them.

LECTURE 21

THE LIFE AND WORK OF DARWIN

HITHERTO we have been concerned with the great theory with which Darwin's name is inseparably connected; we have dealt successively with its birth and maturation; we have tried to form some idea of its wide-reaching influence, and of the effect which it has had, not merely on biological thought, but on other fields of science, literature, and art, and branches of knowledge apparently widely remote. We have seen how this theory has knit together human knowledge, giving the word *history* a new, a wider, a more wonderful significance than was possible before. We have dealt, I admit too briefly, with the main objections to the theory, and have taken a single instance in detail as a test and as an example of methods.

There is no more fitting way of concluding this series of lectures than by giving an outline of the life and work of the man to whom this great advance, this opening up of new fields, this widening of human interests and human posers, is due. Concerning his life, the progress of the events through which such results were obtained, the methods by which success was completed, the successive steps in the development and ripening of the great theory; all these must have much of interest, much that will repay the hearing. Concerning his works, although the "Origin of Species" remains by far his greatest achievement, yet it must not be supposed that Darwin gained one great victory and then rested. No more conscientiously industrious man ever lived; and besides his masterpiece he has left us a great series of books, each dealing with a separate group of problems in animal or in plant life; each based on long-continued and scrupulously exact observations; each breaking entirely new ground; and each contributing powerfully to the advancement and widening of knowledge. While never forgetting that the "Origin of Species" stands foremost, it is well that the other works should not be overlooked; for had the "Origin of Species" never been written, these works—as yet hardly mentioned in our course —would have given Darwin a foremost place among the biologists of all nations and of all ages.

FAMILY HISTORY.

Charles Darwin was born on February 12th, 1809, at Shrewsbury. His mother was a daughter of Josiah Wedgwood, the founder of the great pottery works at Etruria. His father, Robert Waring Darwin, was a physician in large practice at Shrewsbury; a man of marked individuality of character, a quick and acute observer, with a great power of reading character and of winning the confidence of his patients. He was highly esteemed for his skill in diagnosis, but was not a man of special scientific ability. By his large practice he accumulated a considerable fortune, and was able to leave his children in easy circumstances. Darwin's grandfather, Erasmus Darwin, 1731–1802, was a physician of great repute, and the author of "Zoonomia," an ambitious treatise, showing extensive rather than profound acquaintance with natural phenomena; containing many bold, ingenious, and at times fantastic speculations. He was also the author of numerous and voluminous poetical works. He propounded a hypothesis as to the manner in which species of animals and plants have acquired their character, which is identical in principle with that subsequently rendered famous by Lamarck.

Charles Darwin in his childhood and youth gave no indication that he would do anything out of the common. He was a strong, well-grown, active lad, interested keenly in field sports. *"In fact,"* says Huxley, *"the prognostications of the educational authorities into whose hands he first fell were distinctly unfavourable, and they counted the only boy of original genius who is known to have come under their hands as no better than a dunce."*[468] From 1818 to 1825 Darwin was at Shrewsbury School[469] under Dr. Butler, leaving at the age of sixteen. *"Nothing could have been worse,"* he says, *"for the development of my mind than Dr. Butler's school, as it was strictly classical, nothing else being taught except a little ancient geography and history. The school as a means of education to me was simply a blank."* Yet, not incapable of appreciation, he writes: *"The sole pleasure I ever received from such studies was from some of the Odes of Horace, which I admired greatly."* He also says: *"I used to sit for hours reading the historical plays of Shakespeare."*

He was interested in chemistry, and fond of making experiments with his brother in the tool-house at home. He writes: *"The subject interested me greatly, and we often used to go on working till rather late at night."* This became known at the school, and earned for him from his schoolfellows the nickname of "Gas"; and from the head-master a public rebuke for "wasting his time on such useless subjects."

Doing no good at school, he was sent to Edinburgh in 1825, with the intention of studying medicine. This, however, was not much of an improvement, for, as he writes: *"The instruction at Edinburgh was altogether by lectures, and these were intolerably dull."* The Professor of Anatomy made his lectures *"as dull as he was himself"*; and the lectures on Materia Medica were *"something fearful to remember,"* even forty years later. But the climax seems to have been attained by the Professor of Geology and Zoology, whose praelections were so *"incredibly dull"* that they produced in their hearer the determination—fortunately for the world not adhered to—*"never to read a book on geology, or in any way to study the science."* He, however, became acquainted with some good practical naturalists, and got lessons in bird-stuffing, and also became a good shot.

After two sessions at Edinburgh his father decided that he had little or no taste for the life of a physician, and fearing that he might sink into an idle sporting man, proposed that he should go to an English University with the view of becoming a clergyman. So far as the direct results of academic training were concerned, the change of Universities was hardly a success, for he writes *"During the three years which I spent at Cambridge my time was wasted, so far as the academical studies were concerned, as completely as at Edinburgh and at school."*

And yet it would appear that the fault lay rather with the method than the man; for he speaks of Algebra and Euclid as giving him much pleasure. He also

---

468   TH Huxley in his *Obituary* of Darwin (*Obituary Notices of the Proceedings of the Royal Society* 1888; 44) and also in his *Darwiniana*.

469   *"I boarded at this school, so that I had the great advantage of living the life of a true schoolboy; but as the distance was hardly more than a mile to my home, I very often ran there in the longer intervals between the callings over and before locking up at night."*

studied Paley's "Evidences"[470] very thoroughly, and expresses himself as being much delighted with the logic of the book, and charmed by the long line of argumentation. He was fond of out-door sports, especially riding and shooting.

He was devoted to collecting insects, or, as he expresses it, *"mad on beetles."* This was a point of much importance, as it brought him in contact with Henslow, the Professor of Botany,[471] a man of singularly extensive acquirements, who took a keen pleasure in gathering young men around him, and in acting as their counsellor and friend: *"A man of winning and courteous manners; free from every tinge of vanity or other petty feeling."*

At Henslow's advice Darwin was led to break his vow never to touch geology, and through him he obtained permission to accompany Professor Sedgwick on a geological excursion in Wales,[472] by which he gained the practical knowledge which he was so soon to put to the test. It was Henslow who advised him to read Lyell's "Principles of Geology," which had just been published in 1830, advising him, however, on no account to adopt Lyell's general views; a piece of advice which he promptly neglected, for, as we have seen, it was by the unflinching application of Lyell's ideas and methods to Biology that Darwin was led to his greatest results. Finally, it was Henslow who obtained for him the permission to accompany the *Beagle* on her memorable voyage. This, he says, was *"the turning-point of my life."*

The *Voyage of the Beagle* occupied from 1831 to 1836, Darwin being then twenty-two years of age. In the autumn of 1831 it was decided by the Government to send a ten-gun brig of 242 tons burden, under Captain Fitzroy, to complete the unfinished survey of Patagonia and Tierra del Fuego, to map out the shores of Chili and Peru, to visit several of the Pacific archipelagoes, and to carry a chain of chronometrical measurements round the world. This was essentially a scientific expedition, the captain, afterwards Admiral Fitzroy, being himself an accomplished and highly-trained officer, and famous as a meteorologist. Anxious to be accompanied by a competent naturalist, to collect animals and plants, he generously offered to give up part of his own cabin accommodation. On Henslow's recommendation this was offered to Darwin, who was eager to accept it. His father, however, objected strongly, adding: "If you can find any man of common sense who advises you to go, I will give my consent." His uncle, Josiah Wedgwood, strongly urged him to accept, whereupon his father gave his consent. He, however, narrowly escaped rejection, Fitzroy doubting whether a man with such a shaped nose could possess sufficient energy and determination for the voyage!

This voyage, originally intended to last two years, and ultimately extended to five years, started from Devonport on the 27th of December 1831, returning to Falmouth on October 2nd, 1836. Darwin writes *"This was by far the most important event in my life, and has determined my whole career."* It was during this time that he acquired habits of energetic industry and concentrated attention. He collected largely, and from his collections laid the foundation of his great work.

---

470 William Paley (1743–1805) *Evidences of Christianity.* Darwin also studied the same author's *Moral Philosophy* and *Natural Theology* at this time.

471 John Stevens Henslow (1796–1861).

472 Llangollen, Conway, Bangor, and Capel Curig.

It is necessary to bear in mind that this was essentially a surveying voyage, and the bulk of the time was occupied in a detailed survey of the east, south, and to a less extent the west coast of South America; involving slow work and repetition of much of it, the ground having to be covered more than once in difficult places.

The *Beagle* left Devonport on December 27th, 1831, calling at the Cape Verde Islands and St. Paul's Rocks, and reached Bahia on February 29th, 1832. After a short stay she proceeded southwards to Rio and Monte Video. The next three years were spent in the special work of surveying: nearly two years on the east coast off Tierra del Fuego and the Falkland Islands, and rather more than a year on the western coast. Darwin went partly with the ship, but spent long periods on shore, travelling over much country, collecting, observing, and thinking.

Darwin's most important overland journey was in 1833 from Rio Negro to Bahia Blanca, and thence five hundred miles further on to Buenos Ayres. During this journey over Pampas he discovered the remains of vast numbers of extinct animals, some of enormous dimensions, such as the *Megatherium* and the *Mylodon*, which were found in gravel fifteen to twenty feet above the sea level near Bahia Blanca. It was here that Darwin was much struck with the relations between living and extinct forms.

"*This wonderful relationship,*" he writes, "*in the same continent between the dead and the living, will, I do not doubt, hereafter throw more light on the appearance of organic beings on our earth, and their disappearance from it, than any other class of facts.*" He noticed the replacement of huge extinct forms by unlike yet allied forms, and that species are replaced, but by allied species. Darwin was impressed rather with their resemblances than their differences, whereas Cuvier, we saw, was most struck with the differences. He was also much impressed with the evidence of changes in the land in recent times, a point which laid the foundation of his theory of coral reefs. At Tierra del Fuego he was struck with the characters of savage races, and noted the almost entire absence of everything which we regard as characteristically human.

The Galapagos Islands, five or six hundred miles west of America, on the equator, are all volcanic, and therefore presumably recent in a geological sense. Nearly all the animals found here are peculiar to the islands, and different islands have their own fauna; yet these are more nearly akin to those of South America than to any other forms. In 1835, Darwin visited these islands, and set himself to discover the reason of this. Writing in 1837, after his return, he notes: "*In July opened first note-book on Transmutation of Species. Had been greatly struck from about the month of previous March on character of South American fossils, and species on Galapagos archipelago. These facts (especially latter) origin of all my views.*"

From 1836 to 1842, the first six years after his return from the *Beagle* voyage, much time was spent at first in unpacking specimens and distributing the collections among specialists. The geological specimens he reserved for his own share. The publication of detailed results occupied much time, and in 1839 the "Naturalists' Voyage" was published, the second edition appearing in 1845 as the "Narration of the Voyage of the *Beagle*." The third edition in its present form appeared in 1860.

The geology of the voyage was written by Darwin himself, and consisted of three parts, the best known of which dealt with the *structure and distribution of coral reefs.* "*No other work of mine,*" he says, "*was begun in so deductive a spirit as this, for the whole theory was thought out on the west coast of South America, before I had seen a true coral reef.*"

The years between 1836 and 1842 were marked by the gradual appearance of that weakness of health which ultimately forced him to leave London, and take up his abode for the rest of his life in a quiet country-house. This greatly discouraged him, and threatened thus early to become permanent. In 1839 he married his cousin Emma Wedgwood, and lived for three years in Upper Gower Street, then suffering much from illness. In 1842 he purchased a house at Down, near Beckenham, where the rest of his life was spent, and the greater part of his work accomplished.

The "Origin of Species" was his first and greatest work, the *"chief work of my life."* We have already sketched the history of the development of the theory, and may now consider it from the more personal point of view. During the voyage of the *Beagle,* Darwin was led to think much on the subject; and on his return, as soon as he had leisure, returned to it. During the voyage he *"believed in the permanence of species,"* though experiencing occasionally *"vague doubts."* On his return in 1836, while preparing his journal, he noted how many facts indicated the common descent of species.[473]

On the 1st of July 1837 he opened his first notebook to record any facts bearing on the transmutation of species, and *"did not become convinced that species were mutable until I think two or three years had elapsed."*

Darwin was most impressed by the South American fossils, and by geographical distribution, especially in the Galapagos Islands. The problems which occurred to him were: *"Why are the animals of the latest geological epoch in South America similar in facies to those which exist in the same region at the present day, and yet specifically and generically distinct?"* And, *"Why are the animals and plants of the Galapagos Archipelago so like those of South America, and yet different from them? Why are those of the several islets more or less different from one another?"*

These problems were only explicable on the assumption of modification, and in order to explain the cause of these modifications he turned to the only certainly known examples of descent with modification—viz., those presented by domestic animals and cultivated plants. The details of these he worked up and experimented upon in a much more thorough manner than his predecessors, especially in regard to pigeons. He soon perceived *"that selection was the keystone of the main success in making useful races of animals and plants"*; but says: *"how selection could be applied to organisms living in a state of nature remained for some time a mystery to me."*

"In October 1838," he writes, *"that is, fifteen months after I had begun my systematic inquiry, I happened to read for amusement, 'Malthus on Population,' and being well prepared to appreciate the struggle for existence which everywhere goes on from long-continued observations of the habits of animals and plants, it at once struck me that under these circumstances favourable variations would tend to be preserved, and unfavourable ones to be destroyed. The result of this would be the formation of new species. Here, then, I had at last got a theory by which to work; but I was so anxious to avoid prejudice*

---

473   *Permanence* here has its antithesis in *transmutation*, the changing of one species into another. Darwin's statement reflects the early–mid 19th-century debate about which was the true picture in relation to organism diversity. August Weismann and, more relevantly in the present context, Alfred Russel Wallace, came, as did Darwin, to believe strongly in transmutation. Darwin and Wallace's great stride forward was to provide a plausible explanation of how it might happen.

that I determined not for some time to write even the briefest sketch of it. In June 1842, I first allowed myself the satisfaction of writing a very brief abstract of my theory in pencil, in thirty-five pages, and this was enlarged during the summer of 1844 into one of 230 pages."

Not till 1858 was the theory published, and then only on pressure of the strongest character being brought to bear on him. On June 18th, 1858, Darwin, having convinced himself and accumulated the evidence and proofs he wanted, was at last at work on the book, when he received, most unexpectedly, Wallace's MS. from Ternate, in which the Theory of Natural Selection was set forth clearly and decisively, almost in Darwin's own words. Darwin wished to publish Wallace's paper without reference to his own work; but at the urgent solicitation of Lyell and Hooker he consented to allow extracts from his own MS. of 1844, together with a letter to Asa Gray of 1857, to be read before the Linnean Society on July 1st, 1858.

Darwin was at this time forty-nine years of age, and Wallace thirty-five. On Darwin's part this publication was the result of twenty-one years of deliberate work; and of views formed nineteen years beforehand, and actually written out in MS. of 230 pages fourteen years previously. The dual authorship of the "Theory," and its simultaneous announcement from opposite sides of the world, were causes for sincere congratulation. It was fortunate for Darwin, in causing him to publish his views more speedily, and in a more condensed and attractive form than he originally purposed, and as leaving him at liberty for further work. It was also fortunate for Wallace in securing cordial and sympathetic recognition in the most gratifying manner of his independent discovery. Finally, it was fortunate for the world, and a lesson for all time to come, of how an emergency, involving the tenderest susceptibilities of scientific reputation, can be treated so as to redound to the infinite and lasting credit of all concerned.

With regard to Wallace, it is interesting to know the immediate causes which suggested the theory, especially in view of Darwin's history, and I cannot do better than quote Wallace's own words *"In February 1858 I was suffering from a rather severe attack of intermittent fever at Ternate, in the Moluccas; and one day, while lying on my bed during the cold fit, wrapped in blankets, though the thermometer was at 88° Fahr., the problem again presented itself to me, and something led me to think of the 'positive checks' described by Malthus in his 'Essay on Population,' a work I had read several years before, and which had made a deep and permanent impression on my mind. These checks—war, disease, famine, and the like—must, it occurred to me, act on animals as well as man. Then I thought of the enormously rapid multiplication of animals, causing these checks to be much more effective in them than in the case of man; and while pondering vaguely on this fact there suddenly flashed upon me the idea of the survival of the fittest—that the individuals removed by these checks must be on the whole inferior to those that survived. In the two hours that elapsed before my ague fit was over, I had thought out almost the whole of the theory; and the same evening I sketched the draft of my paper, and in the two succeeding evenings wrote it out in full, and sent it by the next post to Mr. Darwin."*

Thus we see that Malthus was the turning-point with both authors; and all historically inclined will place this famous essay on their shelves, alongside the works of Darwin and Wallace. To me it has always been a striking fact, that with both authors, and perfectly independently, the turning-point in the argument should have been the application to the whole animal world of principles already established and accepted in regard to man. This is curiously significant in view of the objection, so often and so ignorantly brought against Darwinism, that it is degrading, because it applies to man laws governing the structure and habits of animals.

From 1846 to 1854 Darwin devoted himself to a monograph on Barnacles, consisting of four large volumes, two on recent and two on fossil species: a heavy task which he was led to undertake, largely, through *"a sense of presumption in accumulating facts and speculating on the subject of variation without having worked out my due share of species."* *"No one,"* he says, *"has a right to examine the question of species who has not minutely described many."*

Subsequent to the publication of the "Origin of Species" much time was taken up by successive editions of the work, and much lost through ill-health. Darwin himself described his books as the *"milestones of my life"*; and they fall into two great groups— those completing the "Origin of Species," and those on more or less independent lines, such as the great Botanical Series.

I. WORKS COMPLETING THE "ORIGIN OF SPECIES."

This work was itself an "abstract," and this was the title he proposed to give it, having originally designed it to be of much greater length. However, yielding to the publishers, he produced it in its present form.

In 1868 appeared "Variations of Animals and Plants under Domestication," in which a detailed account was given of artificial breeding; and this was intended to be followed by two other works dealing with variation, heredity, embryology, geographical distribution, &c., in similar detail; but these were never written.

The detailed application of the theory to man was inevitable; and in the first edition of the "Origin of Species" he says: *"In the distant future I see open fields for far more important researches. Psychology will be based on a new foundation, that of the necessary acquirement of each mental power and capacity by graduation. Light will be thrown on the origin of man and his history."* In 1871 the "Descent of Man" was published, and in 1872 "The Expression of the Emotions," which was originally intended to be only a chapter in the "Descent of Man."

II. THE SERIES OF BOTANICAL WORKS.

These dealt with the development of special questions and problems arising in direct connection with the "Origin of Species."

In 1862 appeared the "Various Contrivances by which Orchids are Fertilised by Insects," of which Professor Huxley says:[474] *"Whether we regard its theoretical significance, the excellence of the observations, and the ingenuity of the reasonings which it records, or the prodigious mass of subsequent investigation of which it has been the parent, it has no superior in point of importance."*

From the first, Darwin was convinced that no theory could be satisfactory which did not explain the way in which mechanisms, involving adaptation of structure and function to the performance of certain operations, had come about. In 1793 Sprengel had established the fact that in a large number of cases a flower is a piece of mechanism, the object of which is to convert insect visitors into agents of fertilisation. What Sprengel did not do was to show that plants provided with such flowers

---

474 *Obituary.*

gained any advantage thereby. Darwin worked at this subject for many years, from 1839 onwards, and showed cross-fertilisation to be favourable, and in many cases essential, to the fertility of the plant and the vigour of the offspring; and that all mechanisms favouring cross, and hindering self-fertilisation, give an advantage, and are hence preserved and improved through Natural Selection.

Orchids form an excellent group for the study of these mechanisms of cross-fertilisation, extraordinary modifications of which are found. The flowers are large and conspicuous, and many of them of singular form. The method of cross-fertilisation in *Orchis mascula*[475] we considered in detail in a former lecture.[476]

We must remember that all these details were worked out by Darwin before he had himself seen any insects visit the particular orchids which he described, the necessary confirmation being supplied in 1873 by Hermann Müller in his work on the "Fertilisation of Flowers."

In fine, an extraordinary diversity of devices, of marvellous interest, for ensuring cross-fertilisation, is met with among orchids. In some the pollen cases explode on being touched, and shoot their pollen at the insect. Some flowers are adapted for particular insects, and failure to become acclimatised is now recognised in many cases as being due, not to the plants being unable to live, but to the absence of the proper insects which alone can effect fertilisation, these as a rule being bees and butterflies. This subject may well be described as the Romance of Natural History, and Darwin says: *"I never was more interested in any subject in all my life than in this of orchids."*

In 1876 appeared "Cross and Self-Fertilisation in Plants," the outcome of a great series of laborious and difficult experiments on the fertilisation of plants, which occupied him for eleven years. The book on orchids showed how perfect are the means for ensuring cross-fertilisation; the new book demonstrated how important are the results. The investigations of which this book was the outcome were commenced with a simple experiment made for quite another purpose. Darwin raised two large beds, close together, of cross-fertilised and self-fertilised seedlings of *Linaria vulgaris*, the common toad-flax, and his attention was aroused by the fact that the cross-fertilised plants, when fully grown, were plainly taller and more vigorous than the self-fertilised ones.

In 1877 the "Forms of Flowers" was written, being the development of a short paper read before the Linnean Society in 1862 on the two forms, or dimorphic conditions, of *Primula:* one of which has short stamens situated in the middle of the tube of the corolla, and a long style, the stigma of which is on a level with the open flower. The other form has long stamens reaching to the centre of the flower, while the style is short and the stigma half-way down the corolla, at the same level as the stamens of the other form. (See Fig. 36.) Darwin showed that these flowers are barren if insects are prevented from visiting them, and further, that each form is almost sterile when fertilised by its own pollen, but each is fertile when fertilised with the pollen of the other. By referring to the figures it will be seen that insects visiting the flowers will carry pollen from the long anthers of the short-styled form to the stigma of the long-styled form, but would not reach the stigma of the short-styled form. Darwin showed

---

475   Early Purple Orchid.
476   {M} Lecture 18, Figure 27.

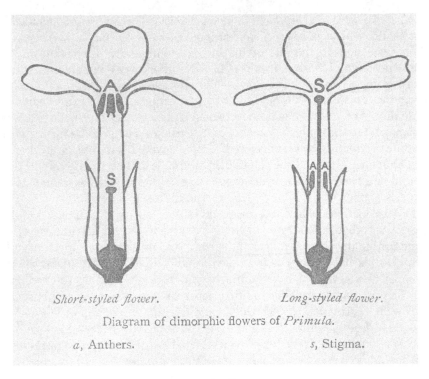

Short-styled flower.   Long-styled flower.

Diagram of dimorphic flowers of *Primula*.

a, Anthers.   s, Stigma.

FIG. 36.

that in this way more seeds were produced than by any other of the four possible unions. Thus, although both forms are hermaphrodite, they bear the same relation to each other as do the two sexes of an ordinary animal.

A few more botanical works must be mentioned, published in 1875, each of which was the outcome of many years of laborious observations and experiments; each breaking new ground and giving results of great importance and interest.

In "Climbing Plants," the spontaneous revolutions of tendrils and of the stems of climbing plants were investigated, and the causes producing them reduced to simple laws.

"Insectivorous Plants" was a typical piece of Darwinian work. "*In 1860,*" he says, "*I was idling and resting near Hartfield, where two species of Drosera*[477] *abound, and I noticed that numerous insects had been entrapped by the leaves. I carried home some plants, and on giving them some insects, saw the movements of the tentacles, and this made me think it possible that the insects were caught for some special purpose. Fortunately, a crucial test occurred to me, that of placing a large number of leaves in various nitrogenous and non-nitrogenous fluids of equal density; and as soon as I found that the former alone excited energetic movements, it was obvious that here was a fine new field for investigation.*"

These researches showed that the plants secreted a digestive fluid like that of animals, and that insects were actually used as food.

The "Power of Movement in Plants," a tough piece of work, was published in 1880. "*In accordance with the principles of Evolution it was impossible to account for*

---

477 Sundew.

climbing plants having been developed in so many widely different groups, unless all kinds of plants possess some slight power of movement of an analogous kind."

We come now to the last of his books, a singularly interesting and most characteristic piece of work, On the 1st of November 1837—*i.e.*, about a year after his return from the voyage of the *Beagle*—Darwin read a paper before the Geological Society on the "Formation of Mould," in which he called attention to the characters of vegetable mould or earth; to its homogeneous nature, whatever was the character of the subsoil; and to the uniform fineness of its particles. He gave the history of some fields, some of which a few years before had been covered with lime, and others with burnt marl and cinders, and showed that on cutting into the soil, the cinders or lime were found, in a fairly uniform layer, some inches below the turf—as though, to quote the farmer's opinion, the fragments had worked themselves down.

The explanation he offered, originally suggested by his uncle, Mr. Josiah Wedgwood, was that earthworms, by bringing earth to the surface in their castings, must undermine any objects lying on the surface, and thus cause an apparent sinking. According to his habit, Darwin continued steadily to accumulate observations, to devise experiments, and to collect information from all possible sources. Finally, in 1881, or forty-four years after his first paper on the subject was published, his last book—the "Formation of Vegetable Mould through the Action of Worms"—was published. He describes himself as getting almost foolishly interested in it; and when the book was published, it was received with what he describes as *"almost laughable enthusiasm,"* 8000 copies being sold in about three months.

Earthworms can live anywhere in a layer of earth, even if thin, provided it retains moisture; dry air is fatal to them, but they can live submerged in water for nearly four months. They live chiefly in superficial mould, from four to ten inches in thickness, but may burrow into the subsoil to a much greater depth. The burrows are effected partly by pushing away the earth on all sides, and partly by swallowing it. In cold weather they may burrow to a depth of from three to eight feet. The burrows do not branch, and are lined by a layer of earth voided by the worm; they end in slightly enlarged chambers in which the worm can turn round.

Earthworms are nocturnal, remaining in the burrows all day and corning out to feed at night. They usually keep their tails in the burrows while feeding, but may leave them entirely, and can crawl backwards or forwards. They are very sensitive to vibrations in the earth, a fact which can be shown by placing a pot of earth containing them on a piano. They do not, however, seem to hear, but can distinguish light from darkness.

Their food consists of leaves or fresh raw meat, especially fat. The worms drag the leaves into their burrows, using their lip to lay hold of them, and showing judgment as to which end to draw them in by. They swallow enormous quantities of earth, which they void in spiral heaps, forming worm castings. By means of the gizzard the earth is ground and mixed with vegetable matter, and the castings are hence of a black colour. They are in this way always bringing earth from below and depositing it on the surface.

Darwin showed that a layer of coal cinders, spread over a field, had sunk an inch in about five years, or one-fifth of an inch per year. In the case of a stony field with very scanty vegetation, and covered thickly with flints, some half the size of a child's head—all these had disappeared in thirty years.

The weight of the castings, collected and dried, which were ejected on a square yard of ground on Leith Hill Common during a year, amounted to over 7 lbs., or 16 tons per acre.

At Stonehenge some of the outer ring of stones have fallen, and are now partially buried in the soil, owing to the action of worms. The specific gravity of objects does not affect their rate of sinking. In fact, "worms have played a more important part in the history of the world than most persons would at first suppose."

Hensen placed two worms in a vessel eighteen inches in diameter, which was filled with sand on which fallen leaves were strewed: these were soon dragged into burrows to a depth of three inches. After about six weeks an almost uniform layer of sand, nearly half an inch thick, was converted into mould, from having passed through the alimentary canals of these two worms.

It is not difficult to account for the success which this work achieved. *"In the eyes of most men the earthworm is a mere blind, dumb, senseless, and unpleasantly slimy annelid. Mr. Darwin undertakes to rehabilitate his character, and the earthworm steps forth at once as an intelligent and beneficent personage, a worker of vast geological changes, a planer down of mountain-sides, a friend of man."*[478]

DARWIN'S METHOD OF WORK.

On this subject his own estimate and account are most helpful. His method was eminently *inductive,* consisting in the accumulation of large numbers of facts, which he compelled to yield up their secrets. He had the greatest distrust of deductive reasoning, and objected to his grandfather Erasmus Darwin's "Zoonomia," because of the undue proportion of speculation to the facts given. He observes: *"From my early youth I have had the strongest desire to understand or explain whatever I observed; that is, to group all facts under some general laws. These causes combined have given me the patience to reflect or ponder for any number of years over any unexplained problem."* Professor Huxley notes how *"that imperious necessity of seeking for causes, which Nature had laid upon him, impelled, indeed compelled him, to inquire the how and the why of the facts, and their bearing on his general views."*[479]

His patience was extraordinary, and the dogged insistence with which he stuck to a subject and followed a clue wherever it led, was wonderful. He spent eleven years over his investigations on orchids; sixteen years over those on the insectivorous plants; the "Origin of Species" was the result of twenty-one years' work; and "Earthworms" that of forty-four years of observations and experiments. He had an unusual power of noticing things which easily escape attention, and of observing them carefully.

"I had," he says, *"during many years followed a golden rule, namely, that whenever a published fact, a new observation or thought came across me, which was opposed to my general results, to make a memorandum of it without fail and at once; for I had found by experience that such facts and thoughts were far more apt to escape from the memory than favourable ones. Owing to this habit, very few objections were raised against my views ('Origin of Species') which I had not at least noticed and attempted to answer."*

---

478 By an unknown author reviewing Darwin's *Formation of vegetable Mould Through the Action of Earthworms*; quoted in Francis Darwin (1880) *The Life and Letters of Charles Darwin*, Vol. 2, Chap. 9.

479 *Obituary.*

"I gained much by my delay in publishing, from about 1839, when the theory was clearly conceived, to 1859; and I lost nothing by it."

PERSONAL CHARACTERISTICS.

Darwin was about six feet high, of active habit, but with no natural grace or neatness of movement. He was awkward with his hands, and unable to draw at all well. He had a full beard and grey eyes, overhung by extraordinarily prominent and bushy eyebrows. In manner, he was bright, animated, and cheerful; a delightfully considerate host; a man of natural and never-failing courtesy—leading him to reply at length to letters from anybody, and sometimes of a most foolish kind. He was fond of animals, and had a great power of stealing the affections of other people's pets.

His private life was burdened by almost constant illness, which rendered him incapable of work for weeks, or even months at a time; and in later years these long periods of suffering often made work of any kind an impossibility. *"For nearly forty years he never knew one day of the health of ordinary men, and his life was one long struggle against the weariness and strain of sickness."*

Under such conditions absolute regularity of routine was essential, and the day's work was carefully planned out. At his best he had three periods of work: from 8 to 9.30; from 10.30 to 12.15; and from 4.30 to 6. Each period being under two hours' duration.

Darwin was a man greatly loved and respected by all who knew him. There was a peculiar charm about his manner, a constant deference to others, and a faculty of seeing the best side of everything and everybody.

The striking characteristic of his manner of work was his respect for time. His natural tendency was to use simple methods and few instruments; little odds and ends were saved for the chance of their proving useful. One quality of mind, which seemed to be of special and extreme advantage in leading him to make discoveries, was the power of never letting exceptions pass unnoticed. He enjoyed experimenting much more than work which only entailed reasoning.

For books he had no respect, regarding them merely as tools to be worked with, and he did not hesitate to cut a heavy book in half, to make it more convenient to hold. He marked the passages bearing on his work, and made an index at the end of the volume. Like many eminent people, he experienced the greatest difficulty in writing intelligible English, and took much pains to accomplish this. *"There seems,"* he says, *"to be a sort of fatality in my mind leading me to put at first my statements or propositions in a wrong or awkward form."* His tone of writing was courteous and conciliatory, and he deliberately avoided controversy.

The closing scene in Darwin's life was in the early months of the year 1882, when his health underwent a change for the worse, and on the 19th of April he died. On the 24th he was buried in Westminster Abbey, in accordance with the general feeling that such a man should not go to the grave without public recognition of the greatness of his work.

And of that work, how shall we estimate its value? To form any notion, however inadequate, we must try to realise the world into which he was born; to picture to ourselves what naturalists up to his time were doing, and what were their aims, ambitions and methods.

From the time of Linnaeus the majority of naturalists were devoted to classifying, naming, and labelling animals, and then leaving them. Others went further and studied more deeply. Increase of knowledge led to constantly increasing specialisation and division of labour; each worker coming to look on his own department as more or less isolated and independent. There was no bond of union between these men, no system, and no true basis of classification.

What Darwin did was to put the backbone into the whole structure; to knit together knowledge from all sources; to point out clearly what was the real nature of these mysterious affinities between the animals now living; to render possible the conception of Natural History as one coherent whole; to show that the real bond was one of blood-relationship, and that the differences between fossil and living animals, which so impressed Cuvier, were not difficulties in the way of such blood-relationship, but necessary consequences of it; to show men that there was no need for them to invoke mysterious agencies to effect they hardly knew what; to show them that all they had to do was to look about, to follow Lyell's method, and see what was now happening around them, in order to get the clue to the past.

Not merely has he changed the whole aspect of biological science, giving it new aims and new methods; but the influence of his work has spread far beyond its original limits. Principles and laws, first established by him for biology, are now recognised as applying to all departments of science, indeed to all departments of knowledge; and it is to him that the phrase the "Unity of History"[480] owes its real significance.

And if we are struck with the importance and grandeur of the results obtained, so are we equally impressed with the simplicity of the means by which they were achieved. The lesson to be derived from Darwin's life and work cannot be better expressed than as the *cumulative importance of infinitely little things*.

Such was the man whom we revere and marvel at, for the greatness of his services to mankind and his contributions to human knowledge; and love for the truthfulness, the patient endurance in suffering, and the gentle courtesy of his life.

---

480 Promoted by English historian Edward Augustus Freeman (1823–1892), also known for the statement (1886): "History is past politics, and politics is present history."

LIST OF AUTHORITIES

Individuals named in Marshall's Lectures, alphabetically by surname, with dates and a brief guide to their importance in the history of biology.

Referred to in Lecture(s):	Surname	Full name	Dates	Contribution to biology
10, 13	**Agassiz (elder)**	(Jean) Louis Rodolphe Agassiz	1807–1873	Swiss, then American, biologist and geologist; reluctant to accept Darwinism due to incompatibility with his religious beliefs, preferring orthogenesis which was a popular alternative between 1870 and 1930; controversial due to supposed racist implications of some of his views of human development; founded Museum of Comparative Zoology, Harvard University; father of Alexander Agassiz.
10, 13	**Agassiz (younger)**	Alexander Emmanuel Rodolphe Agassiz	1835–1910	Swiss, then American scientist, engineer and fish scientist; supported father's work and wrote on marine biology with stepmother Elizabeth Cary Agassiz; son of Louis Agassiz.
15	**Aldrovandi**	Ulisse Aldrovandi (Aldrovandus; Aldroandi)	1522–1605	Italian naturalist and physician, collector of plant and animal specimens; compiled a pharmacopaea; established botanical gardens in Bologna; recognized for his pioneering biological work by Linnaeus and Buffon; first recorded the deliberate breeding of pigeons for visually desirable traits, as noted by Darwin.
15	**Allen**	Joel Asaph Allen	1838–1921	American ornithologist, zoologist and specimen collector; student of L. Agassiz; author of *On the Mammals and Winter Birds of East Florida: With an Examination of Certain Assumed Specific Characters in Birds, and a Sketch of the Bird Faunae of Eastern North America* (Cambridge University Press, 1871).
14	**Aristotle**	Aristotle	384–322 BC	Ancient Greek philosopher, logician, scientist and polymath; a student of Plato but rejected his idea that abstract qualities exist independently from the objects to which they relate; took an empirical approach and separated sense perception from reason and understanding; classified human knowledge into separate disciplines, many of them used today.
9	**Auerbach**	Leopold Auerbach	1828–1897	German neuroanatomist; discovered the eponymous nerve plexus which regulates the gut.

Referred to in Lecture(s):	Surname	Full name	Dates	Contribution to biology
15	**Babington**	Charles Cardale Babington FRS	1808–1895	British field botanist, entomologist, taxonomist and archaeologist; correspondent of Darwin to whom he sent seeds.
18	**Baker**	Sir Samuel White Baker FRS	1821–1893	British explorer, geographer, engineer and politician; sought the source of the Nile and contributed to an understanding of its geography, although beaten to the primary goal by others.
9, 12	**Balbiani**	Édouard-Gérard Balbiani	1823–1899	French microbiologist and embryologist who described the autonomy of germ cells; described structural elements (= chromosomes) in nucleus prior to division.
10, 13, 17	**Balfour**	Francis (Frank) Maitland Balfour FRS	1851–1882	British biologist and embryologist, younger brother of Arthur Balfour (UK Prime Minister); made FRS at the age of 27 and awarded its medal three years later; given University of Cambridge Chair of Animal Morphology in 1882; strong supporter of Darwin's theories; friend and colleague of AMM; made two visits to the Naples Zoological Station, the second in 1875 with AMM; died while attempting to climb Mont Blanc; one of the Naples Station's small steamboats was named after him.
10	**Barrande**	Joachim Barrande	1799–1883	French geologist and palaeontologist; noted for work on trilobites, molluscs and fish; promoted the idea of catastrophes, as represented in the fossil record, rather than the more gradual evolution indicated by Darwin's theory.
14	**Bates**	Henry Walter Bates FRS	1825–1892	English naturalist, entomologist and explorer; friend, colleague and correspondent of Wallace with whom he travelled to the Amazon basin to collect insect specimens; studied animal mimicry, explained its protective potential ("Batesian mimicry") and demonstrated the role of natural selection in its development.
18	**Bateson**	William Bateson	1861–1926	British biologist, evolutionist, traveller and specimen collector; early proponent of Mendel's laws following the rediscovery of the latter's works in 1900; pupil of Balfour and colleague of WK Brooks; influenced by Galton and Lankester and controversially advocated discontinuous alongside continuous variation in the origin of new forms and its species; originated the use of "genetics" to describe the science of inheritance and its mechanisms, although the word was in use a good deal earlier (see AMM's Lecture 1).

Referred to in Lecture(s):	Surname	Full name	Dates	Contribution to biology
3	Bell	Sir Charles Bell FRS FRSE	1774–1842	Scottish anatomist, surgeon and theologian who established the difference between motor and sensory nerves in the spinal cord.
18	Belt	Thomas Belt	1832–1878	British naturalist, explorer, geologist and mining engineer who worked in Wales, Australia, Russia, various parts of North America and Nicaragua.
15	Bentham	George Bentham FRS	1800–1884	British botanist who laid the foundations for the classification of seed plants; donated his herbarium of over 100,000 plants to Kew Gardens; correspondent of Darwin; nephew of the philosopher Jeremy Bentham.
13	Bischoff	Theodor Ludwig Wilhelm Bischoff	1807–1882	German biologist, embryologist and physician; described ovum development in rabbit, dog, guinea pig and deer.
13	Bles	Edward Jeremiah Bles FRSE	1864–1926	British zoologist and author, educated in Hanover and Owens College, Manchester (where he worked with AMM); Junior Demonstrator at Owens College in 1892; sometime Secretary of the Manchester Microscopical Society; later director of the Marine Biological Association at Plymouth; spent time at the Trieste laboratory of the Naples Zoological Station and at Glasgow University; subsequently took degrees in Cambridge and in 1904 became FRSE and DSc, London University; material he collected on *Xenopus* morphology was posthumously edited and published by K Peter in the *Journal of Linnaean Society, Zoology* (1931).
9	Blochmann	Friedrich Blochmann	1858–1931	German zoologist and comparative anatomist; specialist in marine and freshwater organisms; student of Bütschli.
14	Boerhaave	Herman(n) Boerhaave	1668–1738	Dutch botanist, physician, physiologist and renowned clinical teacher, noted for practical, observational approaches to medicine grounded in chemistry and physics.
7	Bonnet	Charles Bonnet	1720 (1729?)–1793	French (Geneva) naturalist and metaphysician noted for propounding mind-body dualism as well as physiological and developmental investigations in animals; described natural parthenogenesis; thought to be the first person to use the term "evolution" in relation to living organisms.
13	Boulenger	George Albert Boulenger	1858–1937	Belgian herpetologist; curator of reptiles at the British Museum.

Referred to in Lecture(s):	Surname	Full name	Dates	Contribution to biology
9	**Boveri**	Theodor Boveri	1862–1915	German zoologist who studied the embryological development of sea urchin eggs; described meiosis; student of R Hertwig.
10	**Boyd Dawkins**	Sir William Boyd Dawkins FRS	1837–1929	British geologist and archaeologist; Curator of the Manchester Museum and Professor of Geology at Owens College, Manchester, contemporary with AAM; noted for human archaeology, especially at sites in UK.
11, 13	**Brooks**	William Keith Brooks	1849–?	American embryologist, studied with L Agassiz; thought that variation arose from external causes and was transmitted through gemmules, which might retain the characteristics of ancestors; made pioneering studies of reproduction in oysters, including description of free-swimming larvae; mentor of EB Wilson.
9	**Brown**	Robert Brown	1773–1858	Scottish botanist, naturalist and explorer who gave the first detailed description of the cell nucleus (1833), later incorporated into the cell theory, and cytoplasmic streaming.
14	**Buffon**	Georges-Louis Leclerc, count de Buffon	1707–1788	French naturalist, plant physiologist, physician and mathematician; undertook the first attempt to produce a comprehensive and systematic account of all existing biological knowledge.
9	**Bütschli**	Otto Bütschli	1848–1920	Heidelberg zoologist; published a renowned description of cell division, egg maturation and conjugation in infusoria; proposed a theory of protozoan phylogeny based on the primitivity of the flagellum.
9	**Carnoy**	Jean Baptiste Carnoy	1836–1899	Belgian priest and cell scientist; described (1886) the halving of chromosome number during "reduction division" (meiosis) in Ascaris; devised an eponymous fixative used in histology and pathology.
10, 13, 15	**Carpenter**	William Benjamin Carpenter FRS	1813–1885	British physician, zoologist and physiologist, noted for studies on marine invertebrates and other work associated with the Challenger expedition.
11	**Chabry**	Laurent Chabry	1855–1894	French biologist and embryologist.
1, 13, 14, 16, 19, 21	**Cuvier**	George Cuvier	1769–1832	French naturalist who recognized the importance of fossils as representing the history of life on Earth; sometimes called the "father" of palaeontology.

Referred to in Lecture(s):	Surname	Full name	Dates	Contribution to biology
1, 3, 4, 5, 7, 10, 13, 14, 15, 16, 18, 19, 20, 21	**Darwin**	Charles Robert Darwin FRS	1809–1882	Traveller, naturalist, scientist, author; originator, with Wallace, of a successful theory of evolution by natural selection; believed in the inheritance of acquired characteristics; *On the Origin of Species by Means of Natural Selection, or the Preservation of Favoured Races in the Struggle for Life* was published in 1859; *The Variation of Animals and Plants Under Domestication* was published in 1868.
13	**Dendy**	Arthur Dendy FRS	1865–1925	British zoologist, one of AMM's first zoology students at Owens College (honours degree in zoology 1885, MSc 1887; and later DSc 1891); worked at Millport Biological Station, Scotland, then studied sponges and other invertebrates while demonstrator and assistant lecturer in Melbourne, Australia, describing many new species; subsequently held chairs in New Zealand, South Africa and King's College London.
4, 10, 13, 19, 20	**Dohrn**	(Felix) Anton Dohrn	1840–1909	German (Prussian) evolutionist and advocate of Darwinism; student of, and later assistant to, Haeckel; lifelong friend of Weismann; believed that vertebrates originated from annelids; founded the zoological research institute in Naples, Italy, the most influential marine zoological station of its day, where AMM studied for a period after graduation (and also visited briefly 9 years later).
11	**Driesch**	Hans Adolf Eduard Driesch	1867–1941	German philosopher, zoologist and embryologist; studied with Haeckel and Weismann; originally a proponent of "vitalism" and believed that development was driven by internal forces; worked with sea urchin larvae and disputed Roux's theory of mosaicism.
9	**Dujardin**	Felix Dujardin	1801–1860	French invertebrate biologist; studied *inter alia* protozoans, echinoderms and insects; described protoplasm as "sarcode".
9	**Dutrochet**	René Joachim Henri Dutrochet	1776–1847	French botanist, physician and physiologist; carried out early research on osmosis and embryology and made contributions to the cell theory.
18	**Eisig**	Hugo Eisig	1847–1920	German marine biologist; Dohrn's first assistant at the Naples Zoological Station and later its assistant director; student of Haeckel; expert on polychaete worms.
9	**Flemming**	Walther Flemming	1843–1905	German biologist and microscopist who identified "chromatin" as the material in the nuclei of cells which stained with basophilic dyes; called indirect cell division "mitosis".
9	**Fol**	Herman Fol	1845–1892	Swiss microscopist and marine explorer; described the nuclear spindle (1870).

Referred to in Lecture(s):	Surname	Full name	Dates	Contribution to biology
13	**Fürbringer**	Max Carl Fürbringer	1846–1920	German comparative anatomist; student of Gegenbauer whom he eventually succeeded as professor in Heidelberg.
7, 8, 15, 18	**Galton**	Sir Francis Galton FRS	1822–1911	English explorer, statistician, eugenicist and polymath who applied numerical principles to observations on inheritance; understood evolution to occur by discontinuous steps; suggested the continuity of the germplasm in 1872, about a decade before Weismann; half cousin of Charles Darwin (through Erasmus Darwin).
10	**Gaudry**	Jean Albert Gaudry	1827–1908	French geologist and palaeontologist; research on fossil mammals supported Darwin's theory of evolution.
13	**Gegenbaur**	Carl (Karl) Gegenbaur	1826–1903	German comparative anatomist; promotor of Darwin's theories; student of Kölliker, Leydig and F Müller; teacher of Haeckel and Fürbringer; described the migration of germ cells between tissue layers in early embryos of hydroids.
1, 14	**Goethe**	Johann Wolfgang von Goethe	1749–1832	German poet, writer, naturalist and philosopher.
9	**Goodsir**	John Goodsir FRS FRSE	1814–1867	Scottish surgeon and anatomist who carried out early research on cells and tissue organization.
12	**Götte**	Alexander Wilhelm von Götte	1840–1922	German zoologist and teacher, noted for studies on animal development; wrote a lavishly illustrated study of the development of the toad which went against Haeckel's embryological interpretations (and was aggressively criticized by him). For discussion of this: Nyhart LK, *Biology Takes Form. Animal Morphology and the German Universities* (Chicago, University of Chicago Press, 1995).
20	**Grant**	Robert Edmond Grant FRS	1793–1874	Scottish anatomist, naturalist and marine zoologist, eventually Professor of Anatomy at University College London; friend and mentor of Darwin during his abortive medical studies in Edinburgh, advising him on specimen preparation prior to the *Beagle* voyage; anti-clerical radical; follower of the evolutionary theories of Lamarck and Geoffroy St Etienne with whom he corresponded.
14, 21	**Gray**	Asa Gray	1810–1888	American botanist and plant collector; established a taxonomy of North American flora; supporter and frequent correspondent of Darwin.

Referred to in Lecture(s):	Surname	Full name	Dates	Contribution to biology
6	Gunther	Albert Karl (Charles) Ludwig (Lewis) Gotthilf Günther	1830–1914	German-born, English zoologist; keeper of the zoology department at the British Museum; identified and named large numbers of fish, reptiles and amphibians.
7, 9, 10, 11, 13	Haeckel	Ernst Heinrich Philipp August Haeckel	1834–1919	German (Prussian) biologist, embryologist, traveller, polymath and artist; promulgated Darwin's theories, sometimes controversially, especially as applied to human societies; described several thousands of new species, especially marine invertebrates; named plastids in plant cells; proposed that nucleus contains the matter of heredity while cytoplasm is concerned with the cell's adaptation to environment; promulgated the recapitulation theory.
9	Harvey	William Harvey	1578–1657	English physician, anatomist and physiologist, noted for describing the circulation of the blood but also for opposition to the idea that life is spontaneously generated.
10, 13	Hatschek	Berthold Hatschek	1854–1941	Austrian zoologist, embryologist and invertebrate morphologists; noted for describing the eponymous "pit", a secretory site at the back of the mouth in Amphioxus.
9, 21	Hensen	Christian Andreas Viktor (Victor) Hensen	1835–1924	German zoologist and embryologist; led a high-profile marine research voyage in 1889 and coined the term plankton; defined an eponymous "node" which initiates the primitive streak in early development of avian embryo; published observations on earthworms and their behaviour, including calculations of their density in soil which Darwin made use of in his later studies of earthworms.
11	Herbst	Curt Alfred Herbst	1866–1946	German zoologist; studied with Haeckel; investigated the influence of ions on morphogenesis.
9	Hertwig	Richard (Wilhelm Karl Theodor Ritter von) Hertwig	1849–1922	German zoologist; with his brother Otto (Oscar), both students of Haeckel, proposed a role for the nucleus in heredity and deduced that fertilization involves combination of egg and sperm; mentor of Boveri.

Referred to in Lecture(s):	Surname	Full name	Dates	Contribution to biology
11	His	Wilhelm His, Sr	1831–1904	Swiss anatomist and embryologist, known for establishing the first objectively assembled reference sequence of developing human embryos and for inventing the microtome; believed that human embryos developed in a unique manner, rather than as an elaboration of other vertebrate embryos; father of Wilhem His, Jr (1863–1934), eponymous discoverer of a cardiac conduction pathway. On 13 March 1892, AMM wrote to Sir Edward Sharpey-Shäfer recommending His Sr for election to foreign membership of the Physiological Society in preference to another candidate.
14, 21	Hooker	Sir Joseph Dalton Hooker	1817–1911	English botanist, traveller and specimen collector; assistant director and then director of Kew Gardens; published, with Bentham, a world flora of over 90,000 species; president of the Royal Society; close friend of Darwin and champion of his theories, realizing their applicability to the plant world.
13	Hunter	John Hunter FRS	1728–1793	Scottish surgeon and anatomist; proponent of the scientific method based on careful observation and its application in medicine; with brother William established large collections of medical and biological specimens, now on display in museums in London and Glasgow.
Prefaces	Hurst	(Charles) Herbert	1856–1898	Lancashire-born biologist, anatomist and teacher; student of Huxley; close colleague of AMM as Demonstrator and Assistant Lecturer in Zoology, Owens College (1881–1892); with AMM wrote A Junior Course of Practical Zoology (3rd edn published in 1892); given leave of absence by AMM in 1889 to complete a doctorate in Leipzig, gaining a PhD in 1890 with a thesis on "The Pupal Stages of Culex"; failed to succeed to AMM's Chair in 1894 but took up a similar position in Dublin in 1895; died of blood poisoning and influenza following tooth extraction.
14	Hutton	James Hutton FRSE	1726–1797	Scottish agriculturalist, industrial chemist and geologist; believed the Earth to be millions of years old rather than the 6000 previously thought, which brought clerical condemnation, and that the structure of the planet was dynamic rather than fixed; considered to be a founder of modern geology.

Referred to in Lecture(s):	Surname	Full name	Dates	Contribution to biology
1, 9, 14, 16, 20, 21	**Huxley**	Thomas Henry Huxley	1825–1895	English biologist, anatomist, evolutionist, educator and agnostic; known as "Darwin's Bulldog" for his defence of Darwin's theories, especially against theologically based criticism from Samuel Wilberforce in 1860.
13	**Jelgersma**	Gerbrandus Jelgersma	1859–1942	Dutch psychiatrist, criminologist and neuroanatomist; initially believed that mental illness could be traced to specific structures in the brain, observable post mortem by thin slicing and staining; subsequently a follower of Freudian psychoanalysis.
11, 13	**Kleinenberg**	Nicholaus Kleinenberg	1842–1897	German zoologist and evolutionary anatomist; student of Haeckel; colleague of Dohrn and helped him to found the Naples Zoological Station; investigated the migration of germ cells in *Eudendrium* (a cnidarian).
9	**Kölliker**	Albert von Kölliker	1817–1905	Swiss comparative histologist and anatomist; investigated the metamorphosis of axolotl; described ova and sperm as single cells (1845); made extensive descriptions of tissue structure in several different organs; believed (with others) that the evolution of new forms was teleologically driven by an innate force; his hand, with ringed finger, was famously X-rayed by Röntgen in 1896.
10, 13	**Kowalevsky**	Alexander Onufrievich Kovalevsky	1840–1901	Polish/Russian embryologist and microscopist; studied under Leydig and collaborated with Metschnikoff on methods for fixing marine specimens for histological sectioning; discovered that tunicates (sea squirts) are chordates rather than molluscs; showed that all animals undergo gastrulation during development; showed that invertebrates, like vertebrates, were built of germ layers originating in the embryo.
11	**Krause**	Johann Friedrich Wilhelm Krause	1833–1910	German anatomist, known for neuroanatomical discoveries as well as pioneering embryological work; famously presented a controversial embryo specimen as human, despite anatomical anomalies which gave it greater resemblance to that of a bird.
13	**Kupffer**	Karl Wilhelm von Kupffer	1829–1902	Russian (Latvian) embryologist and anatomist; studied (*inter alia*) the development of mesoderm and formation of the central nervous system; remembered for describing the eponymous phagocytic cells of endothelium; student of F Müller.

Referred to in Lecture(s):	Surname	Full name	Dates	Contribution to biology
1, 14, 21	**Lamarck**	Jean-Baptiste Lamarck	1744–1829	French botanist, largely remembered for his theory (greatly exaggerated and oversimplified by history) that characteristics acquired during life can be passed by organisms to their offspring; also suggested the importance of cells in living organisms.
10, 13, 20, Prefaces	**Lankester**	Sir E(dwin) Ray Lankester FRS	1847–1929	British zoologist and evolutionary biologist; protégé of Huxley and advocate of theories of Weismann and Darwin; like several others (Dohrn, Balfour, Hertwig brothers), tried to establish phylogeny of various species on the basis of germ layer embryology and recapitulation; Linacre Professor of Comparative Anatomy at Oxford; third Director of Natural History Museum, London; son of Edwin Lankester (1814–1874) who was a close friend of Darwin.
9	**Leydig**	Franz von Leydig	1821–1908	German comparative biologist and human histologist; remembered principally for identifying the eponymous interstitial cells in seminiferous tubules.
1, 14, 15, 21	**Linnaeus**	Carl Linnaeus	1707–1778	Also called Carl von Linné, Swedish botanist, physician and zoologist who established the system of binomial nomenclature for living organisms in general use today.
11	**Loeb**	Jaques Loeb	1859–1924	German biologist and experimental physiologist; demonstrated parthenogenetic activation of sea urchin eggs.
1, 14, 21	**Lyell**	Sir Charles Lyell FRS	1797–1875	Scottish lawyer and influential geologist who provided the first generally accepted explanation of sequential geological strata, with its implications for the age and evolution of life forms.
1	**Malpighi**	Marcello Malpighi	1628–1694	Italian physician and biologist who made significant advances in microscopic anatomy; discovered pulmonary capillaries and gave his name to the excretory tubules of insects.
8, 16	**Marsh**	Othneil Charles Marsh	1831–1899	American palaeontologist; discovered the first pterodactyl in the US and a thousand other fossil vertebrates; wrote about toothed birds, fossil horses and gigantic horned mammals as well as dinosaurs; defined the type species of *Brontosaurus* in 1879.
12	**Maupas**	Émile Maupas	1842–1916	French zoologist; gave the first identification of the nematode *Caenorhabditis elegans* and some other invertebrates.
13	**Maurer**	Friedrich von Maurer	(dates unknown)	Author of several papers (1887–1890) on the anatomy of the frog including descriptions of the spleen, arteries, gills, thyroid gland and thymus.

Referred to in Lecture(s):	Surname	Full name	Dates	Contribution to biology
4	**Mivart**	St George Jackson Mivart FRS	1827–1900	English biologist, noted for his teleological understanding of organismal variation and for trying but failing to reconcile the theory of natural selection with his Catholic beliefs; initiated a theory of evolutionary change of function of anatomical structures, subsequently developed by others including Dohrn.
9	**Möbius**	Karl August Möbius	1825–1908	German marine zoologist and ecologist who described several species of marine invertebrates.
19	**Morse**	Edward Sylvester Morse	1838–1925	American zoologist, malacologist and traveller; described several new species of snail; correspondent of Darwin; assistant to L Agassiz at the Museum of Comparative Zoology, Harvard University.
1, 10, 13, 18	**Müller**	Fritz (Johann Friedrich Theodor) Müller	1821–1897	German biologist, atheist, traveller; brother of Hermann; early promotor of Darwinian theories and (with von Baer and others) of a teleomechanistic approach (deterministic but reducible to physical and chemical processes) to understanding life; expounded on embryonic recapitulation in a scientific conference paper *Für Darwin* (1864) illustrated by the early stages of development in crustaceans; used Schwann's cell theory to explain cancer; studied inter-species mimicry in butterflies with Van Beneden; believed that bird aversion to unpalatable caterpillars was learned rather than innate; helped to establish marine zoology as a subject for practical and academic study.
18, 21	**Müller**	(Heinrich Ludwig) Hermann Müller	1829–1883	German botanist, brother of Fritz, who studied insect behaviour and mechanisms of pollination in flowers; corresponded with Darwin, providing evidence for his theory of evolution.
18	**Murray**	Andrew Dickson Murray FRSE	1812–1878	Scottish entomologist, botanist and naturalist; Scientific Director of the Royal Horticultural Society.
9	**Nägeli**	Carl Wilhelm von Nägeli	1817–1891	Swiss botanist who made (disputed) contributions on plant cell structure and cell division; believed that an internal "phyletic force" drove evolutionary development towards a state of perfection; did not accept Darwin's theories.
10	**Neumayr**	Melchior Neumayr	1845–1890	Austrian palaeontologist and geologist.

Referred to in Lecture(s):	Surname	Full name	Dates	Contribution to biology
7	**Paget**	Sir James Paget	1814–1899	English physician, surgeon and pathologist, known for work on tumour treatment and for an eponymous bone/soft tissue disease; friend and correspondent of Darwin and Huxley.
13	**Parker**	William Kitchen Parker FRS FRMS	1823–1890	English anatomist, physician and self-taught zoologist; an authority on foraminifera and birds; noted for remarkably accurate drawings of skeleton and skull; close associate of Huxley; using anatomical analysis, successfully disputed Owen's theory of archaetypal forms; father of anatomist and zoologist William Newton Parker (1856/7–1923), who studied with Huxley and Balfour and was a close associate and son-in-law of Weismann, and of zoologist Thomas Jeffrey Parker FRS (1850–1897).
9	**Pfitzner**	Wilhelm Pfitzner	(Dates unknown)	Described "parachromatin", the non-staining material of the nucleus (1883); discovered chromomeres on chromosomes (1880).
15	**Poulton**	Sir Edward Bagnall Poulton FRS	1856–1943	British evolutionary biologist, zoologist and entomologist; strong supporter of Darwin's theory of evolution (although ambivalent about the importance of sexual selection) and Wallace's recognition of external ornaments and colours as the subject of selection; promotor of Weismann's concept of the germ-plasm; his influential book *The Colours of Animals* (1890), which described frequency-dependent selection, was praised by Huxley.
9	**Rabl**	Carl Rabl	1853–1917	German anatomist and microscopist; described chromosomes and their dissolution during the cell cycle and suggested them as hereditary material.
9	**Ranvier**	Louis-Antoine Ranvier	1835–1922	French anatomist and histologist, known for identifying eponymous nodes along the length of myelinated nerves.
9	**Remak**	Robert Remak	1815–1865	German (Prussian) neurologist and physiologist, student of F Müller; with Franz Unger and Rudolf Virchow gave first descriptions of binary fission and concluded that life has continuity through cells (*omni cellule e cellule* or "all cells come from cells"); noted for identifying the egg as a cell and for rationalizing the description of tissue structure into three cell layers: ectoderm, mesoderm, endoderm.

Referred to in Lecture(s):	Surname	Full name	Dates	Contribution to biology
11	**Roux**	Wilhelm Roux	1850–1924	German experimental embryologist; promoted, with others including Weismann, the mosaic theory of development.
11	**Salensky**	W Salensky (VV Zalenskii)	1847–1919	Russian embryologist; friend of Kowalevsky.
13	**Sarasin F**	Karl Friedrich (Fritz) Sarasin	1859–1942	Swiss zoologist and anthropologist; cousin of Paul Sarasin with whom he travelled extensively (1883–1912) in the East Indies and Indian Ocean studying native societies and natural fauna; they published multi-volume monographs on Ceylon (now Sri Lanka) and the Celebe Islands in the Moluccas; Weismann used their description of *Formanketten* ("chains of forms"—a continuous line of development through variants, with no distinct species) amongst snails to inform his evolutionary ideas; the Sarasins accepted neither Darwinian natural selection nor Lamarckian inheritance but offered no alternative explanation for the variety of forms they observed.
13	**Sarasin P**	Paul Benedict Sarasin	1856–1929	See Sarasin F
9	**Schewiakoff**	Vladimir Timofeyevich Shevyakov	1859–1930	Russian protozoan biologist.
9	**Schleiden**	Mattias Jakob Schleiden	1804–1881	German botanist and botanist, early acceptor of Darwinian theory; co-founder, with Schwann and Virchow, of the cell theory, importantly helping to unify ideas about the nature of plants and animals; described nucleoli (1838); thought that cells could arise in three ways: endogenously, exogenously and by direct division.
9	**Schultze M**	Max Schultze	1825–1874	German microscopist, known for identifying rods and cones in the eye retina; described protoplasm, cell membrane and nucleus; father of Oscar Schultze.
9	**Schultze O**	Oskar Max Sigismund Schultze	1859–1920	German anatomist and developmental biologist; son of Max Schultze.

Referred to in Lecture(s):	Surname	Full name	Dates	Contribution to biology
9	**Schulze**	Franz Eilhard Schulze	1840–1921	German zoologist and comparative anatomist, noted for research on sponges; described division of amoeba and defined the type specimen of *Amoeba polypodia* (now *Amoeba vitrea*); long-time friend of Weismann.
9	**Schwann**	Theodor Schwann	1810–1882	German (French Empire) biologist and physiologist; student of F Müller; noted for identifying peripheral nerve cells and discovering pepsin; originator, with Schleiden and Virchow, of the cell theory, applying it to animals; described several intracellular processes including the cyclical disappearance of the nucleus and nucleolus, nucleolar condensation, and cell wall formation.
10, 13, 21	**Sedgwick**	Adam Sedgwick FRS	1854–1913	British zoologist, embryologist and comparative anatomist; student and colleague of Balfour; held chairs in zoology in Cambridge and London; great-nephew of Darwin's mentor the geologist Adam Sedgwick (1785–1873).
13	**Selenka**	Emil Selenka	1842–1902	German comparative zoologist and explorer; contributed to the taxonomy and embryology of echinoderms and other marine invertebrates.
6, 13	**Sollas**	William Johnson Sollas FRS	1849–1936	British geologist, zoologist and anthropologist; wrote a series of papers (1885–1894) describing sponges, echinoderms and other marine organisms.
7	**Spallanzani**	Lazzaro Spallanzani	1729–1799	Italian physiologist and Catholic priest, noted for describing the gamete-based processes of reproduction; reportedly the first to carry out artificial insemination (dogs) and fertilization in vitro (frogs).
2, 8, 13, 14	**Spencer**	Herbert Spencer	1820–1903	English biologist, philosopher and political theorist; developed his own ideas about gradual evolution by natural selection a few years before the publication of *Origin of Species* but acknowledged Darwin's theories and is said to have coined the phrase "the survival of the fittest" (1864) to encapsulate them; criticized Weismann's germ-plasm theory and a long, acrimonious debate ensued between them about the role of selection in the determination of structure.

Referred to in Lecture(s):	Surname	Full name	Dates	Contribution to biology
21	**Sprengel**	Christian Konrad Sprengel	1750–1816	German theologian and naturalist; amongst the first to recognize the importance of insects in cross-pollination and the role of flowers in facilitating this; his mechanistic details of flower structure were rediscovered and confirmed by Darwin.
1, 14	**St Hilaire**	Etienne Geoffroy St Hilaire	1772–1844	French biologist who agreed with Lamarck that species were mutable and that acquired characteristics could be inherited.
13	**Stannius**	Hermann Friedrich Stannius	1808–1883	German anatomist and pathologist; investigated vertebrate neural system and marine organisms; gave his name to calcium-regulating endocrine cells in teleost fish kidneys and to experimental ligatures used to demonstrate conduction of heart contractile impulse.
1	**Swammerdam**	Jan Swammerdam	1637–1680	Dutch biologist and microscopist; showed that insects go through life stages of egg, larva, pupa and imago.
13	**Thomas**	(Michael Rogers) Oldfield Thomas FRS	1858–1929	British zoologist, specimen collector and mammalogist; worked at the Natural History Museum; described some 2000 species and subspecies.
9	**Van Beneden**	Edouard Joseph Louis Marie Van Beneden	1846–1910	Belgian biologist and early cell biologist; contemporaneously with others, described the centriole and the organization of chromosomes during mitosis and meiosis; showed that gametes have half the number of chromosomes as somatic cells (reduction division); son of Belgian biologist Pierre Joseph Van Beneden (1809–1894) who was a pupil of Cuvier and who is considered to have established the first biological field research station (at Ostend in 1843).
1, 3, 9, 10, 11, 13, 17	**von Baer**	Carl Ernst von Baer	1792–1876	Russian (Estonian) naturalist, explorer and geographer; identified the mammalian egg cell; pioneer of embryology; developed "Laws" of embryological development which disputed the recapitulation theory of Haeckel and others; opponent of Darwin's theory, preferring a teleological approach.
9	**Von Mohl**	Hugo von Mohl	1805–1872	German botanist; described the wall of plant cells and coined the term protoplasm for cell contents; with Bartélemy Charles Dumortier described cell fission in plant meristems.

Referred to in Lecture(s):	Surname	Full name	Dates	Contribution to biology
9	**Waldeyer**	Heinrich Wilhelm Gottfried von Waldeyer-Hartz	1836–1921	German anatomist and embryologist, known for naming the chromosome (in 1888) and for describing the role of the germinal epithelium in gonadal development.
1, 5, 14, 15, 16, 18, 21	**Wallace**	Alfred Russel Wallace FRS	1823–1931	British explorer, naturalist, specimen collector and co-originator with Darwin of the doctrine of evolution by natural selection; realized that there were non-blending forms of polymorphisms, thus presaging Mendelian genetic variation; noted for studies on geographical distribution of species, mimicry and camouflage, seasonal dimorphism and other phenomena.
5	**Weir**	John Jenner Weir	1822–1894	English amateur naturalist who corresponded with both Darwin and Wallace, providing the first evidence of warning coloration in caterpillars.
2, 6, 7, 10, 11, 12, 13	**Weismann**	August Friedrich Leopold Weismann	1834–1914	German biologist, embryologist and evolutionary theorist, whose analyses represent a link between the 19th-century embryological/anatomical understanding of evolution and the Mendelian/gene-based/neo-Darwinian interpretations of the 20th century; adopted Darwinian evolutionary theory and rejected Lamarckian and teleological explanations of change of form; appreciated (pre-Mendel) that the inheritance of characteristics depended on the "amphimixis" of particulate units, rather than blending, but saw these as working at a series of different levels (ids, idants and determinants); proposed "germ-plasm" (1885) as a testable, conceptual model for the non-somatic material through which characteristics are passed on through generations; studied germ cells (eggs and sperm) and chromosomes and emphasized the importance of reduction division in maintaining the correct amount of inherited material during sexual reproduction; in the first decade of the 20th century produced a landmark course of lectures and textbooks on evolution.

Referred to in Lecture(s):	Surname	Full name	Dates	Contribution to biology
1	**Whewell**	William Whewell FRS	1794–1866	English polymath and philosopher; besides a lengthy term as Master of Trinity College, Cambridge, was the University's Vice-Chancellor on two occasions; co-founder of the British Association for the Advancement of Science; said to have coined the word "scientist"; published *The Philosophy of the Inductive Sciences, Founded upon their History* in 1840.
10, 11, 13	**Wilson**	Edmund Beecher Wilson	1856–1939	American zoologist and geneticist; wrote a monograph on the development of *Renilla* in 1882; later determined the XY chromosome basis of sex.
11	**Wolff**	Caspar Friedrich Wolff	1733–1794	German (Prussian/Russian) biologist, physician, physiologist and early embryologist; champion of epigenesis (in its 19th-century sense, that the living form of organisms arises during development rather than being "preformed"); described the mesonephric or "Wolffian" ducts of the developing vertebrate kidney/urinogenital system.
10, 13	**Würtenberger**	Leopold Würtenberger	1846–1886	German/Swiss palaeontologist and geologist; champion of the principle of condensation—a development of the recapitulation theory in which beneficial adaptations are inherited and passed back so that they appear in the embryology of later generation; noted for work on Jurassic ammonites and for pestering Darwin for recognition and financial assistance.

ARTHUR MILNES MARSHALL AND HIS FAMILY

The Marshall Family in Birmingham

AMM was the second of five sons born to William Prime Marshall (1816–1906; WPM) and his wife Laura (née Stark, 1827–?). The 1851 census shows William and Laura living, with their newborn first son William Bayley Marshall (WBM) and three female servants (nursemaid, cook and housemaid), at 54 Newhall Street, Birmingham. This property, where AMM was born in 1852, still stands: a four-storey, double-fronted mansion with an imposing portico, now with added rooms in the roof and occupied by a firm of solicitors.

WPM and Laura also had two daughters who died in early infancy, Mary Laura (1849–1850) and Alice Laura (1856). There was evidently a further child, sex and dates unknown, because the 1911 census records the 85-year-old Laura Marshall, five years widowed and living in London with her youngest son Charles Frederic Marshall (CFM) and his wife, as being the mother of eight offspring, four of whom were alive at that time: WBM, Frank Herbert Marshall, Percy Edward Marshall, and CFM. (The *Dictionary of National Biography* entry for AMM, published in 1901, describes him as WPM's third son, but this cannot be correct; the *Oxford DNB* entry of 2016 correctly says second son.)

Mary Laura and Alice Laura were buried, as eventually were WPM and AMM, in the family grave in Key Hill Cemetery, Icknield Street, Birmingham. Cemetery records state that the grave (N241) contained six burials at the time of WPM's interment, so it may be that the unknown child was among those laid to rest there. Unfortunately, the gravestone was removed by the council in the 1990s and the plot is now flat and unmarked.

By 1861, the expanding family had moved to an address in Portland Road, Edgbaston, where they remained at least until 1871. *Grace's Guide to British Industrial History* for 1880 gives the address of WPM and WBM as 14 Augustus Road, Birmingham. Censuses for 1891 and 1901 show that WPM and Laura were by then living at "Struan", Richmond Hill Road, Edgbaston, an address that WBM also used for engineering correspondence (held in the archive of the Royal Institution of Mechanical Engineers).

Laura (b 22 May 1826) was the fifth of at least eight children (including a pair of twin girls) born between 1819 and 1835 to Mary (b 1794) and Mark/William Stark (b 1798) of St George Plain, St Michael Coslany, Norwich, Norfolk. The Royal College of Surgeon's biography of CFM (*Plarr's Lives of the Fellows*) states that Laura was "a niece of William Stark, the artist" but the allusion is unclear and no artist of that name has been traced. The first of Laura's siblings was also called William (as was the fashion), so this information may not be correct. In the *Edgbastonia* obituary of AMM, Laura is described as an accomplished amateur musician.

Public records show that the third of the eight Stark siblings Jannet [*sic*] (b 1824) married one Alfred Bowles, a dancing teacher from Ipswich, and gave birth to a daughter, Gertrude Elinor Dale Bowles, in 1856. The 1861 Census lists Gertrude, age 5, as staying with the Marshall family, including her cousins AMM (age 8), FHM (5) and PEM (2), at the Portland Road address. Gertrude married in 1886 but her subsequent fate is not known, and other Stark family progeny have not been traced.

William Prime Marshall (1818–1906)

AMM's father William Prime was a railway engineer and for many years secretary of the Institution of Mechanical Engineers. He was born on 28 February in St Albans, Herts, to the Revd William Marshall and his wife Mary Bayley, but baptized in Birmingham on 11 June. Thus the subsequent Marshall family history in that city may stem from a move during the first few months of WPM's life.

Grace's Guide records that following home education by his father, WPM studied engineering at King's College London with Edward Cowper, and from 1835 was employed in the office of Robert Stephenson (son of railway pioneer George Stephenson). He worked on projects including the London and Birmingham Railway, the Great Western Railway and the North Midland Railway, designing stations, experimenting with new methods of rail traction, and assisting the development and implementation of a uniform track gauge. He established an engineering consulting practice in Birmingham in 1848, engaged in material inspection and in rolling stock design, and worked for railways in India and Africa as well as in Britain.

In 1849, he was involved in the Birmingham Exhibition of Manufactures and Art, part of the British Association for the Advancement of Science meeting, and he served again as Secretary for a further meeting in Birmingham in 1865. He subsequently helped to establish the Birmingham and Midland Institute, created for the 'Diffusion and Advancement of Science, Literature and Art amongst all Classes of Persons resident in Birmingham and Midland Counties'. He was associated with the Birmingham Natural History Society and became its secretary in 1887. Obituaries of AMM describe his father as a keen naturalist and accomplished microscopist; indeed, the pair authored a joint paper in 1881.

William Prime died at the Portland Road address on 27 March 1906. In the 1901 census, the household included a nursemaid and a nurse attendant, besides other servants. He left his estate of £28,666 2s 1d to WBM and PEM. Why his other living sons CFM and FHM, and indeed Laura, were excluded from his will is not clear.

William Bayley Marshall (1850–1912)

WBM, AMM's elder brother, was born on 17 November 1850. He followed his father into railway and civil engineering and in 1877 became, like his father, a member of the Institution of Mechanical Engineers. He was born in Birmingham (not Norwich, as stated in Grace's Guide) and remained at the Marshall family addresses for most of his childhood and working life. He trained as a locomotive engineer and became manager of the Staffordshire Wheel and Axle Company. He joined his father's consulting practice, William P Marshall and Son, in 1882 and was associated with the development of railways in South Africa.

In 1890, WBM delivered three papers (on servo tubes, brakes and safety factors) to the Mechanical Science Section of the British Association for the Advancement of Science meeting in Leeds. This was the same meeting at which AMM was President of the Biological Section (giving the Address which forms Lecture 13).

WBM was Secretary of the BAAS Mechanical Science Section for the 1890 meeting and for the previous four meetings.

WBM never married. Towards the end of his life he moved to Malvern, Worcester-shire, and is recorded in the 1911 census at the Imperial Hotel, Avenue Road. He died in Malvern on 23 July 1912, leaving his estate of £9863 14s 11d to his brother PEM.

Frank Herbert Marshall (1854–?)

FHM is also listed in *Grace's Guide*, being the manager of Tees Ironworks in Ormsby, Northumberland. In 1890, at the age of 36, he too became a member of the Institute of Mechanical Engineers. He was also a member of the Chemical Society and the Cleveland Institution of Engineers.

FHM was one of the two Marshall brothers to marry and have children. He married Lucy Harriett Jordison of Coatham, daughter of Joseph Jordison, a post-master, on 2 May 1883, in Kirk Leatham. His brothers WBM and AMM, along with their father WPM, were witnesses to the marriage. His occupation at the time is shown as "Secretary", so the date when he took over the ironworks is uncertain.

The 1891 census records two sons, Laurence Herbert Marshall (b 1885) and Gerald Struan Marshall (b 1890), and shows the family, including two of Lucy's sisters and a servant, living at Blenheim Villas, Grove Hill, Marton, Middlesbrough. Of Laurence Herbert, nothing more is known. Gerald Struan, whose second name recalls that of the final Marshall family home in Edgbaston, had an exceptionally distinguished military and medical career (outlined below). FHM's date of death is not recorded.

Percy Edward Marshall (1859–1947)

PEM became a solicitor and is notable in AMM's story for the role he played at the latter's inquest. CFM also acknowledges PEM's help as proof-reader for the first volume of the collected lectures.

PEM's law practice, Hensman & Marshall, was established in London at 25 College Hill and later at 35 Bedford Row. He was sometime Director of the Law Association, a legal charity. His long-term residence and address at death was Netley Cottage, The Grove, Hampstead, and he is shown as living there, with a housekeeper, in the 1911 census. Netley Cottage, originally built in 1779 and extended during PEM's time, is currently a Grade II listed building and has historical associations with Chief Justice Coleridge and Robert Louis Stevenson.

PEM died on 7 September 1947, leaving his effects (£5725 9s 8d) to "Gerald Struan Marshall group captain R.A.F. [his nephew] and Hugh Waldron Dallas solicitor". There is no evidence that he married.

Charles Frederic(k) Marshall (1864–1940)

AMM's youngest brother, CFM, plays a crucial part in our narrative, for it was he who posthumously assembled the lectures on which this book is based. The essence of his professional life was summarized in a *British Medical Journal* obituary (8 June 1940):

> Mr. CHARLES FREDERIC MARSHALL, M.D., F.R.C.S., who died on May 22 at Golders Green, N.W., was the joint author with E. G. Ffrench of Syphilis and Venereal Diseases, which reached its fourth edition in 1920, and of a small book, Syphilis and Gonorrhoea, published in 1906. He was born in Birmingham in 1864,

and studied at Owens College and the Victoria University of Manchester, taking the B.Sc. degree in 1883, the M.Sc. in 1886, the M.B., Ch.B. in 1889. and the M.D. in 1890. After being admitted a Fellow of the Royal College of Surgeons of England in 1893 he held for a year the post of surgical registrar at the Hospital for Sick Children, Great Ormond Street, and was for six years assistant surgeon to the Hospital for Diseases of the Skin, Blackfriars. During the last war and for two years after the armistice he was a civilian medical officer attached to the R.A.M.C. He had been a member of the British Medical Association from 1895 until his retirement at the end of 1937.

... but we should note a few more details of his interesting life.

CFM was born on 13 February 1864, while the family lived at the Portland Road address. His early education is not recorded but he entered Owens College Manchester in 1879, the year his brother AMM became Professor of Zoology. He was a Committee Member of the Biological Society in the 1881/2 and 1882/3 academic sessions and gave a lecture on *Fruits* during the latter session.

The Wasdale Inn visitors' book for 1884 has an intriguing entry by "Chas. F. Marshall, Owens Coll. Manchester" and "Walter Hurst, Owens Coll Manchester" on 8 April (the Tuesday before Easter). Their activities there are not stated but this was three years before the first recorded visit of AMM to the area.

CFM gained a BSc in 1884, produced an academic paper on lobster nerves during the 1884/5 session (listed in a statement of publications by AMM's laboratory, held in the Manchester University Archives) and gained an MSc in 1887. (The dates of these degrees, and of the later MD, are as recorded in the Victoria University Graduate Register and are each a year later than stated in the *BMJ* obituary quoted above.) According to a Royal College of Surgeons biography (*Plarr's Lives of the Fellows*) he was Dauntes medical scholar, Platt physiological scholar, Dalton natural history prizeman, and senior physiological exhibitioner. In addition to the medical qualifications listed by the *BMJ* obituary, he became a Licentiate of the Society of the Apothecaries in 1888 and gained MB BCh and MRCS in 1889.

Following the formal completion of his studies he published three research papers:

> Observations on the structure and distribution of striped and unstriped muscle in the animal kingdom, and a theory of muscle contraction. *Quarterly Journal of Microscopical Science (Journal of Cell Science)* 1887; 28: 75–105.
>
> Further observations on the histology of striped muscle. *Quarterly Journal of Microscopical Science (Journal of Cell Science)* 1890; 31: 65–82.
>
> The thyro-glossal duct or "canal of His". *Journal of Anatomy and Physiology* 1891; 26(1): 94–99.

In each, he acknowledges the assistance of AMM; the conclusion of the third paper says: "*I express thanks to my brother, Professor Milnes Marshall, for much help in connection with the embryological bearings of the subject.*" The muscle work was "*carried on in the Physiological Laboratory of Owens College during the winter of 1887. My thanks are due to the Council of the College for a special grant to enable me to carry on the research.*"

CFM's active medical career began as house surgeon at the North Eastern Hospital for Children, London, before taking the Great Ormond Street position in the same year that his collections of AMM's lectures were published (1894). Subsequent surgical appointments were at the London Lock Hospital (for venereal diseases; Soho), the British Skin Hospital (Euston Road) and, from 1908 to 1914, at the Blackfriars Hospital for Diseases of the Skin. In the same year as starting the Blackfriars position he married, in London, Blanche Fanny Emmett (b 1884), daughter of William Henry and Frances Emily Emmett of Bristol, Gloucestershire. On a later census entry, her occupation was given as "Dispenser".

The Marshall Family after AMM

AMM (d. 1893) was survived by his father WPM and his mother Laura, as well as by his four brothers WBM, FHM, PEM and CFM. At various times, most of the brothers, including AMM, are listed as shareholders in the Great Western Railway and of a new iron company set up in 1890 by the Stirling Phosphate and Mining Company of Kingston. WBM, AMM and PEM died without issue, so the Marshall family story continues through CFM and FHM.

CFM's medical career continued after the First World War until 1937. He published research papers and letters on venereal diseases, including on its vertical transmission (*British Medical Journal* 1908; 1(2463): 656), and on the origin, diagnosis and radiological treatment of cancer. He was evidently experimental in his treatments, sometimes using controversial or unproven methods which attracted the criticism of colleagues.

In letters to *The Lancet* (25 February, p 501) and the *British Medical Journal* (28 February, p 401) in 1911, he criticized the introduction of salvarsan (dioxy-diamidoarse-nobenzol, also known as "606", recently invented by Paul Ehrlich and marketed by Hoechst AG), as a treatment for syphilis in place of the traditional mercury and iodides then in widespread use. He implied that the new drug was being promoted for commercial reasons rather than on the basis of proven efficacy and safety. In subsequent correspondence (*The Lancet*, 18 March 1911, p 724) its advocates disputed CFM's authority to opine on the matter and the debate grew professionally ill-tempered.

As it turned out, salvarsan became the treatment of choice for syphilis (and for trypanosomiasis) and is credited with saving many lives. CFM's resistance may reflect the mood of the times: salvarsan represented a new type of chemotherapy, originating from directed research within a fledgling pharmaceutical industry rather than from the pragmatic experience of experimentally inclined physicians.

CFM and his wife Blanche had a son Robert Michael, born at a nursing home in Ealing, London on 4 July 1914. The family home at the time was 55 Acacia Road, St John's Wood, which is where mother Laura was living at the time of the 1911 census. Robert's subsequent story is uncertain but it is possible that he became a ship's broker, living in Thorpe Bay, Essex and dying at a nursing home in Westcliffe-on-Sea in October 1995, survived by wife Amy Ethel and son Jonathan Michael Marshall.

CFM himself died on 22 May 1940 at his wife Blanche's address (69 The Drive, Golders Green, Hendon) although his own address at the time was 68 Crowstone Road, Westcliffe-on-Sea, Essex. (Whether the similar location to his son's place of death half a century later holds any significance is not known.) His estate, left to Blanche, amounted to just £60 8s 8d.

FHM's second son Gerald Struan Marshall (16 July 1889–25 December 1967) came to London and qualified as a dentist (LDS) in 1913 and as a doctor (MRCS) in

1915, before being commissioned into the Royal Army Medical Corps.[*] He had a son, John Lawrence, b 1922, who is listed in the Epsom College Register as LMSSA (an apothecary qualification, then giving licence for medical practice) and living in New South Wales, Australia.

Arthur Milnes Marshall

AMM was born at the Newhall Street address on 8 June 1852. Little is known of his early life but an extensive, multi-authored obituary in *Edgbastonia* (reprint held in the University of Manchester Library) has him attending *"the well-known preparatory school conducted by the two sisters of the late Mr Arthur Ryland"*. This school and its matriarchs cannot now be traced although Ryland was a prominent local politician and benefactor. Subsequently *"he was sent to the Rev. D. Davies, Lancaster, and afterwards to Mr. J. Sibree, Stroud"*; both of these gentlemen are similarly now lost from record.

The 1871 census has AMM living at the family's Portland Road address and described as a "Graduate from University". He had obtained an external BA degree from London University that year, aged 18, *"whilst still a schoolboy ... as soon, in fact, as it was possible for him to do so"* (*Edgbastonia* obituary). It is difficult to interpret this undoubted prodigality using the modern idea of a university degree (which requires an extended period of registered study), although it came with a scholarship in Animal Physiology. He entered St John's College, Cambridge, on the Natural Sciences Tripos, studying at the newly established School of Biology.

Cambridge and Naples

It was probably at Cambridge that he first encountered his near contemporary, the even more prodigiously gifted Francis Maitland Balfour (b 10 November 1851: seven months before AMM). Balfour had entered Trinity College in 1870, gained a Natural Science scholarship in his second year (1871) and graduated second in the Tripos in 1873. AMM was elected as a College Scholar in 1873 and graduated Senior in the Tripos in 1874.

---

[*]   He achieved rapid promotion to captain and served for two years in Mesopotamia, being three times mentioned in despatches. He received a military OBE at the end of the war, at which point he gave up his army commission and joined the Air Force as a temporary captain. He became squadron leader in 1922 and worked for four years as one of the first RAF medical officers at the Chemical Warfare Experimental Establishment at Porton. He served in Iraq for a year and was promoted to wing commander in 1932. Prior to the Second World War he was Chief Assistant to the RAF's Director of Medical Research, conducting experimental research in several areas including the development of the first pressurized suits for high-altitude flying. According to his obituary in the *British Medical Journal* (28 January 1968, p 256), he was noted for "his skill and dexterity in laboratory work". Concurrently, he obtained a diploma in tropical medicine and hygiene, then moved to studies on vision at the Central Medical Establishment. He was promoted to group captain in 1936 and made Honorary Physician to the King in 1939. He was Deputy Principal Medical Officer for Coastal Command and again mentioned in despatches in 1942. Following retirement from the Air Force in 1944 he became a Home Office inspector under the Cruelty to Animals Act and was made FRSE in 1947.

Both men studied under the renowned physiologist Michael Foster, whose courses in basic biology followed the approach of TH Huxley, the eminent advocate of Darwin. The emphasis was on comparative anatomy, based on practical dissection; animals were grouped according to their similarities and types, using this to understand their relationships. (Blackman H, *Journal of the History of Biology* 2007; **40**: 71–108 gives a detailed history of the Cambridge School and the development of its biological courses.)

On the strength of his Tripos performance, AMM was awarded a travel scholarship to the Zoological Station in Naples, recently opened by Anton Dohrn. The Station operated through temporary occupancy of "tables" by visiting researchers and students. Cambridge University had right of allocation of two of these tables: one went to AMM and the other to Balfour, for whom it was a second visit. AMM was there from 19 February to 5 June 1875. It is not clear whether he travelled to Naples with Balfour or met him there but they certainly returned to England together. There is no direct information about what AMM studied at the Station although Balfour was investigating the development of elasmobranchs (Blackman H, *Studies in the History and Philosophy of Biology & Biomedicine* 2004; **35**: 93–117).

It is intriguing to speculate on the friendship between AMM and Balfour. The latter was of aristocratic stock, well used to networking amongst the higher echelons of society, and using his influence and contacts to promote his interests. AMM's origins, as we have seen, were firmly grounded in the technical and economic practicalities of Victorian industry. Despite the contrasting backgrounds, both came from intellectual families who placed great value on education. Their inherited work ethics were undoubtedly similar too: a meticulous focus on detail and accuracy and a fierce belief in the value of hard work, perhaps to the level of obsession.

Their friendship, underpinned by the shared Cambridge approach to the biological world, evidently endured. It is likely that Balfour's embryological skill and his rapidly developing reputation as an academic leader played a large part in stimulating and legitimizing AMM's later interests. For both, comparative embryology, morphology and evolution were intertwined through recapitulation.

Obituaries of the two men emphasize the friendliness and genial charm of their personalities, coupled with a desire to encourage others in an uncompromising pursuit of truth about the natural world. Both were keen sportsmen (Balfour especially enjoyed football). Thus we may imagine they felt comfortable in each other's company at several levels.

On their return from Naples, Balfour gained the lectureship at Trinity College which was to open the door to the rest of his short but remarkably productive academic life (including FRS at the age of 27 and a personal Chair in Animal Morphology five years later). He established an elementary course on Embryology and employed AMM during the 1875–76 season to teach its Practical Morphology component, evidently to large classes in cramped conditions (Blackman, 2007).

In 1876, AMM gained an open entrance natural science scholarship to St Bartholomew's Hospital and began a course of study in medicine. Obituaries suggest that this was a move of last resort, implying either that he found full-time service teaching unfulfilling or that he saw few openings for the sort of experimental research he wanted to do. Whatever the reason for the change of direction, and whether or not his heart was really set on becoming a physician, he obtained a London University DSc and a Cambridge University MB in the remarkably short time of two years, including sitting exams for the two awards on consecutive days. In 1877 he was also elected a Fellow of St John's College. He never practised as a doctor, for his next move was to Manchester.

Manchester

The detailed story of AMM's appointment to the new Chair of Zoology at Owens College, Manchester in 1879 (aged 27) is told by Kraft & Alberti (*Studies in the History and Philosophy of Biology & Biomedicine* 2003; **34**: 203–26). They explain the significance of the position in the development of biological research and teaching in England, particularly the emergence of institutional and public education in the north of the country. Biology was moving out of museums into observational and eventually experimental laboratories. AMM, with his formal Cambridge training and steeped in the Huxley tradition, was the perfect embodiment of the new approach, although his selection after competition against a candidate with museum-based credentials was evidently close-run.

It is not difficult to see how this opportunity to resume active research and combine it with teaching appealed to AMM. It was an established position which came with the expectation that he would develop a productive laboratory and a leading centre of biology education. The lectures in this volume, together with the much admired textbooks he created and his research papers (Publication List), are testament to his achievements on both fronts and also to the close connection he saw between teaching and research.

Papers in the Manchester University Archives record his tireless efforts to develop an academic environment of the highest quality, including appeals to College authorities for funds, equipment and staff and for the support of the College's museum for which he also had responsibility. His papers for the 1880–81 session include a scale drawing of a laboratory, planned for 38 men. Amongst early purchases were microscopes for teaching the foundations of biology. The relentless demand for such instruments continues to feature in his yearly cost estimates and laboratory receipts.

He continued to correspond collegially with Balfour. In a letter of 13 January 1880 (*Darwin Correspondence Project*, letter DCP-LETT-12421) he gave his forthright opinion on Lawson Tait, whom Balfour was considering as a research associate, asked Balfour for a supply of *Amphioxus* for teaching, mentioned plans for a Marine Station in the Menai Straits, gave a progress report on illustrations for a publication (not identified), bemoaning the lack of time available to draw them, described a supply of starfish and dogfish eggs secured from Southport Aquarium, praised the "men" of his team and their fair prospects, and invited Balfour to visit should he be "coming this way at any time".

Coverage of the cost (£22 19s 3d) of AMM's second visit to the Naples Station (5–16 April 1884) appears amongst his departmental accounts for the 1883–84 academic year. The following year's papers include a set of beautifully fine and detailed lithographs, prepared for the 1884 paper on *Antedon* which emerged from that visit. Printers' estimates and samples show the immense care taken to ensure high-quality reproduction of the drawings. Five hundred copies of the 1882 paper on cranial nerves were also prepared for distribution. AMM was made an FRS in 1885.

Beside advocacy of his own department, AMM evidently had a flair for organization and was appointed to Senate and other decision-making bodies in Owens College. In 1887, the College opened the Beyer Laboratories using funds from a

Publications by AMM
(in addition to the Lectures and book review in this volume)

On the mode of oviposition of Amphioxus. *Journal of Anatomy and Physiology* 1876; **10**(3): 502–505

Note on the early stages of development of nerves in the chick. *Proceedings of the Royal Society of London* 1877; **26**: 47–50.

The early stages of development of the nerves of birds. *Journal of Anatomy and Physiology* 1877; **11**(3): 491–515.

The development of the cranial nerves in the chick. *Quarterly Journal of Microscopical Science (Journal of Cell Science)* 1878; **18**: 10–40 and illustrations at 103–108.

The morphology of the vertebrate olfactory organ. *Quarterly Journal of Microscopical Science (Journal of Cell Science)* 1879; **19**: 300–340 and illustrations at 438–448.

On the habits and life-history of Lepidoptera Hyalina. 1880 [Publication details unknown]

On the head-cavities and associated nerves of elasmobranchs. *Quarterly Journal of Microscopical Science (Journal of Cell Science)* 1881; **21**: 72–97.

[With his pupil W Baldwin Spencer] Observations on the cranial nerves of Scyllium. *Quarterly Journal of Microscopical Science (Journal of Cell Science)* 1881; **21**: 469–499.

[With his father WP Marshall] Report on the Pennatulida, collected in the Oban Dredging Excursion of the Birmingham Natural History and Microscopical Society. 1881.

[With Dr GH Fowler] Report on the Pennatulida, dredged by HMS *Triton* and HMS *Porcupine*. *Transactions of the Royal Society of Edinburgh*, Vols 32 and 33.

The segmental value of the cranial nerves. *Journal of Anatomy and Physiology* 1882; **16**(3): 305–354.1.

Certain abnormal conditions of the reproductive organs in the frog. *Journal of Anatomy and Physiology* 1884; **18**(2): 1–144.

The Frog: An Introduction to Anatomy, Histology and Embryology (Manchester: JE Cornish, 1st edn 1882; 5th edn 1894).

The nervous system of *Antedon rosaceus*. *Quarterly Journal of Microscopical Science (Journal of Cell Science)* 1884; **24**: 507–548. [Following his second visit to Naples]

On shallow water faunas. In: *The First Report upon the Fauna of Liverpool Bay and the Neighbouring Seas. Liverpool Marine Biology Committee Report No 1*, ed. WA Herdman (London: Longmans, Green & Co, 1886), pp. 32–41.

[With C Herbert Hurst, PhD, Demonstrator and Assistant-Lecturer in Zoology at Owens College] *A Junior Course of Practical Zoology*. 1st edn 1887; 3rd edn. 1892.

[With his pupil E J Bles] On the development of the kidneys and fat bodies in the frog. 1890.

[With E J Bles] On the development of the blood vessels in the frog. 1890.

[With E J Bles] On variability in development. Paper to the British Association for the Advancement of Science, Biology Section, Friday, 5 September 1890.

Vertebrate Embryology: A text-book for Students and Practitioners (London: Smith, Elder & Co, 1st edn, 1893).

generous endowment by a Manchester locomotive engineer, Charles Frederick Beyer (1814–1876), a former Governor of the College. It is likely that AMM used his persuasive powers to direct funds from the legacy to this purpose. The laboratories became his realm. Over the next few years, various evening events and presentations at the College included tours of the Beyer Laboratory and the Museum. A photograph taken in about 1890 (reproduced in Kraft and Alberti, 2003) shows a set of Italian embryological models for teaching the development of *Amphioxus*.

Throughout his time at Owens College, AMM lived close to his work. His addresses on announcements of Senate meetings are shown as (1883/84) 131 Cecil Street, Greenhays and (1886/87–1890/91) Moss Grove Villa, Moss Lane East. These locations are within a mile or two of the College, although the actual buildings have long since disappeared in redevelopments.

The Menai Marine Biological Station was a worthy but ill-fated project (Baker, *Archives of Natural History* 1994; **21**: 217–224). It was established on Puffin Island/ Priestholme/Ynys Seiriol, a small limestone promontory at the north-eastern end of the Menai Straits. AMM was a founding member of the Liverpool Marine Biology Committee (est. 1885) which collaborated with University College Liverpool to set up the station in 1887. The Committee's objects were *"to make arrangements for 1st, organising dredging, tow-netting, and other collecting expeditions; 2nd, the examination and description of the specimens obtained; and, 3rd, the publication of the results."*

The initiative was stimulated by a growing awareness of the diversity of marine species, the success of the *Challenger* expedition, and the recent formation of the Marine Biological Association and its research station at Plymouth. AMM's enthusiasm for the venture probably stemmed from his visits to the Naples Station, his interests in marine invertebrates and fish, and his experience of dredging (an excursion in 1887 to Arran with the Birmingham Natural History and Microscopical Society.)

The station took over some abandoned signalling buildings on the island. An abundance of wildlife, especially seabird colonies, coupled with ready access to the waters of the Mersey estuary and North Wales coast, made it seem the ideal location. Alas, active fieldwork failed after a few seasons, confounded by a combination of bad weather, the tidally restricted accessibility of the island, its lack of fresh water and the relentless need to manhandle supplies to and from the mainland. It was finally abandoned in 1892 when the Committee was disbanded and a more practical centre established on the Isle of Man. Its legacy consists in five annual reports from the Committee, to the first of which (1886) AMM contributed a substantial paper on "Shallow Water Faunas".

Student societies and sport

In the year of his appointment to Owens, AMM founded a Biological Society. He became its first President and was Vice-President in subsequent years. The Society came under the aegis of the College Union in 1881 and continued to present a full annual programme of lectures by prominent College staff including AMM. Besides the talk in 1883 by CFM (a committee member), the Society's syllabuses until at least 1893 included talks by AMM's colleagues EJ Bles and CH Hurst. Other societies with which he was closely associated were the College Union Debating Society, the Medical Students' Debating Society and the Philosophical Society.

He was equally if not more involved with sporting and athletic pursuits at the College. Fixture lists amongst his papers reveal presidency, vice-presidency or other roles in the football, athletics, cricket, lawn tennis, bicycle and lacrosse (*"All players are expected to turn up whether wet or fine"*) clubs. He was heavily involved with the College's annual Assault-at-Arms (a semi-competitive, public demonstration of gymnastic prowess): in his first two years at Owens he participated in parallel bars, dumb bells, ring, vaulting bars, horizontal bar, club, rapier fencing and salute, and in subsequent years he acted as a judge. He regularly took on the role of judge for the College Athletic Sports and his papers include entry tables and handwritten lists of competitors' initials, times and placings.

Taken together, this is evidence of AMM's keen participatory interest in sport of all kinds, coupled with an enthusiasm for promoting it amongst students. Perhaps it is also evidence that the student body learned to exploit his inability to say "No". He was evidently in demand as an advocate for the improvement of sports facilities, appending his signature to letters to College authorities, managing estimates and invoices for new equipment, and generally oiling the wheels of recreational development across the institution. Papers associated with his 1885 presidency of the College Athletic Union include a long memo to the Chairman of Council requesting a loan of £20 for ground improvements, incorporating drainage and levelling, together with technical plans and estimates for a new cricket pavilion and a heavy roller.

This influence on College sport was posthumously commemorated by a Medal and a Challenge cup. The "Marshall Gold Medal" was established with £102 8s 6d from a memorial fund set up by colleagues and students. (The remainder of the fund, £650, was invested and used to maintain a Marshall Biological Library given to Owens College by relatives, now apparently absorbed into the Manchester University Library.) The Medal and Cup were awarded variously for athletics and all-round performance in other sports until at least 1939. Considerable status evidently attended these accolades: some recipients who fell in the Great War had their awards mentioned in press obituaries.

In 1891, AMM was guest of honour at the annual prize-giving of the Lower Mosely-Street Schools, a centre of working class education, originally set up by the Cross Street Chapel in a deprived area of industrial central Manchester. In his speech (reported at length in the *Manchester Guardian*, 21 September, p 6) he praised academic achievement, especially in the natural sciences, and discussed the value of studying biology as a hobby, for its own sake, as well as for its potential in understanding the basis of disease. He reminded the audience that *"what was known as the 'the survival of the fittest' ... was often determined through the value of little things"* and that *"competition was keenest between those who most nearly resembled each other"*. He presented awards to successful students but was at pains to encourage those who had not yet achieved success: *"very few people who were really worth anything ... did not occasionally come to grief. Such little misfortunes should, rather than being a discouragement, be an incentive to further effort."*

Public education

Programmes and reports for the variety of public lectures with which AMM was associated appear amongst his papers from 1879 (the 14 October lecture to the Leeds Philosophical and Literary Society on *"The Modern Study of Biology"*: see Lecture 1) and every year thereafter. The commentaries on the Lectures describe his involvement with Ancoats and other altruistic organizations as well as with courses and lectures for the public based at Owens College. In 1893, as a member of the Executive Committee of the Lancashire and Cheshire Association for the Extension of University Teaching, he led a deputation to the Technical Instruction Committee of Lancashire County Council to seek renewal of grants and to ensure freedom in managing programmes of lectures across the County (*Manchester Guardian*, 7 July, p 3). Public education was clearly something about which he was passionate.

CH Hurst talks in his obituary of AMM's *"unflinching loyalty to the College, to the new University, and to the cause of higher education generally"*. The "new University" was the Victoria University, set up as an amalgamation of several colleges in the north of England, including Owens. In relation to educational outreach, Hurst explains: *"The success of the Victoria University Extension movement has been due largely if not mainly to his efforts and his great power of organisation, and especially to his tact in adjusting conflicting interests. He thoroughly believed in the usefulness of the extension scheme, and threw himself heartily into the work. ... Many of the zoologists who knew him ... looked with disapproval upon this expenditure of his time—the energy did not much matter, for he seemed to have an unlimited supply of it."*

Mr Sadler, Secretary of the Oxford University Extension Delegacy, told the *Manchester Guardian* that AMM had *"given cordial ascent to a policy which secured good feeling and friendly emulation ... among authorities engaged in University Extension in the North of England"*, referring to his *"energy, influence and foresight"*. Regarding AMM's seven-year presidency of the Manchester Microscopical Society, Hurst says: *"wealthy mill-owners and justices of the peace sit in the meetings of the society, unconscious of social inequality, cheek-by-jowl with poor men who work for a weekly wage."*

Death in the mountains

It is not clear when AMM's enthusiasm for mountaineering began, although his first recorded exploits in the Lake District were in 1887. It is entirely possible that he was inspired or even encouraged by Balfour, but the latter's death in the Alps in 1882 made him initially reluctant to take up the sport. He became especially fond of Wasdale but also made several visits (uncatalogued) to Wales and to the Alps, and was a member of the Alpine Club. (There seems to be no direct link between his Alpine adventures and the visits to Naples.)

The origin of his climbing skill, assuming he was other than self-taught, is not recorded. Witnesses at the inquest described him as "an expert mountaineer, but also one of the most careful". Indeed, confirmation of his methodical, safe approach supported the verdict of accidental death and contributed to the conclusion that his fall was not the result of personal "rashness" or negligence.

The story of his final visit to Scafell is told in newspaper reports of the inquest (*Manchester Guardian*, 3 January 1894, p 8) and in the *Edgbastonia* obituary. At 10 am on Sunday, 31 December 1893 he set out with seven companions from the Wastwater Hotel (now the Wasdale Head Inn) with the aim of scaling Scawfell (Scafell), making sketches and taking photographs. The party divided just below Brown Tongue. AMM, Owen Glynne Jones and Joseph Collier roped themselves together and ascended the mountain via Steep Ghyll and the Low Man.

Moving rapidly, they crossed the Scafell pinnacle and reached the head of Deep Ghyll before the rest of the party had made their ascent by some alternative route. They made *"measurements of various distances important to climbers"* and *"had a short lunch at the top of the crags"*. They then took an "easy descent" by Great Ghyll and Lord's Rake to the foot of Deep Ghyll where they unroped and rested for more lunch. It was now 2.15 pm.

Arthur Milnes Marshall: recorded visits to Wastdale

Date	Activity/location	Companions
1887, September 10–13	Wastdale area	C Hopkinson
1889, April 19–23	Wastdale area: fourth ascent of Curt's Gully, Great Gable	JE Keocher (New College, Oxford), OJ Koecher
1889, December 24–January 8	Wastdale area	F Wellford, CF Clay, GH Rendall, HB Dixon, OJ Koecher, GJ Koecher, JP Gilson, RC Gilson
1890, March 29–31	Wastdale area: Deep Ghyll	JP Gilson, OJ Koecher, GJ Koecher, Dora Koecher, HB Dixon, Olive Dixon, R Adamson.
1890, September 28–30	Wastdale area	Mr & Mrs HB Dixon, C Hopkinson
1890, December 27–January 3	Wastdale area: first ascent of Grainy Ghyll right-hand branch	R Adamson, HB Dixon, Ellis Carr, C Hopkinson, RC Gilson
1891, April 1–3	Wastdale area; first ascent of Central Gulley, Great Gable; first ascent of Professor's Chimney, Scafell	HB Dixon, WI Beaumont, A G[?]
1892, December 30–January 5	Wastdale area: Moss Ghyll	R Adamson, H B Dixon, J Collier
1893, April 2-8	Wastdale area: Deep Ghyll	J Hopkinson, B Hopkinson, A Hopkinson, S Hopkinson, SD Weaver[?], J Collier
1893, December 28–31	Wastdale area: Great End; Pillar Rock, Ennerdale; Steep Ghyll, Scafell	Mr Owen Glynne Jones, Dr Joseph Collier, Professor Dixon, Mrs Dixon, Mrs Dixon's brother, Mr Otto Koecher, Mr Koecher's nephew

Sources: Wastdale Climbing Book, Wasdale Head Inn Visitors' Books, records of the Fell and Rock Climbing Club, Owens College Obituary, Death Inquest Report.

According to the records of the Fell and Rock Climbing Club and the Owens College obituary, Marshall spent "part of each autumn" climbing in the Alps, was a Member of the Alpine club, climbed in the Mont Blanc chain, climbed in Switzerland and spent one season in the Tyrol and the Dolomites. In 1893 he climbed the Matterhorn (with Owen Glynne Jones), Grand and Petit Dru, and Mont Blanc "by a variation of one of the known routes".

Amongst the companions listed, J Collier and OG Jones were witnesses at Marshall's inquest, while HB Dixon supplied information for the 1901 entry in the *DNB*. Further information on the climbing companions, including the Hopkinson brothers, is available in Cocker M, *Wasdale Climbing Book: A History of Early Rock Climbing in the Lake District Based on Contemporary Accounts from the Wastwater Hotel, 1863–1919* (Ernest Press, 2006).

AMM moved away from his companions to seek a vantage point from which to photograph Deep Ghyll. Having found a suitable spot, he called for Collier to bring his camera. Progressing a little higher up, he then called to Jones that he had found a location from which a sketch could be made. Collier immediately heard the sound of falling stone and a rock "about two feet by one foot" rolled past him, followed by the body of AMM.

Collier and Jones rushed down to the place where they estimated the fall would be arrested and found Marshall there, head downward and quite dead. Being unable to attract the attention of the other members of the party, they secured the body in a rock cavity and Jones hurried back to the Hotel for assistance. Marshall's body was brought down and laid out in the front room.

The inquest, held before a jury at the Hotel on 3 January, was attended by AMM's brothers WBM and PEM, the latter acting as the family's legal representative. The district coroner, John Webster, who was sadly used to investigating the deaths of young climbers in the locality,[*] concluded from the witness statements that AMM lost his footing, most probably because the rock he was standing on gave way, causing him to fall backwards. However, he believed that the precise circumstances of the fall would have to remain a matter of conjecture. It was noted that AMM had been "in good bodily trim" and "well equipped", and that his boots (produced for inspection) were "strong and well nailed".

Further details emerge from a letter sent as a factual corrective to the *Manchester Guardian* (10 January 1894, p, 5) by solicitors Hensman & Marshall (PEM and his partner):

> "*The medical examination of the body revealed the following points: 1. Death was instantaneous and due to extensive comminuted fracture of nearly all the bones of the skull and laceration of the brain. 2. The blow which caused the injury was due to a heavy rock striking the front of the cranium while the head was against another rock (or the ground) behind. Whether the rock which caused the injury was one falling from above, or whether it was one on which he was sitting or standing which fell upon his head after he had been stunned and knocked out by a blow from smaller stone, is a matter of conjecture. 3. There were no other injuries of any magnitude to any other part of the body or limbs. This fact makes it almost impossible for death to have been due to a fall, because a fall which could cause such injuries to the head would almost certainly have injured other parts of the body. This fact also shows that the crush occurred without any warning, otherwise the hands and arms would have been injured in the instinctive effort to ward off the blow. There was no such injury.*"

The letter concludes, vehemently: "*We wish to point out emphatically ... that death was a pure accident ... [it] was not in any sense a climbing accident.*"

The body remained at the Hotel until the Wednesday after the inquest, whereupon it was taken back to Birmingham, first to the family home at Richmond Hill and then for burial in the family plot. Those attending the funeral included AMM's

---

[*] The graveyard of St Olaf's church in Wasdale bears witness. Note that the cross chiselled into the rockface at the base of Lord's Rake commemorates a later death in that location.

father WPM and three of his brothers (WBM, FHM, CFM), representatives of Owens College Council, several professors, lecturers and other academic associates, H Dixon, Dr Collier, Edgar Koecher, Dr Hurst, and Mr J Collier (*Manchester Guardian*, 5 January, p 8). Probate records show that AMM's estate of £4576 12s 10d went to "William Prime Marshall, civil engineer".

Neither the Wasdale Inn Visitors' Log nor the Wasdale Head Climbing Book for 1893/94 mentions the death. However, a poignant entry in the Visitors' Log for 28 March–2 April 1894 records a subsequent visit by Mr & Mrs Harold Dixon, Miss Ella Hepworth Dixon, and Roy Dixon:

> "Climbed the small needle on Stirrup Crag (Yew Barrow). Since Jan 1st the East side of the top has fallen, forming a bridge across the east gulley. The shock has loosened rocks on both sides of the needle, so that we did not care to ascend the ordinary route on the left (shown us by Mr J W Robinson) but went straight up the face.
> Examined the scene of the accident in Lord's Rake.
> Ascended the Pillar Rock. R.D. (aged 7) found no great difficulty in scrambling up."

Marshall personified

And what of AMM's personality? It is biographically frustrating that the Manchester University archives contain only material related to the academic and professional sides of his life. There are no personal letters, diaries or notes from which one might divine something of his character. Public records enable details of his family, life and achievements to be pieced together but say nothing about him as a human being.

The *Edgbastonia* obituary and its accompanying "personal reminiscences" are fulsome in praising him as a colleague, teacher, mentor, friend and athlete, as are the eulogies written by the Owens College staff and climbing companions who mourned his premature passing. Such documents tend inevitably to be one-sided, so perhaps we should be impressed as much by their extent as their content: criticism by those who knew him well would surely have been indicated more by reticence than candour.

He was clearly a man of enormous energy and drive who came from a family of high achievers. From his father's time and earlier, the Marshalls represented the progressive values of Victorian science and engineering. AMM directed this heritage into biology rather than railways, but it is easy to see parallels between the Industrial Revolution and the intellectual revolution represented by Darwin, Wallace, Huxley, Lyell and so many others. The family culture was grounded in education and enquiry, no matter that it led to technology, biology, medicine, law or elsewhere. There was also an artistic heritage from his mother's side, although this has been more difficult to identify.

We can be certain that he saw his professional academic role as extending beyond the confines of the classroom and laboratory. He had a strong social, even altruistic sense and it is tempting to describe him politically as a socialist (small p, small s) who might have felt at home in the emerging Fabian movement. This energy was directed into public education as much as into the advancement of his beloved biology and the development of facilities for his students. The two great cities in which he lived, Birmingham and Manchester, evinced strong traditions of social advancement among working people and in the use of industrial profits to create better living conditions. Thus he came from and moved into cultural environments entirely consistent with his personal motivations.

Whether his social perspective was grounded in any kind of religious conviction is harder to say. Public records of births, baptisms, marriages and deaths, through several generations and across the connected families, portray a non-conformist Christian background but there is nothing to indicate the strength of any beliefs he or his family may have held. The apparent absence of a memorial service following his death may be significant in this regard. The content of the Lectures steers firmly clear of theistic or moral debate.

The more personal reminiscences among the obituaries seem to describe an individual with what has more latterly become known as charisma. His intellectual and leadership qualities were evidently balanced by good humour, encouragement and indefatigable optimism. He was also noted for his tact and cheerfulness. Students, colleagues and friends found these characteristics attractive and engaging. He evidently inspired complete trust, whether for his biological insight and methods, for the integrity and openness of his leadership, or for his physical fitness and reliable climbing technique.

Perhaps we can do no better than to allow him to rest beneath the admiring description of Charles Darwin which concludes the final Lecture (stolen, presumptuously, for this book's dedication).

INDEX

In the original publication of Marshall's lectures, only the second volume had an index. The following index is based on that, extended to include the first volume and the book review (Interlude). In the original text and index, italicization of organism names was inconsistent; here I have italicized the names, both common and systematic, of all organisms and groups. The names of scientists appearing in the List of Authorities are not included: they can be located by reference to the chapter numbers in the List.

CPSIA information can be obtained
at www.ICGtesting.com
Printed in the USA
BVHW04*0805140618
519011BV00003B/4/P